T0073585

Madness and Enterprise

Madness and Enterprise

Psychiatry, Economic Reason, and the Emergence of Pathological Value

NIMA BASSIRI

The University of Chicago Press
Chicago and London

The University of Chicago Press, Chicago 60637
The University of Chicago Press, Ltd., London
© 2024 by The University of Chicago
All rights reserved. No part of this book may be used or reproduced in any manner whatsoever without written permission, except in the case of brief quotations in critical articles and reviews. For more information, contact the University of Chicago Press, 1427 E. 60th St., Chicago, IL 60637.
Published 2024
Printed in the United States of America

33 32 31 30 29 28 27 26 25 24   1 2 3 4 5

ISBN-13: 978-0-226-83087-2 (cloth)
ISBN-13: 978-0-226-83089-6 (paper)
ISBN-13: 978-0-226-83088-9 (e-book)
DOI: https://doi.org/10.7208/chicago/9780226830889.001.0001

Library of Congress Cataloging-in-Publication Data

Names: Bassiri, Nima, author.
Title: Madness and enterprise : psychiatry, economic reason, and the emergence of
     pathological value / Nima Bassiri.
Description: Chicago : The University of Chicago Press, 2024. | Includes
     bibliographical references and index.
Identifiers: LCCN 2023023150 | ISBN 9780226830872 (cloth) | ISBN 9780226830896
     (paperback) | ISBN 9780226830889 (ebook)
Subjects: LCSH: Mental illness—Economic aspects. | Mental health—Economic
     aspects. | People with mental disabilities—Economic aspects.
Classification: LCC RC454.4 .B388 2024 | DDC 616.890068—dc23/eng/20230623
LC record available at https://lccn.loc.gov/2023023150

♾ This paper meets the requirements of ANSI/NISO Z39.48-1992 (Permanence of
Paper).

According to a profound analogy that has yet to be illuminated, what is repressed, the sinful idea, is capital itself, which pays interest on the hell of the unconscious.

WALTER BENJAMIN

Contents

This book explores a style of reasoning that suffused transatlantic psychiatric thought across the turn of the nineteenth to the twentieth century—a style that can be characterized as economic in nature. With notable frequency, this book argues, practitioners of psychological medicine made liberal use of informal and sometimes only tacit economic interpretations in order to find meaning in pathological conditions whose ambiguities and enigmatic natures were becoming especially pronounced during the final decades of the nineteenth century. Under the scrutiny of the clinical gaze alone, patients did not always present symptoms that could be neatly partitioned into clear-cut categories of health or disease. One of the most expedient ways clinicians and medical theorists discovered to resolve whether a particular behavior or disposition was diagnosable as morbid was to appraise it according to the financial value it was believed either to possess, generate, or forfeit. The turn to a provisional economic hermeneutic functioned above all as a powerful tool and stratagem for psychiatrists, to adjudicate what an otherwise strict medical assessment could not always. If, as many nineteenth-century clinicians believed, the mad tended to squander their wealth—had they any to begin with—then, what to make of those "eccentrics," for example, who were strongly suspected of madness, but whose fortunes prospered or who remained otherwise quite adept at managing their affairs? Perhaps they were not mad after all. Or perhaps they were indeed mentally ill, but in a manner that was increasingly viewed as tolerable, even venerated—the consequence of a growing psychiatric recognition that mental pathologies were an ever more unavoidable feature of modern society, and that the truth of a person's mental health could not always be deciphered through a medical lens alone.

In this book, I consider an emerging psychiatric conviction that sanity was not, tout court, an essential or indispensable condition for economic life, and that the grounds for economic success were far less yoked to the ideals of health and rationality than otherwise believed. Such a conviction took the form of a diverse set of medical questions and quandaries at around the turn of the century: Was it not possible for the apparently rational behavior of *homo economicus* to be surreptitiously performed and even unconsciously imitated by profoundly pathological individuals? Was the quintessential early twentieth-century entrepreneur actually healthy in a strictly medical sense, or did entrepreneurial virtuosity not in fact depend upon a measure of mental abnormality? Why were the gravest neurological diseases of older, wealthy, and landed patients viewed as compatible with the aims of civil and financial life, in ways that the occupational psychopathologies of industrial workers, deemed perennially costly and fiscally dangerous, could never be? How did medical determinations concerning the underlying sanity of alleged eccentrics come to be so contingent on their financial successes or failures?

These questions, which circulated throughout European and North American psychiatry, did more than simply highlight the extent to which pathological conditions were overlaid with economic meaning. They also showed that within the psychiatric imaginary, an economic lens could reveal the vital truth of health and illness and, perhaps more importantly, uncover the hidden *economic value of pathology itself.* Such a possibility not only raised the questions whether and to what extent remunerative economic conduct could make sense of, supersede, and even redeem pathological behaviors; it also raised the more radical conjecture that *homo pathologicus* might yet have some place in the economic sphere—that modern capitalism could very well make room for, even encourage, the most extreme forms of irrationalism.

Economy and Psychiatry

The relationship between economic thought and psychological science and medicine has conventionally been framed around the ways economists historically drew from and made use of conclusions from modern psychology in order to generate novel economic theories of human behavior. This inclination began in earnest in the 1910s, motivated in large part by the urge to revise a predominantly rationalist model of economic action and *homo economicus.*[1] Psychological doctrines, John Dewey claimed, could make plain that "economic man" was "morally objectionable because the conception of such a being empirically falsifies empirical facts."[2] The economic turn to psychology was rooted in part in marginalist economic theories that rose to prominence after the 1870s and focused on "microeconomic" inquiries into the nature of consumption and the subjective theory of value.[3]

Eventually, some of the most prominent economic theorists found themselves enticed by more radical psychological and psychiatric doctrines.[4] Perhaps most famously, John Maynard Keynes drew upon psychoanalysis in *The General Theory of Employment* in order to flag as morbid some very common economic proclivities and behaviors, from liquidity preferences to extreme short-term speculation.[5] Like Freud, Keynes would characterize the unmitigated "love of money" as a "somewhat disgusting morbidity, one of those semi-criminal and semi-pathological propensities which one hands over with a shudder to the specialists in mental disease."[6] At the same time, Keynes's engines of economic growth—his famous account of the "animal spirits" that incited the "spontaneous urge to action" in the form of enterprise and investment—clearly relied on psychological and psychiatric sources as well, for, as scholars have observed, Keynes appended to his notes on Descartes's theory of animal spirits from a Cambridge course on early modern philosophy the phrase "unconscious mental action."[7] Although the economic interest in psychological and psychiatric theories eventually declined, it gained a second wind with the formalization of behavioral economics after the 1970s, the prominence of which has been amplified particularly by the growing popularity and commercial marketability of the brain and "psy" sciences.[8]

In this book, however, I approach the relationship between economic and psychological-psychiatric thought from the opposite direction, by focusing instead on the use *psychiatrists* have made of forms of economic reasoning in order to resolve diagnostic enigmas with respect to patients whose mental health is either ambiguous or inscrutable. By turning to economic conduct and asking, for example, whether patients appeared capable of managing their affairs, whether they were good with money, or whether ostensibly irrational business ventures ultimately yielded a profit, psychiatrists could often compensate for and circumvent uncertainties with respect to the status of a person's psychiatric health. This exploration of how and why psychiatry came to make use of an economic style of reasoning is in large part prompted by how prominently within the cultural imaginary of the North Atlantic the benchmarks of mental health and illness have been constructed in relation to economic ideals. What links a person's psychiatric status to their economic success? As it turns out, financial prosperity is not always predicated on mental health alone. A seemingly entrenched cultural conviction holds that abnormal and other neurotic states can function as equivalent, if not more potent sources of financial gain.

Keynes's animal spirits were tinged, after all, with an ineffable irrationalism, a precursor to what has been characterized more guardedly as the "irrational exuberance" underwriting economic booms and busts alike.[9] The historian Elizabeth Lunbeck has described how after the 1960s, narcissistic personalities were regarded as economically advantageous states of mind, "perfectly suited for success in the late-capitalist United States," so long as they did not become overly malignant.[10] The anthropologist Emily Martin has observed a similar attribution of economic value to mania as a source of endless economic productivity and value-creation. When manic states yield economic growth, they are revalued as assets rather than disabilities.[11] American economic culture in particular has associated the business prowess of entrepreneurs with quasi-pathological or otherwise unorthodox behaviors and states of mind, the taken-for-granted idea that "an entrepreneur's quirks spur success."[12] In *Trump: The Art of the Deal*, ghostwritten by the journalist Tony Schwartz, Donald Trump extolled the value of "controlled neurosis," a quality he purported to have observed in the most successful entrepreneurs. "They're obsessive," Trump/Schwartz writes, "they're driven, they're single-minded and sometimes they're almost maniacal, but it's all channeled into their work. Where other people are paralyzed by neurosis, the people I'm talking about are actually helped by it."[13]

This somewhat diffuse inclination to imagine that pathological states can yield economic advantages is not an expression of a vague cultural ideology but an explicitly emergent conceptual possibility, which I call *pathological value*. Pathological value refers to a historically circumscribed conviction that pathological states can be appraised in economic terms and can thus be recast not only as value-laden but, in many cases, as value-generating. Pathological value is not limited to acclamations of entrepreneurial eccentricities but has also extended to psychiatric debates on inherited wealth and industrial labor—though, as I discuss later in this book, entrepreneurial eccentricity has been a core iteration of pathological value since the late nineteenth century. The central claim of this book is that the notion of pathological value itself arose as an effect of the psychiatric adoption and deployment of an economic style of reasoning. An economic style of psychiatric reasoning effectively transformed how mental pathologies were conceptualized and measured with respect to their social worth by infusing them with an appraisable degree of economic value.

Why psychiatry would adopt an economic style of reasoning in the first place is a topic I discuss at length in chapters 1 and 2. As we will see, the adoption of this style of reasoning was a consequence of developments internal to psychiatry itself as a system of thought, a result of efforts to resolve diagnostic challenges with respect to changing medical perceptions of mental illness, and so therefore due neither to explicit oversteps among economists nor to an "economic imperialism" seeking to construct a systematic "economic approach to human behavior."[14] In other words, the social and institutional authority of economics, at least in the first instance, was not the reason why economic rationalizations suffused psychiatric thinking in ways that were often quite latent. For that reason, the remit of this book's analysis, which is historical but also decidedly theoretical, is wider than what a narrow history of medical capitalism might otherwise entail, and goes beyond the idea that capitalist structures influenced or were simply entangled with scientific and medical institutions (though those assumptions are not in dispute).[15] Over the course of this book, I attempt instead to unearth a deeper conceptual possibility: that the psychiatric turn to economic reasoning signaled a transformation of the very idea of *value* in the modern social order, and that the differences between the most common forms of social valuation—moral value, medical value, and economic value—were flattened by the start of the twentieth century such that these otherwise disparate expressions of value could be rendered equivalent and even interchangeable.[16] If what was *good* and what was *healthy* were increasingly conflated with what was *remunerative* (and vice versa), then a conceptual space was opened through which madness itself, the most irrational of irrationalisms, could be converted into an economic form and subsequently be redeemed and even revered.[17]

The Moral and the Irrational

Modern social and political thought has held that irrationality is perilously antithetical to the very fabric of society, and psychiatric discourse has played a large part in formalizing the hypothesis that sociopolitical conflicts are rooted in the internal volatilities of the psyche.[18] This hypothesis has taken many forms—from late nineteenth-century claims that hereditary degeneracy catalyzed criminality and facilitated European national decline to Sigmund Freud's apprehensions that our desire for social bonds was incompatible with our morbid drive toward aggressiveness. Frankfurt School researchers like Erich Fromm and Theodor Adorno, and nonaffiliates such as the psychoanalyst Wilhelm Reich warned of the underlying correspondences between psychopathology and political authoritarianism as well as fascistic and other antidemocratic attitudes. This grim modern stance concerning the danger of irrationality has been both fueled and validated by recurrent political violence and international war.[19]

A more careful rendering of this anxiety about irrationality, however, would say that some displays of irrationalism, such as biases, spurious heuristics, and idiosyncratic belief systems, could be countenanced as an unavoidable hallmark of the heterogeneity of the social world in a way that genuinely morbid mental illness could not. Psychopathology has long functioned as the surrogate for social ills—not on the basis of any empirical justifications but because of the inextricable alliance between madness and immorality. In *History of Madness*, Michel Foucault detailed the long history of the moral devaluation of mental illness, due in no small part to the challenges involved in transforming the mentally ill population into a contingent labor force. Madness could never quite become situated within an acceptable moral economy in part because of how long it had served as an emblem of moral shortcomings. By the early nineteenth century, heinous depravity and profound psychopathy became synonyms for one another, and they continue to operate as such despite ongoing efforts to destigmatize mental illness.[20]

What is significant, however, about the moral approbation of mental health and disavowal of madness is that it signaled the degree to which, in the eyes of medical practitioners and incipient social scientists, the human sciences were over the course of the nineteenth century securing the epistemic power to arbitrate a new moral world. As the historian Roger Smith has argued, the human sciences were not just a set of disciplines or institutions but an open-ended historically emergent domain of inquiry spread across numerous fields, from the social sciences and natural sciences to the humanities, consolidated around the charge of elucidating the nature of "human nature."[21] While codes of conduct informed by religious belief, ethical traditions, and conventions of politesse undoubtedly persisted, the emergent sciences of human nature sought to redefine the parameters of moral propriety in ways that cut across the social fabric with a greater degree of authority and veridicality.[22] Morality, wrote Dewey, was not only "the most humane of all subjects" but also the most "ineradicably empirical" insofar as "everything that can be known of the human mind and body in physiology, medicine, anthropology, and psychology is pertinent to moral inquiry."[23]

This moral world of the human sciences was defined not simply by new sites of expertise but also by novel conceptions of human nature and action— theories that would have had little traction in long-established moral traditions that relied on unambiguous notions of agency and deliberative voluntarism. Eventually the moral dimensions of the self would come to be regarded as the means through which individuals could be manipulated and governed, and consequently the medicalization of morality could be one of the central and crucial means by which sociopolitical and economic behaviors and attitudes could be shaped.[24] This is not at all to deny the enduring power of religiously inflected moral norms. The historian Rhodri Hayward has even proposed that the emergence of secular conceptions of the unconscious—and its attendant forms of conduct—was linked to historical religious narratives and "nineteenth-century struggles over the historical status of the supernatural."[25] But while the history of the unconscious, for example, could certainly be told from the standpoint of the intersections of psychiatry and religion, the promulgation of the notion of the unconscious as a potential truth about human nature hinged on its scientific status, even if it remained, to some degree, a medically encoded remnant of religious thought. In any case, what should be emphasized is the noticeably concerted effort on the part of practitioners of the human sciences to move away from religiously informed conceptions of moral conduct. For Émile Durkheim, a "sociology of morality" could demonstrate the extent to which moral norms were neither wholesale religious edicts nor philosophical propositions but ultimately the expression of the dynamic force of social facts; this was why any inquiry into the nature of morality consequently fell within the remit of the human sciences—sociology and psychology, in particular.[26]

In the increasingly secularized and scientifically informed moral world of the North Atlantic, as ethical cultivation was slowly relegated to the domestic and private sphere, the human sciences assumed, in no small measure, the task of filling the vacuum of a distinctly public ethos. One of the ways they did so was by striving to resolve the demands of an increasingly rationalized society against the eruptive irrationalisms that plagued it—from the most ordinary to the most perilous. The sciences of human nature, mind, and behavior were ultimately preoccupied with the challenge of maintaining social coherence in the face of decohering forces. The American sociologist Talcott Parsons famously listed Freud, Durkheim, and Max Weber as members of an early twentieth-century triumvirate of thinkers who, in a historical turning point, most starkly emphasized the role of precisely those decohering influences on ordinary life.[27] Weber's contributions to this topic in particular have established a thesis of modernity that has become almost commonplace. For Weber, the nonrational constituents of social action consisted primarily of affects, traditions, belief systems, and religious orientations.[28] A central task for Weber was to uncover the ethical frameworks that determined the "conduct of life" (*Lebensführung*) in modern, postindustrial Europe, given that premodern spiritual ethics were ostensibly no longer the central guiding constraints on action.[29]

According to Weber, it was capitalism that took up the reins of regulating the ethical heuristics of the modern world. Everyday practical reason was not emptied of meaning so much as it was reimagined through an "ethical 'rationalization' of life conduct [*Lebensführung*]," a newly emergent "lifestyle that stood at the cradle of modern capitalism"[30]—an ethos that while capitalist in nature could function independently of the actual economic systems of capitalism. Human behaviors were subsequently situated along a spectrum of social action, from the most instrumentally rational demonstrations of utility maximization to the most value-laden expressions of traditions and belief systems, which appeared "irrational" in only a comparative sense.[31] However, as Weber admitted, rational conduct was only an idealization and sociological abstraction, a hermeneutic tool.[32] In reality, the vast majority of ordinary conduct was profoundly "irrational," an irrationalism that Weber made sense of in part through the lens of novel psychological doctrines. "In a great majority of cases," he explained, "actual action goes on in a state of inarticulate half-consciousness [Halbbewusstheit] or actual unconsciousness [Unbewusstheit] of its subjective meaning. The actor is more likely to 'be aware' of it in a vague sense than he is to 'know' what he is doing or be explicitly self-conscious about it."[33] Impulses and habits were the common guides of human action, falling often and imperceptibly under the sway of crowds, the contagious effects of social imitation, or the power of charisma.

While economic rationality might have assumed the aim of establishing and reinforcing social cohesion and coherence, it could never fully integrate the fragmentary and diverse expressions of nonrational conduct that pervaded social life. These ordinary irrationalisms most often took the form of allegedly capricious justifications for individual behavior, either prompted by emotions, customs, fallacies, and dogmatisms, or grounded on premises that were "rational" in only an inferred, bounded, biased, or heuristic sense.[34] The economist Joseph Schumpeter claimed that even economic conduct might just as well reflect the behavior of homo religiosus or homo eroticus as it might homo economicus. The dispersion of homines throughout society corroborated the fact that "life is ontologically irrational, at least as much as 'nature.'"[35] The rationalization and so-called disenchantment of the modern world were not successful, certainly not by any empirical measure.[36]

At the same time, however, not all irrationalisms were alike in the dangers they were perceived to pose. Quite likely in light of his own personal preoccupation with financial precarity, Freud concluded that the pursuit of wealth was an inherently neurotic form of behavior, that desire for money was a symptom of psychosexual fixations, anal eroticism, and the symbolic attachment to feces.[37] Different forms of economic behavior could equally portend social peril, particularly given the turn-of-the-century rise of consumerism and mass culture across Europe and the United States—a feature of modern capitalist society that Adorno and Max Horkheimer would implicate in the rise of authoritarianism, and that Herbert Marcuse would dub the most "vexing aspect of advanced industrial civilization," an irrationality that was both "fantastic and insane."[38] Some expressions of nonrationality, in other words, merely signaled the variegations of human nature, the cognitive errors and emotional despondencies and excitements that rippled across the social fabric and that economic norms sought to regulate often in the form of policies and other soft governing techniques intended to steer financial conduct. Other irrationalisms, however, evinced the presence of a veritable *homo pathologicus* who now haunted the new moral world of the human sciences, and who, unlike the other social *homines*, was typically framed as irreconcilably in conflict with, and thus a danger to, the entire socioeconomic domain.

But this supposition—this veritable social myth—that "pathological man" was perilous to the social order in large part because he could not be contained or regulated by the economic reasoning of modern capitalism, or was simply the monstrous creation of it, is precisely the misconception that this book seeks to invert. Madness was not, as I will endeavor to show, in irreconcilable conflict with economic reason, but a condition that was in fact thoroughly amenable to economic rationalization, a possibility that could be reimagined as a source of remuneration or loss, a social fact that could be translated into economic terms and thus revered or renounced on the basis of its financial prospects. One of the central claims of this book is that if madness could be redeemed because it could be regarded as a source of financial profit, or simply tolerated as a state compatible with the management of one's affairs, then its moral and social disapprobation too was staked on its economic promises. If madness was a detriment, it was not because it was unhealthy or immoral but because it could occasion financial losses.

Value and Pathology

In 1900, just months after Freud published *The Interpretation of Dreams*, the sociologist Georg Simmel published his ambitious and unruly magnum opus, *The Philosophy of Money*. Grappling in part with marginalist conceptions of subjective value, Simmel proposed that economic forms encoded modes of action that were not necessarily "economic" in any narrow, that is, financial sense, but were the effects of "more profound valuations and currents of psychological and even metaphysical pre-conditions."[39] For Simmel, the broad concept of value was no more monetary than it was moral and psychological. Ultimately, all social interactions were fundamentally interpretable as forms of exchange. Durkheim, who was hesitant in his review of *The Philosophy of Money* to accept all the book's conclusions, nevertheless conceded how deeply economic forms intertwined with moral life and how profoundly modern conceptions of moral value had come to embody an economic sense.[40] By the end of the nineteenth century, the economic inflections of moral conduct were strongly organized through and around categories like contract freedom, commercial propriety, and the responsibilities associated with appropriate wealth management.[41]

The Philosophy of Money announced a new spirit of modern life and ordinary social relations. When we situate it alongside Freud's contemporaneous appraisals of dreams and the many other psychopathologies of everyday life, we see that, according to the most prominent turn-of-the-century medical and social theorists, the practical dimensions of human behavior had become suffused with both medical and economic meaning.[42] This double medico-economic valuation was crucial in tackling a particularly bleak realization among practitioners and researchers of psychological medicine throughout Europe and the United States. As the German neuropsychiatrist Wilhelm Griesinger lamented in 1867, asylum patients have "fallen victim to an illness that can befall and make a victim of each and every one of us."[43] Mental pathologies were an inevitable fact of social life, which could not in all cases be clearly or unambiguously identified. But even so, this did not mean that madness was in principle always and forever in irreconcilable conflict with the social order. If the specter of madness could be neither eradicated nor simply disavowed, then perhaps it would be more advantageous not to regard mental pathologies according to their hypothetical dangers—that is, to continue to treat them as objects of moral disapprobation. Madness could even be reappraised as a source of social value if it could be shown to have some minimal degree of financial worth.

The concept of value, after all, is not singularly economic. It possesses a strong moral dimension, as Friedrich Nietzsche observed in *On the Genealogy of Morality*, which was not only an inquiry into the origins of the value judgments (*Werturteile*) of good and evil but a critique of the value of those values (the philosophical study of value has since been codified as the subfield of axiology).[44] The concept of value also possesses a robust medical sense, as the historian and philosopher of science Georges Canguilhem argued when he defined living beings (*vivants*) as "organizations whose validity (that is, value [*valeur*]) must be referred to the eventual success [*réussite*] of their life." The *value* of life consists in the sheer vitality of a living being, its capacity to adapt to and endure successfully in its environment. "Therein lies the profound meaning of the identity between value and health," Canguilhem explained, "to which language attests: *valere*, in Latin, means 'to be well.'"[45] One can readily speak of the *moral value* of life, for what this ultimately denotes is the vital success of the living being; to be healthy was in principle good. With the rise of institutionalized modern medicine after the eighteenth century, health continued to acquire not only a distinctly moral sense but, through it, a form of intrinsic economic value as well.[46] Health was not only morally revered but economically valued and esteemed as the source of financial productivity.[47]

As Canguilhem would go on to explain, after the eighteenth century "the truth of the body in an ontological sense" was no longer defined through metaphysical doctrines but, rather, by virtue of the status of the body's health and its capacity to endure and thrive in the context of environmental forces and fluctuations.[48] And yet, to be healthy did not imply imperviousness to injury or disease, since, as Canguilhem would insist, "disease is the risk of living as such."[49] To be healthy means to live in a way that is open to injury and infirmity; health denotes not the avoidance of pathological states but, rather, the capacity to recover from them in order to reclaim one's flourishing vitality. Health was not an attribute a body passively and statically possessed but a condition that needed to be perennially reachieved, reestablished, and renewed. If health was formalized over the course of the nineteenth century as the most fundamental truth *and value* of life, then pathology was not simply error or aberration but itself constitutive of the underlying logic of health, the condition against which health was continuously defined.

In Canguilhem's historical rendering, pathology had become the perpetual shadow of health, the condition upon which the body's vital truth was not only possible, but possible in an adaptive and dynamic way. Given that pathology had transformed into an inescapable reality built into the medicalized essence of human nature, we might begin to understand why *mental* pathology was not merely excised from the social domain, why madness did not simply signify the absence or negation of value and social worth. Madness demanded new forms of positivity, new conditions by which to determine not only its truth and falsity, its legitimacy and acceptability, but also its merits and disadvantages. Why would the attribution of economic value not be precisely one such condition?

Madness, after all, may very well have played a crucial role in the historical "embourgeoisement" of the modern transatlantic world.[50] Scholars have described for some time how psychiatry as a field and institution was co-constituted alongside a burgeoning European liberal order and an industrial-capitalist bourgeois society.[51] Madness became a possibility that was integrated into the social fabric not always as something to be feared and medically sequestered but as an irresolvable and unavoidable fact of the social order—something that needed to be assimilated into the social purview of nineteenth-century economic liberalism and normalized as an inevitable feature of the human populace, the social landscape, and "economic life" in general.[52]

But to suggest that mental pathologies could be economically valuated, that there could be such a thing as pathological value, opposes a standing contention in the history of psychiatry first articulated by the sociologist Andrew Scull. It was, as Scull writes, "a mature capitalist market economy" and the "thoroughgoing commercialization of existence" that prompted in the first instance the "segregative response to madness (and to other forms of deviance)." The mentally ill were excluded from the market economies of the eighteenth and nineteenth centuries because they were deemed to be an even more incorrigible subclass of the population of impoverished masses and so were unable, unlike their indigent counterparts, to be mobilized as an expendable workforce in a newly industrialized labor market.[53]

And yet the exclusion of the mentally ill from the socioeconomic back-drop of nineteenth-century Europe and the United States did not mean that *madness itself* was wholly excised from the entirety of socioeconomic life. Madness inevitably became part and parcel of the ethos of the modern trans-atlantic world—a veritable form of life and "style of existence" that could still be assigned an economic sense, even if the mentally ill themselves were for the most part seen as opposed to the economic order.[54] Drawing in part on what science studies scholar Michelle Murphy has described as the "econo-mization of life," this book seeks to sketch out the *economization of madness*, a task that introduces two scholarly challenges.[55] The first is the counterintui-tive nature of such an argument. The idea that madness could be economi-cally remunerative and thus morally redeemable, even if still an expression of unhealth, opposes the conventional Weberian story (described above) that the strongest forms of irrationalism were opposed to economic appropria-tion. Scholars have admittedly put quite a lot of pressure on this myth of a disenchanted modernity, pointing out how much mysticism and even religi-osity infused and informed many core attributes of modern capitalism, par-ticularly, as the historian Charly Coleman has pointed out, the "enigmatic nature" of the commodity form.[56] But to suggest that capitalism is consistent with magic and religion, which are only "irrational" in a very qualified sense (magic and religion are quite rational doctrines[57]), is one thing; to suggest that capitalism is consistent with *madness*—and not simply insofar as capitalism can induce psychopathologies but that madness is itself consonant with the very operation of modern capital, that madness could even in some form become an asset for capitalism—is quite another task, and precisely what this book contends.

The second and perhaps more significant challenge is that the economiza-tion of madness did not transpire at the level of an explicit institutional de-velopment (and so in that sense it differs from Murphy's treatment of the economization of life), nor was it merely a discursive epiphenomenon, the expression of economic figurations that famously litter psychiatric writings, from the American neurologist George Miller Beard's notion of "nervous bankruptcy" to Freud's "economic" model of the psyche.[58] The economization of madness was not a formal doctrine deliberately contrived by the institutions of psychiatric medicine but the inadvertent consequence of a diffuse and sometimes only latent *style of reasoning* that permeated turn-of-the-century psychiatric thought; and so, to that end, it behooves us to consider what precisely a "style of reasoning" entails.

Styles of Reasoning

As this book contends, economic reasoning was not limited to the domain of economics alone; it was a form of thought adopted by psychiatric clinicians and researchers as well—not, however, in order to make economic knowledge claims but, rather, to facilitate diagnostic assessments about potential psychiatric patients on the basis of their apparent economic behaviors. In that sense, economic interpretations constituted a *style* of psychiatric reasoning. But what precisely is a style of reasoning? A style of thinking or reasoning, an undeniably open-ended term, is intended to designate the organizational coherence of utterances and thought patterns that are situated within the boundaries of a shared historical or epistemic locale, whether it be a discipline, knowledge field, or institution. A style of reasoning demarcates what it is possible to say about a given subject matter. Within the history and philosophy of science, the discipline in which styles of reasoning tend to be most explicitly investigated, the notion of a style of reasoning has been especially popularized by the philosopher Ian Hacking. For Hacking, styles of reasoning stabilize the "objectivity" of novel scientific phenomena, such as emergent objects or types of evidence.[59] Drawing from the historian of science A. C. Crombie, Hacking proposed that such styles could be divided into relatively strict taxonomies, including postulational reasoning, experimental reasoning, and statistical reasoning, among other forms.[60] In that sense, however, Hacking's account of styles of reasoning remains somewhat constrained to a very specific kind of cognitive and epistemic practice, one that is for the most part intentional, deliberate, and institutionally circumscribed.

Does a style of reasoning need to be the expression of explicitly formalized and deliberate forms of cognition? One important feature of the economic style of psychiatric reasoning is that it was not the effect of an institutional impetus, not an explicit program, and in many cases fell on the side of what Michael Polanyi and Harry Collins have called the "tacit" dimensions of knowledge production.[61] And yet, it is still possible to consider tacit forms of thinking as possessing a sufficient degree of systematicity in order to be acceptably dubbed a "style." There have been efforts to reimagine the concept of styles of reasoning in a broader and more encompassing sense, less anchored to notions of deliberative cognition, something akin to a form of shared conceptualization that organizes and introduces moral and epistemological coherence into the world.[62] These more generous accounts of styles of reasoning draw from analogous concepts in the history and philosophy of science that seek to make sense of the nature and preconditions for the very specific form of intellectual practice called scientific and medical reasoning.

One such concept with which styles of reasoning have been associated is Thomas Kuhn's "paradigm." Itself a slippery notion that has never really been fully or finally explicated (by Kuhn or others), a paradigm for the most part denotes a set of shared practices, theories, and exemplars that are embedded within the sociological conventions—that is, the textbooks, training protocols, and experimental procedures—of a scientific community.[63] There is a strong and arguable sense that at least part of the definition of the Kuhnian paradigm entails communal attitudes as well as collective perceptual and behavioral patterns among scientific practitioners, attitudes and patterns that shape knowledge production but do not always amount to self-reflective and intentional forms of cognition.[64]

Another concept that has been helpful in refining what it might mean to think of styles of reasoning in nondeliberative and nonconscious terms has been the very analogous theory of "thought styles" (*Denkstiles*), introduced by the physician and philosopher of science Ludwig Fleck to refer to the semiautomatic patterns of thinking, which in turn generate a "thought collective" that functions as the communal "carrier" for that style of thought. As Fleck insisted, any individual within a given thought collective "is never, or hardly ever, conscious of the prevailing thought style, which almost always exerts an absolutely compulsive force upon his thinking and with which it is not possible to be at variance."[65] Fleck's notion of thought styles is reminiscent of Émile Durkheim's notion of a "social fact," or the "manners of acting or thinking" that are "capable of exercising a coercive influence on the consciousness of individuals."[66] Thought styles were incitements to reason in one way rather than another, and in that sense, as Fleck explained, they amount to "definite constraints on thought," but only to the extent that they function as the "intellectual preparedness or readiness for one particular way of seeing and acting and no other."[67] Thought styles are not, strictly speaking, forms of premeditated cognition but constraints that shape the affective and behavioral dimensions of knowledge production; they consist of a "mood," a "readiness both for selective feeling and for correspondingly directed action" as well as a "performance," or "readiness for directed perception, with corresponding mental and objective assimilation of what has been so perceived."[68] The perceptual, affective, and behavioral dimensions of thought styles imply that a community of scientific practitioners can adopt internally coherent and systematic techniques of reasoning without necessarily being able to reflectively articulate what they are doing or why they are doing it.

When viewed through this Fleckian register, a style of reasoning can denote an inclination to perceive and think about scientific objects and problems in certain ways over others, an incitement that is more *felt* than it is *known*. A style of reasoning as an affective inclination dovetails well with what the historian of science Lorraine Daston has called a "moral economy," or a "web of affect-saturated values," a "balanced system of emotional forces," and a collective affectivity that gives to science "its sources of inspiration, its choice of subject matter and procedures, its sifting of evidence, and its standards of explanation" (and which itself is rooted in the idea posed by the sociologist Robert K. Merton that "sentiments," as historian of science Steven Shapin has explained, are "the ultimate motive forces responsible for social action," including scientific knowledge production, because they are "socially patterned psychic structures that lie behind, give form to, and animate a more or less coherent body of cultural expressions").[69]

But perhaps the most compelling way that styles of reasoning have been disentangled from the idea of conscious and deliberative forms of cognition has been through Foucault's concept of the "episteme." Foucault described an episteme as "something like a worldview . . . common to all branches of knowledge" specific to any given historical and geopolitical context, which had the particular effect of "[imposing] on each one the same norms and postulates, a general stage of reason, a certain structure of thought that the men of a particular period cannot escape."[70] An episteme instigates a style of reasoning by structuring its content and prompting its felt necessity. By itself, Foucault insisted, an episteme is not a form of knowledge or reason but the background precondition for why specific thoughts, utterance, and judgments make sense and appear true and acceptable at certain historical moments. Epistemes furthermore give rise to the unique rules, or what Foucault called the background "schemata" (*schèmes*) or "preconceptual" conditions, that generate the specificity and apparent divisibility of individual knowledge fields, which while seemingly disparate are nevertheless linked according to the episteme that more essentially unites them.[71]

In the history of science, a version of the Foucauldian episteme has been appropriated and translated into the popular concept of historical epistemology, which, depending on whose rendering of that phrase one considers, can feel more indebted to Foucault's early Kantian proclivities than to other traditions of French history and philosophy of science.[72] Daston, for instance, defines historical epistemology as an investigation into the "history of categories that structure our thought, pattern our arguments and proofs, and certify our standards of explanation." Unlike the history of ideas, which seeks to identify and contextualize concepts and practices, historical epistemology seeks instead to illuminate the conditions that underlay and legitimated those concepts and practices in the first place. Historical epistemology "transcends the history of ideas," Daston writes, "by asking the Kantian question about the preconditions that make thinking this or that idea possible." It is "an inquiry into the very general structures in which human beings think" or, to put it another way, an inquiry into the conditions for the emergence of specific *styles of thought and reasoning*.[73]

We can begin to see, then, how and why a style of reasoning is not necessarily a form of deliberative cognition but, rather, the systematic effect of incitements among a community of experts to think and speak in certain ways about certain problems. Styles of reasoning, then, can be understood as latent epistemic patterns that are not always clearly announced by the users of those styles. To uncover and study them, we must perform a kind of excavational labor that we can call historical epistemology, or what Foucault called "archaeology." This specific interpretation of styles of reasoning, with respect to the history of psychiatry, has taken center stage in the work of the philosopher Arnold Davidson. A psychiatric style of reasoning designates the medical judgments and statements about mental health that avail themselves of some degree of positivity—that is, they can be claimed to be either true or false and not, therefore, an expression of nonsense or intellectual obsolescence.[74] The idea that madness is the result of demonic possession, for example, is not within the modern positivity of psychiatric thought; it is not part of psychiatry's modern style of reasoning. On the other hand, the idea that mental illnesses arise in part because of a synaptic misfiring or a chemical imbalance is within the positivity of modern psychiatric thought. Whether mental illness genuinely *is* the effect of either of these purported sources is not the issue; at stake is simply the fact that modern psychiatry possesses the tools, techniques, and styles of thought to make determinations about their truth and falsity in a way that it cannot with respect to spirit possession. However, as the historian of psychiatry John Forrester has noted, "There can be different styles of reasoning operative in different disciplines for different purposes."[75] Modern psychiatric styles of reasoning, and the historical positivity that they presuppose, are not always translatable into other scientific or medical fields. Forrester's paradigmatic example is "reasoning in cases." Although medicine has always relied upon case studies, case-based reasoning was the exemplary form of reasoning for psychoanalysis in particular, and psychiatry more broadly.

Accordingly, when I claim that an economic style of reasoning was specific to turn-of-the-century psychiatric thought, I mean that it was a latent epistemic pattern of thinking shared by a community of experts and fully ensconced within the positivity of modern psychiatry. The economic style refers to a set of discursive and epistemic explanatory norms that were used to make sense of and to adjudicate mental pathologies by virtue of economically informed interpretative standards. Economically oriented judgments about behaviors and states of mind structured part of what it was possible to say about mental health and illness, and as such, economic reasoning possessed a strong veridical force; it could in many instances credibly reveal the truth about madness. The economic style of psychiatric reasoning, however, did not amount to a deliberate or intentional form of intellectual calculation. It is not at all clear that psychiatric clinicians would have admitted or even realized that they were supplementing their diagnostic judgments with an economically inflected hermeneutic. But, as we have seen, it is quite possible to conclude that styles of reasoning possess degrees of latency, that they take place at different enunciative levels; sometimes they are overt, but very often they remain embedded and even concealed within formal diagnostic assessments and are therefore detectible only at a distance, through what historian Camille Robcis—invoking cultural theorist Walter Benjamin—has called a "constellational" approach to psychiatric thought, that is, an interpretative viewpoint through which "unexpected relationships between characters, context, and ideas—often seemingly fragmentary or tangential—can emerge."[76] This does not imply that an economic style necessarily represented psychiatry's "unconscious" or ideological underbelly (an interpretation that I do not foreclose), but simply that psychiatric discourse should be understood as occupying multiple levels of meaning, with economic reasoning being among the more latent measures through which madness was conceptualized.[77]

The economic style of psychiatric reasoning was, furthermore, neither unified nor monolithic but had several permutations, and the chapters of this book highlight its vicissitudes by describing some (certainly not all) of the most prominent and palpable iterations of psychiatry's economic style of thought. It is important here to note that reasoning styles typically cut across institutional and national boundaries; while there are styles undoubtedly affixed to specific sites, such as nations, cities, hospitals, schools, and laboratories, this book seeks to extrapolate an economic style in the broadest and most general sense. And so, for that reason, psychiatry will be treated here as a *system of thought*, a conceptual phenomenon, a set of possible utterances and ways of thinking—and therefore not solely as an institution or profession. This book's objectives are, in that regard, limited to an initial exhumation of psychiatry's economic style of reasoning, an effort to confirm its existence in the first place, and to offer an initial division of some of its most prominent iterations, which, as we will see, enjoyed a degree of movement across borders and transatlantic settings.

As I discuss at length below and in the first two chapters, the psychiatric turn to an economic style emerged as a reaction to a problem—namely, a hermeneutic crisis internal to psychiatry with respect to the discernibility of mental illnesses. What this means is that psychiatry's economic style was not the effect of an explicit movement of ideas from the field of economics to the field of psychological medicine, and on that account this book does not attempt to reproduce a story about how ideas circulated through and across institutional settings, though aspects of that kind of methodological approach are made use of when appropriate.[78] The sociologist Elizabeth Popp Berman has exemplified such an approach in her analysis of how an "economic style of reasoning" was adopted by US legislators after the 1970s to address a broad range of policy decisions according to the metric of "efficiency." It was a style, Berman writes, that emerged in the disciplinary context of economic departments before eventually proliferating in the US political establishment, where it was institutionalized and even naturalized as a policy norm.[79] This kind of historical sociology of knowledge, however, is less suitable for this book for several reasons: the economic style of psychiatric reasoning did not emerge from a cross-pollination of economic ideas into psychiatric institutions; the economic style was expansive and internationally diffuse, exceeding any one institutional context; and economic reasoning seemed to operate tacitly (we could say "ideologically") rather than as a formal diagnostic doctrine.

It is precisely the latency of psychiatry's economic style, that it seems to operate at the level of surreptitious notional norms and intellectual dispositions—configurations of thought and mental attitudes whose conditions are not necessarily set and foreclosed by institutional or national boundaries, interests, and pressures—that signals why an investigation into the circulation of ideas across epistemic contexts is not entirely apposite with respect to the aims of this book. To put it another way, if styles of reasoning are not specific to the standard contexts by which we typically historicize ideas—nations, institutions, disciplines, societies, publics, and so on—then we may be forced to admit that psychiatry's economic style of reasoning may not be reducible to a context in the conventional sense of that term, as what the historian Peter Gordon has dubbed an "ideal of containment." Gordon has exhorted historical researchers to adopt a more "qualified allegiance" to contextualism, especially if efforts to contextualize ultimately hinder the "exercise of critique."[80] Psychiatry's economic style of reasoning is a good example of an otherwise historically situated conceptual phenomenon that is not reducible to strict context-specificity (save in the broadest sense, to psychiatry as a *system of thought*) but slips beyond the bounds of contextualization in the narrow sense.

This is not at all to say that the economic style of psychiatric reasoning is without context, or that styles are historically autonomous phenomena; in fact, I recount many different institutional and national contexts in subsequent chapters. It simply means that the "context" of a style of reasoning may not have the sort of sociological familiarity—organized around specific events, periods, and settings—that we are accustomed to encountering in historical analyses, and that we may need to consider broadening our understanding of "context" to include other sorts of "preconditions" that make thinking and reasoning in certain ways possible. To that end, this book is not just a history—whether it be a sociological history of the movement of ideas or an intellectual history of the contextualization of concepts—but a history *and philosophy* of psychiatry, in the sense that I approach psychiatry as a historically situated system of thought, from which I seek to extract a pattern of discursive and epistemic possibilities.

Methods, Objectives, and Orientations

The story that this book tells commences in a period that the Marxist historian Eric Hobsbawm dubbed the "age of capital."[81] The subject matter of this book, however, does not entirely correspond to a straightforward history of the political economy of psychiatry, and it would be worth briefly explaining why that is. At stake in this book is an effort to uncover how psychiatric thought was shot through with an economic rationality and, to put it in the strongest way possible, that the very *thinkability* of madness was tinged with economic valuations in ways that we have not sufficiently appreciated. This is a different—though not at all antithetical—form of analysis than an investigation into psychiatry as a profession, therapeutic institution, and industry organized around distinct markets, commodities, and labor practices. That being said, a critical history of the political economy of psychiatry may very well presuppose that practices of psychiatric care can be severed from the vicious capitalist logics on which they have been historically structured, in order to be resituated in more equitable "moral economies," in E. P. Thompson's sense.[82] This, then, presumes that medical norms and economic logics are only contingently linked at the level of the institution or the market, and that since psychiatry has been progressively financialized over the course of the twentieth century, it can in theory be critically de-financialized as well.

But what this book ultimately suggests is that modern psychiatric concepts may be contaminated with an economic rationality at their very core, irrespective of the markets and industries that surround the institutions of psychiatric care. And for that reason, freeing psychiatry from its capitalist imperatives may not actually purge psychiatry of the economic logic that underlies and animates many of its concepts. For if madness is already an economic form, as I will argue throughout this book it has become, then modifying the economic structures of psychiatric institutions may not necessarily alter that fact. The story of the economization of madness that I tell here is not about how an institution was co-opted by market imperatives, but, instead, how a medical concept attained a new veridical status only by being recast in economic terms, that economic value came to comprise part of the very ontology of madness. These two stories are not opposed and are indeed profoundly imbricated; the version of the story that this book presents differs only with respect to what it seeks to emphasize: that the difference between moral, medical, and economic value has been altogether flattened to a degree of which we are, even now, not altogether aware.

Another way of putting this is to say that this book tells a *philosophical* story about the *history* of psychiatric reasoning. It is a story that begins with a conceptual impasse that emerged within psychiatric thought, and one that was resolved through the deployments of an economic style of reasoning. But these deployments resulted in their own repercussions to the extent that madness, rather than mental health, was consequently infused with an assortment of economic ideals. Forms of monetary value were in different ways attributed to irrational states of mind and behavior, a revaluation that subsequently ascribed moral worth to pathological states, a phenomenon I describe as the emergence of pathological value. This philosophical story, furthermore, espouses a very specific understanding of madness, one that draws upon the work of Emily Martin. For Martin, the *reality* of mental pathologies is defined by the sociocultural meanings that stabilize them and give them their sense of aberrancy and irrationality. The status of madness is wrapped up with what Martin calls the "performative" dimensions of behavior, an approach she derives from the philosopher Judith Butler (which in turn was drawn from the work of J. L. Austin and Jacques Derrida).[83]

A "performative" theory of mental illness reveals the dynamic nature of the social hermeneutics that frame and actualize the meaningfulness of mental illness. It is rarely simply the case that a person either is or is not mentally ill in some definitive or unambiguous sense. Such a dichotomous view does not do justice to the interpretative complexities and impasses that often accompany the question of mental and behavioral normalcy. As Martin explains, my social performances are the means through which others come to recognize the status of my health or unhealth; but performances also enable those "cast into the category of the irrational to comment *on* their putative 'irrationality.' In so doing, they demonstrate that they *are* rational."[84] Mental illness, then, while a feature of *interiority*, an expression of the disease of mind and brain, is fundamentally reliant on *external* expressions in order to become meaningful. It is precisely in the context of social performances that the presence of mental pathologies seems to matter most acutely. This is partially why, as Elizabeth Lunbeck has shown, psychiatrists by the start of the twentieth century understood their remit as extending into the hygienic management of "life's normal, routine aspects."[85] Andrew Scull has similarly observed that psychiatrists "negotiate reality on behalf of the rest of society," that psychiatry "is preeminently a moral enterprise, involved in the creation and application of social meanings to particular segments of everyday life."[86]

Social performances, then, are not only the primary means by which a person is rendered legible as either healthy or pathological; performances also provide an opportunity for people to stylize themselves in relation to the loose boundary that separates the normal from the pathological. As Martin explains, "Situations where people labeled 'irrational' perform 'rationally' throw into light both the arbitrariness of the categories rational and irrational and the fuzziness of the line between them."[87] In this book, I adopt a similarly performative view of madness, that is, an understanding of madness as a matter of social performances. This is not to suggest that madness is not a mental disorder or brain disease, but simply that madness becomes a matter of concern when it is adjudicated at the level of the social. One of the most crucial claims of this book is that while some behaviors ignite suspicions of mental illness, other behaviors (particularly of the economic sort) have the power to mitigate and counteract those suspicions, or even transform an alleged mental pathology from a liability into a redemptive advantage, that is, a socially valuable style of self-presentation. Such an analytic possibility can only arise when we adopt a performative view of madness as a matter of social concern.

But, along those lines, another core assertion that this book makes is that not all seemingly irrational or morbid performances can be offset or redeemed by the pathological value that they are regarded as possessing or generating. Pathological value is a form of privilege, a performative entitlement largely organized around a very specific set of gender and racial boundaries. As I describe in detail, the emergence of pathological value across the North Atlantic hinged on the modern categories of whiteness and masculinity. Beginning in chapter 3 and drawing on the legal and critical race theorist Cheryl Harris's work on the underlying economic nature of whiteness as a form of property, I propose that through white masculine encodements, certain pathological states were able to circulate throughout the social body as acceptable and even idealized expressions of creative irrationality, lauded precisely for their apparent financial lucrativeness. This conversely meant that even the most virulent attributes of masculine whiteness could be recoded as socially beneficial attributes and thereby vaunted according to the remunerative pathological value they were believed to possess.

Masculinity and whiteness were the privileged affordances that enabled some performances of irrationality to be regarded as integral and profitable instruments for financial enterprise; but in the absence of those affordances, that is among Black and other nonwhite economic actors who did not possess the entitlements to pathological value, madness became little more than a financial liability or the effect of an economic traumatism. The discussions of whiteness in particular with respect to pathological value, which are especially prominent in chapters 3 and 6, draw on the work of feminist theorist Sara Ahmed and her analysis of whiteness as a form of experience, an effort to understand precisely how "whiteness is 'real,' material and lived."[88] The experience of whiteness, its lived reality, is not only economically overdetermined—as cultural theorist Stuart Hall has asserted, race is "the modality in which class is 'lived,' the medium through which class relations are experienced"[89]—but it is also, as I propose, especially punctuated by the privilege of pathological value. It would also be worth adding that the concept of pathological value, as a markedly racialized privilege specific to the North Atlantic, potentially introduces a novel analytic perspective with which to reimagine the logics of colonial medicine as well (a topic admittedly beyond the remit of this book). There is a tendency to view the difference between colonizer and colonized, or the very operation of colonization, as an effort to establish a dichotomy between political idealizations of health and unhealth.[90] As the psychiatrist and decolonial political theorist Frantz Fanon has so notably emphasized, pathologization has been one of the most integral instruments of colonial violence that the colonized subject "will become abnormal on the slightest contact with the white world."[91] But what this book suggests is that the financial valorizations of pathology introduce a slightly different sort of dichotomy, not between health and illness, but between *two different styles of pathology*—one that is regarded as valued and lucrative, and another that is perceived as pernicious and costly. Broadly stated, pathological value is a concept through which the history of madness converges with the history of racial capitalism, across the North Atlantic and beyond.

Outline of the Book

The chapters of this book are organized in a cumulative fashion. The first two chapters lay out the explanatory groundwork for why psychiatry would come to deploy an economic style of reasoning in the first place—what the emergent problems were that demanded a turn to economic rationalizations of mental pathologies. And the chapters that follow showcase key implementations of psychiatry's economic style of thought.

Chapter 1, "Boundaries of the Legible" introduces what I describe as a pivotal hermeneutic crisis at the heart of psychiatric diagnosis. The chapter argues that over the course of the nineteenth century, psychiatry inaugurated a medical typology—what I call the borderland typology—that was defined in large part by the fact that borderland figures, the eccentric being among the most prominent, resisted the interpretative expectations of medical clinicians as well as lay observers. Borderland figures were neither discernibly healthy nor insane, and consequently, while they may have been a necessary professional artifact (for if mental pathology were self-evident, psychiatric diagnosis would be redundant), they nevertheless troubled a core social imperative: that members of a social body must be legible to one another, that motives and states of mind must be decipherable, that hermeneutic accessibility was a necessary precondition for a coherent social order. By examining the history of the conceptualization of the borderland—the medical and psychophysiological architecture through which the typology was fashioned—this chapter argues that with respect to such figures, the operative clinical question was not always "Are these people insane?" but, rather, "Am I able, in the first instance, to unerringly interpret the conduct of others?"

In chapter 2, "What Conduct Reveals," I consider how psychiatry discovered a potential solution to the hermeneutic problem described in chapter 1. The solution lay in a clinical shift, away from trying to ascertain the medical status of the *mind* and turning instead toward efforts to evaluate the propriety of *conduct* and asking simply whether behaviors were appropriate with respect to the social contexts in which they were performed. The turn to conduct represented a particular style of reasoning about madness, one that could bypass diagnostic judgments about mental truth by appraising and valuating the merit of external performances. For some notable psychiatrists, the quality of certain social performances could in many cases mitigate the presence of mental abnormalities. Madness, in other words, could be recast as a pathology of conduct alone. As the chapter goes on to explain, economic behaviors became one of the most salient social benchmarks for assessing the propriety and indeed the *value* of any given social performance, for unlike moral or medical valuations, which were not always clear-cut and which could not always bypass the diagnostic indecipherability of the borderland typology, economic conduct was ultimately reducible to an arithmetic measure of worth: Does the behavior in question result in a gain or loss of monetary value? This chapter sets the stage for what follows: a series of in-depth analyses of specific iterations of the economic style of psychiatric reasoning.

Chapter 3, "From Disorders of Enterprise to Entrepreneurial Madness," explores how business enterprise and financial conduct in general, including commerce, speculative ventures, and wealth management, were imagined alongside diverse psychiatric appraisals to such a degree that they could function as benchmarks for mental health and pathology. The chapter investigates three interwoven psychiatric suppositions with respect to the relationship between business enterprise and madness. The first supposition was the common medical refrain that business and commercial endeavors were often severely pathogenic, that the wear and tear of economic life was enough to drive people mad. The second supposition, which offsets the first to some degree, was that financial enterprises did not necessarily demand a sane or rational economic actor, that as far as some psychiatric clinicians were concerned, the actions of *homo economicus* could be replicated by *homo pathologicus*. And the third supposition, which draws upon and augments the first two, was that a degree of madness was actually crucial for business success. This last supposition leads to an extended discussion of the concept of the entrepreneur and of the purported irrational creativity of entrepreneurial enterprise. Entrepreneurial irrationality, this chapter argues, was as much the expression of a psychiatric legacy as it was economic theory and must be understood as an exemplary form of pathological value—a *lucrative* pathology bound up with strongly gendered and racial inflections.

Chapter 4, "The Aphasic's Will, or Dispensations for the Propertied," turns to a very different intersection: the discourse of inherited wealth and the neuropathology of speech disorders. This chapter examines a prominent but transient concern in the history of the study of aphasia as to whether aphasics could write wills and effectively bequeath their wealth. The chapter highlights the diagnostic tendency among aphasiologists to unfailingly defend the testamentary capacities of aphasics no matter how dire the state of language loss was. The psychiatric tendency to rationalize aphasic testation was underwritten, I argue, by a significant economic imperative: that the inherited accumulation of generational wealth was the prevailing method of amassing capital in late nineteenth-century Europe and the United States. Aphasics were medically perceived as value-generating to the extent that they contributed to one of the most remunerative forms of economic activity at the end of the nineteenth century. As such, the psychiatric vindication of aphasic testation was not a defense of patients' civil rights so much as it was a way to protect generational wealth accumulation.

Chapter 5, "The Pathology of Work," focuses on the relationship between industrial labor, traumatic occupational psychopathologies, and the politics of accident insurance legislation. The chapter highlights a short-lived French disorder, dubbed *sinistrose*, which was intended to characterize industrial workers in France who morbidly believed that they had been injured and could not return to work. *Sinistrose* was classified as an authentic and genuine disease in its own right but one that was not caused by workplace accidents. The disease's true source, or so clinicians averred, was the French workplace accident insurance law of 1898, and economic legislation was regarded as the pathogenic source of an economically inflected psychiatric disease. The chapter explores how *sinistrose* was an exemplary case of the economization of madness, a pathology of work indistinguishable from a sense of financial loss for the state and insurance industry. The economization of *sinistrose* was mirrored, as the chapter illustrates, by a comparable pathologization of economic policies related to workers' compensation laws. In contrast to the previous two chapters, which demonstrate how some pathological states can be regarded as value-generating, this chapter illustrates the inverse, that some pathologies can be regarded as inherently value-forfeiting, that pathological affordances granted to mad enterprisers and propertied aphasics simply did not translate to industrial laborers. While some pathological or otherwise abnormal conditions could at times be regarded as a privilege capable of being (economically) defended and vindicated, this chapter shows how and why other purportedly non-normal conditions could not only be discredited but, on the basis of an economic rationality, exaggeratedly disavowed.

In chapter 6, "Appraising Eccentricity," I return to the figure of the eccentric, the emblematic borderland persona who set the book's entire analysis in motion. This chapter presents a case study of eccentricity, one that exemplifies in the life of a single individual many of the core themes presented throughout the book. The eccentric in question was John Chanler, a turn-of-the-century New York businessman and heir to a vast real-estate fortune who believed himself to be a genuine prophet and medium, in possession of an oracular business acumen. As the chapter shows, Chanler, who renamed himself "Chaloner," harnessed the value of both his inherited wealth and his white racial identity in order to recast what was likely an incipient mental disorder into a purportedly remunerative psychological asset. His behaviors and style of life consisted of performances of pathological privilege that played out, and were repeatedly justified, in economic terms. In this chapter, I describe what it might mean for an individual to adopt and internalize an economic style of psychiatric reasoning, and for such a style to function as a hermeneutic strategy for self-intelligibility. "Chaloner" was a man who recognized himself as psychologically abnormal, but who interpreted and vaunted his abnormality as a lucrative financial asset. Ultimately what this chapter confirms is that psychiatry's economic style of reasoning was not limited to the *institution* of psychiatry alone, that it was not contained to clinical deployments, but that it could operate as a diffuse interpretative grid to be utilized as a framework for subjectivity through which some individuals (a privileged few) could practice, stylize, and experience themselves and their apparent eccentricities. This is part of what it means to consider psychiatry as a *system of thought*—namely, that styles of psychiatric reasoning could be adopted by the lay populace just as readily as by medical experts. Madness, this chapter shows, was quite generally thinkable in economic terms, and the emergence of pathological value amounted to a norm for social performances and social legibility that exceeded the strict remit of psychiatric practice.

The book's conclusion, "The Economic Reason of Madness," is a provocation of sorts, an effort to draw out the ramifications of the preceding analyses. What the conclusion underscores above all is a series of questions that, while drawn from traditions of social and critical theory, are intended to be of interest not only to historians and philosophers of the human sciences but to historians and critics of capitalism as well. If madness is consonant with capitalism, as this book proposes, then what does this mean for critiques of capital that look to its irrationalisms, contradictions, and internal paradoxes as the dialectical source of its undoing? What does the idea of pathological value mean with respect to aspirations for a rational society, given that the social order is not only replete with irrationality but with forms of irrationality that are objects of financial valuation? And lastly, how far into ostensibly noneconomic spheres should we consider structures of capital as having proliferated? In other words, if psychiatric styles of thought encode economic logics, if our modern self-stylizations remain affixed to formulations of pathological value, if the idea of medical and even moral value has been imbued and perhaps overridden with a tacit sense of economic worth, then what facets of the "historical ontology of ourselves" are genuinely free from modern economistic appraisals?[92]

Boundaries of the Legible

Eccentrics on the Borderland

Over the course of the nineteenth century, psychiatrists across Europe and North America found a common threat in the figure of the eccentric. From as early as the 1830s, prominent clinicians in England and the United States detailed how extensively eccentric behaviors and dispositions smacked of incipient madness and how thin the line separating eccentricity from insanity truly was.[1] Eccentrics revealed themselves to be morbid just as soon as they became "complete pests to society," which, it was believed, they often were.[2] Such a medical attitude stood in stark contrast to the veneration of eccentricity by prominent and historically well-known figures, such as John Stuart Mill, for whom the eccentric was the paragon of virtuous individualism, and who, in bucking society's overarching commitment to custom, exhibited "genius, mental vigour, and moral courage." Mill was not alone in viewing eccentricity as either a dispositional mark of eminence or extraordinary unconventionality. Eccentricity was the object of a modest nineteenth-century tradition of curiosity, often in the form of compendiums that exhibited human anomalies and deviations both marvelous and ignoble, differences not merely reducible to pathological classification.[3] Still, even Mill could not deny the perils eccentrics often faced in being deemed mentally unsound and "having their property taken away from them."[4]

Mill's apprehensions concerning the medical and psychiatric bad faith regarding eccentric conduct could not have been more justified. For in the ensuing decades, the proximity of eccentricity and insanity only became clearer in the minds of clinicians, as the "line of demarcation" slowly became increasingly "impossible" to detect.[5] It was soon customary to view eccentric conduct as among the "premonitory symptoms of insanity."[6] German, Austrian, and French psychiatrists espoused similar conclusions as their Anglo-American counterparts. Eccentricity (*Exzentrizität*) was located along "a borderland [*Mittelgebiet*] of disordered states," neither normal nor entirely pathological.[7] The eccentric was believed to possess a hereditarily morbid disposition, to share with the madman a common pathological origin, a deep familial affinity.[8] The transatlantic psychiatric aim with respect to eccentricity was less a task of understanding than it was a project of vilification. For some physicians, the malignancy of the eccentric was evident in any "irrational departure from the conventional usage of society, or the violation of some received canon of social or personal propriety."[9] For others, the eccentric was simply the one who "neglects his relationship to his fellow men and to the society and social position in which he was born."[10]

For the most part, the presumed "threats" eccentrics posed to civil society were simply coded expressions of moral disapprobation with respect to divergent, though not necessarily dangerous conduct. At other times, eccentric divergency was aligned with forms of political radicalism, an emerging medical effort to pathologize left-wing and revolutionary action. Writing several years after the Paris Commune, the Austrian psychiatrist Richard von Krafft-Ebing insisted that revolutionary leaders "fall ill comparatively often," for the simple reason that "it is frequently those with hereditarily problematic natures and eccentric minds [*exzentrische Köpfe*] who are at the forefront of such movements."[11]

Ultimately, by the end of the century, medical experts in Europe and North America were fairly unified in a belief that the English psychiatrist Daniel Hack Tuke summed up in his entry on "eccentricity" for his international *Dictionary of Psychological Medicine*, a vast encyclopedic compendium that, in the spirit of its intended internationalism, employed the expertise of an array of European and American researchers and typically elaborated on the international purchase of most of the dictionary's entries.[12] "We should be disposed to hold," he wrote, "that much of what is understood as eccentricity at the present day . . . is suspicious as indicating mental peculiarity," and "in proportion as mental disorder is scientifically understood, the domain of eccentricity will contract, and that of insanity expand."[13]

What the largely negative medical assessments of eccentricity never explicitly disclosed, however, was why the category posed such a problem for psychological medicine in the first place. On the one hand, the turn to the eccentric was linked to the project of brandishing the disciplinary force of psychiatric power beyond the remit of the asylum or the hospital, toward the hygienic normalization of everyday life.[14] But beneath this institutional endeavor lay a deeper conundrum, one that Tuke himself some years earlier made plain. The eccentric fit the mold of a troubling disposition increasingly observed within modern society. There were, Tuke argued, an inevitable population of individuals who were "always on the *border-land between sanity and insanity*," indeed to such a degree that "their friends are sometimes tempted to wish that they would actually cross the line, and save them from constant harass." Only then, through the stamp of genuine insanity, could their "vagaries" be made sense of and subsequently endured.[15]

Eccentrics and other "borderland" personalities provoked a deep frustration in response to conduct that was neither normative enough to denote health nor sufficiently abnormal to unmistakably signal disease. Whatever problems *recognizably* insane individuals posed to the social order, the interpretability of their morbid condition was not usually one of them. What the eccentric broadly introduced, however, was an altogether different challenge, one that concerned the very legibility of social conduct. For at stake in the medical anxieties over eccentricity were essentially performances that could not always, or at least not easily, be deciphered as something to be either tolerated or disavowed. Eccentricity was not, therefore, a predicament strictly limited to psychiatric medicine; it signaled a potential crisis within the very coherence of the social body.

The sociologist Norbert Elias years ago outlined in the pages of *The Civilizing Process* an innovative historical and sociological hypothesis, now somewhat neglected, concerning the nature of modern social organization, of the dynamics that constitute a "good society" in the modern era—a hypothesis that I take to be integral to understanding the social predicament that eccentricity introduced.[16] In the absence of overt forms of violence and domination that characterized feudal social forms, and as modern civil societies were increasingly concentrated into population centers and saturated with new economic and legal imperatives and incentives, novel forms of social interdependency emerged that were organized around interpersonal norms of reliance, pressure, and predictability. Membership in the modern social body effectively demanded an essential stipulation, Elias suggests—to present behavior that could be intelligible to other social actors, decipherable in such a way as to signal an underlying state of mind, desire, intentionality, and, above all, self-control. The legibility of interpersonal social conduct is a first-order imperative, a sine qua non for the very existence of modern social coherence. Elias provided a sociohistorical narrative for what the philosopher Charles Taylor has in more theoretical terms described as the hermeneutic requirements surrounding social behaviors and performances, that "on the phenomenological level . . . a certain notion of meaning has an essential place in the characterization of human behavior."[17]

In this way, eccentricity posed a social problem different from recognizable insanity, for neither the sane *nor the determinably insane* incited the sorts of interpretative frustrations that eccentrics provoked. Eccentricity introduced the unsettling reality that in the context of modern transatlantic social life, defined by the necessary interdependence of concentrated populations, there remained some whose conduct simply troubled the normative benchmarks typically deployed to determine clear acceptance or exclusion from the social body. The eccentric, then, was merely an exemplary case of a broader typology of troublingly "borderland" individuals, supposedly observed among transatlantic psychiatric practitioners, who were always on the "frontières" of disease, on the "region situated on the border of reason and madness,"[18] an "intermediary zone" of medical ambiguity.[19] Such individuals were not *clearly* insane, even as they exhibited "peculiarities of thought, feeling and character,"[20] a vague likeness to or mere "analogies of insanity" (*die Analogien des Irreseins*),[21] which only ever insinuated but never definitely proved the existence of "latent brain disease in the community."[22] To that end, individuals positioned on the borderland of madness differed from the more familiar psychiatric patients to the extent that they were not confined to asylums or hospitals but were instead scattered and deposited throughout the social body.

The borderland typology was a clinical hypothesis increasingly adopted by nineteenth-century psychiatrists; it was a hypothesis organized around no particularly unifying feature save that the individuals who fit the bill were neither assuredly sound nor unquestionably insane. Eccentricity was simply the most familiar illustration of the type, the easiest example to denounce, to fret over, and to warn against. One noticeable attribute, however, of the borderland typology—what did ultimately give it some degree of consistency as a medical idea—was that its very existence was grounded on a growing assumption among clinicians and medical writers that the boundary separating health from disease, the normal from the pathological, was in the final analysis indeterminable, permeable, and perhaps even nonexistent.[23] "There is probably not one of us entirely free from some more or less insane peculiarity," wrote one physician just at the turn of the century.[24] The supposed indistinction between mental health and pathology, the idea that sanity and insanity were progressively continuous and not starkly divided, was perhaps no more concisely nor emblematically expressed than in the work of Sigmund Freud; in some respects, the entirety of the psychoanalytic project could well be summed by the idea that "we are all a little neurotic [*wenig nervös*]."[25]

At the same time, the borderland typology introduced an element of incongruity into the dynamics that otherwise organized and governed the modern social body. By the end of the nineteenth century, many prominent clinicians and medical writers agreed that one crucial function psychiatric medicine fulfilled was the management and safeguarding of the social order. When the British psychiatrist Henry Maudsley defined insanity in 1895, he described it as the mental and affective "derangement" that incapacitates a person from "doing the duties of the social body in, for, and which he lives." Insanity was—not incidentally, but essentially—"a want of harmony between the individual and his social medium," a mental failing that "prevents him from living and working among his kind in the social organization." The insane were those who were simply "out of tune" with the rest, in a state of persistent "social discord."[26]

In trying to remediate the afflictions that gave rise to social disharmony, in trying to root out even the most imperceptible kernels of mental disease, which medical practitioners routinely declared to be continuously on the rise, psychiatry cast the widest net possible. "Since mental distortion may be of every degree and kind," Maudsley explained, "so slight as to be no more than eccentricity that is consistent with sanity or so great as to go beyond the bounds of reason, it is not possible to draw a distinct line of division between sanity and insanity."[27] But in casting such a wide net for the sake of managing the social body—in order for the psychiatrist to become what Elizabeth Lunbeck has called the preeminent "philosopher of social life"—psychiatry inadvertently introduced a troubling degree of ambiguity into it.[28] For without a separating boundary between health and unhealth, psychiatrists effectively affirmed that an alarming number of individuals conducted themselves in ways that fell outside the boundaries of clear and decisive social legibility, through behaviors and performances that either were not transparent or lent themselves to being interpreted in precisely the wrong ways. While psychiatrists promised that they alone possessed the expert acuity to draw the lines of distinction between normalcy and pathology, in order to make such a claim, they needed to presuppose that the social body was infiltrated with indecipherable forms of conduct, in the guise of eccentrics and other borderland types. And whether psychiatrists *actually could* draw the lines separating the normal from the pathological, and to do so according to the constraints of a psychiatric rationality alone, remains the very question of this book, for as later chapters propose, psychiatrists indeed *could not*—at least not without deploying a supplementary (economic) style of reasoning. Psychiatry inaugurated a foundational irresolvability within the very fabric of the social order, one that it could not reconcile on its own terms.

This chapter, then, explores how the borderland typology was fashioned by clinicians and medical theorists in such a way as to inadvertently frustrate conventional assumptions concerning the transparency and legibility

of social conduct. In order to construct this borderland phenomenon, psychiatrists relied upon a crucial medical concept, the core psychiatric contribution to turn-of-the-century social thought, namely, the unconscious—or,
what should be better called a *version* of the unconscious. For as historians of
psychiatry have for some time documented, there have been many notions of
the unconscious even within the domain of psychological medicine, most
of which have performed to varying degrees a similar tumultuous function:
to invert core Enlightenment assumptions about personhood and subjectivity, to "very effectively [undermine] the bases of the classic representation of
the conscious subject and their voluntary power," and to thwart the idea that
human beings are transparent and fully self-possessed.[29]

At the same time, however, if these disparate notions of the unconscious
in fact introduced a certain degree of upheaval, there is no reason to assume
that such disruptions were strictly limited to the category of the individual
alone. To that end, this chapter considers how, within the psychiatric imaginary, a particular notion of the unconscious subtly instituted an unsettling
dilemma within the social body itself. Before we can explore the different
permutations of an economic style of reasoning that suffused turn-of-the-
century transatlantic psychiatry, we must be able to appreciate the conditions
that made that style of thought necessary, warranted, and thus viable in the
first place; it is the task of this and the next chapter to articulate just those
conditions.

The Concept of Automatism

One important variation on the notion of the unconscious that dominated
turn-of-the-century transatlantic psychiatry was the concept of automatism,
perhaps the single most heavily deployed concept to underwrite the theorization of the borderland typology. Automatism represented a diffuse medical
concept that eventually came to possess a remarkable degree of traction for
clinicians and researchers throughout Europe and the United States. While
not an entirely univocal concept, there were efforts to define it. Tuke himself
penned an entry on "automatism" for his *Dictionary of Psychological Medicine*,
calling it a "state in which a series of actions are performed without cerebral
action or conscious will, as during reverie or in certain morbid conditions."[30]
For Tuke and many other clinicians, automatisms were involuntary and unconscious performances that could often amount to incredibly complex behaviors following the onset of mental disorder.

Automatism was not a disease category in any strict sense. While perhaps
most recognized through experiments with automatic writing and drawing

popular among French, English, and American psychologists and psychical researchers,[31] automatism was a prevalent if only loosely unified explanation of the physiopathology and broad symptomatology of some of the most dominant disorders and disease states of the late nineteenth century, including hysteria, epilepsy, general paresis, and traumatic neurosis. It was a way of describing what in fact happens to patients when they succumb to severe mental illnesses—that one way or another, across numerous disease states, mentally ill patients lose some degree of consciousness and voluntary self-control and surrender to more automatic, and thus more unconscious, forms of behavior.

In the early 1930s, the French psychiatrist Henri Ey somewhat retrospectively identified what made the concept of automatism so striking and expedient for psychiatric theory—and for French psychological medicine in particular, for which automatism played a crucial conceptual and clinical role from as early as the 1880s, in the work particularly of Jean-Martin Charcot, Théodule Ribot, and Pierre Janet, among others.[32] What Ey stressed most about automatism was its "essential ambiguity."[33] Automatisms simultaneously denoted normal physiological processes related to reflex, habit, and routine as well as unconscious pathological upsurges characteristic of mental illness.[34] Indeed, some degree of automaticity persists across the entire spectrum of human conduct, from the most voluntary and healthy actions (since even willful acts depend upon automatically coordinated movements) to the most involuntary and pathological. With automatism, Ey maintained, "the distinction between the normal and the pathological collapses at the same time as the distinction between the voluntary and the automatic."[35] The challenge the concept of automatism most urgently posed arose specifically in the context of mental disorders, for in such states clinicians believed that with the higher mental capacities debilitated, patients surrendered to otherwise automatic states of functioning. At the same time, however, such automatic states often amounted to highly routinized and ingrained performances, the effect of years of behavioral acclimatization, habituation, and attunement to the tacit norms of social conduct. It was therefore one of the disquieting assumptions of late nineteenth-century transatlantic psychiatry that pathologically unconscious expressions could replicate, and easily be mistaken for, entirely normal behaviors such as everyday routines, ordinary customs, and quotidian practices.

A person might, for instance, walk the streets of a city, enter a shop, order food, pay the bill—all of these familiar and customary acts could be the expression of willful deliberation on the one hand or the absent-minded execution of routine on the other; or, as will be discussed further below, such performances could very well be the manifestation of a state of undetectable though nevertheless pathological and unconscious automatism, and an observer would likely never know the difference. If eccentricity was the exemplary case of the borderland typology, then automatism was the conceptual device through which the ambiguity of the borderland was most effectively articulated by medical writers. With respect to the legibility of social performances, eccentricity and automatism shared a striking resemblance. For just as eccentricity was the *appearance of oddity* that could conceal healthy idiosyncrasy just as easily as incipient madness, automatism was the *appearance of normalcy* that could conceal deliberation and habit just as easily as severe mental pathology. What psychiatric thought tacitly purported was that neither the bizarreries nor the regularities of conduct could unambiguously signal the status of an individual's mental health, certainly not in any unequivocal way. The question that eccentricity, automatism, and the borderland typology collectively underscored was not simply the status of health and disease—Are these people insane?—but, rather, a more immediate and pressing worry—Am I able to unerringly interpret the conduct of others?

Somatic Entwinements

A crucial reason why automatism became an especially persuasive doctrine for psychiatric practitioners was the cogency with which it could present a highly technical account of the radical ambiguity of human conduct. It was, furthermore, an account that subtended the competing and factious discourses on subconscious and unconscious personalities and impulses that so dominated both the medical and popular imaginary for much of the early twentieth century. The technicality of automatism was a direct feature of the fact that it was first and foremost a *somatic* theory, which is why it could, as a physiological precept, come eventually to possess a substantial degree of veridical authority. Admittedly, however, many of the core suppositions underlying the concept of automatism had been articulated many times over through the earlier notion of *habit*. By the early nineteenth century, habit was particularly esteemed for its capacity to blur classical dualist boundaries. "Habit erases the line of demarcation between voluntary and involuntary acts, between experiential acquisitions and the operations of instinct," wrote the French ideologue Pierre Maine de Biran.[36] As one writer put it, it was the concept of habit that first signaled "the mysterious border-land between the conscious and unconscious functions."[37]

Where automatism differed from habit, however, was around the question of pathology and the relationship between health and disease. A morbid habit typically denoted little more than the habituation of vice—a way of encoding physiology with moral normativity. With respect to "habitual criminals," the Italian physician Cesare Lombroso would declare, misconduct is reinforced so deeply that "crime becomes an organic phenomenon, flesh of their own flesh."[38] The concept of habit alone could never quite entwine the normal and pathological dimensions of human behavior into a unified physiopathology, free of any overt moralism, like the doctrine of automatism could. It was not, therefore, the legacy of habit that gave to automatism its radical potential but, rather, the theory of neurophysiological reflex.[39] After the 1830s, physiologists began to reform and expand the long-standing theory of the mechanical reflex arc by recognizing the role the brain and "cerebral reflexes" played in the complex sensorimotor dynamics of the human body.[40] Even the highest cerebral and mental functions were understood to be reflexive in some broad way, sharing with peripheral nervous activity a functional uniformity.

This idea was most emphatically elaborated by the midcentury British physiologists Thomas Laycock and, most especially, William Carpenter.[41] Laycock relied on a Romantic theory of organic unity to describe the "reflex function" of the brain, a way of proposing that the highest voluntary and intellectual activities were constitutively buttressed by and entangled with an assortment of "reflex, automatic, unconscious, and instinctive" processes.[42] Despite expressing a more theologically motivated commitment to freedom of the will, Carpenter (whom Laycock greeted as both colleague and professional rival) also agreed that no voluntary action was ever *fully* voluntary through and through, that is, ever free from some degree of automatism.[43] In all voluntary action, the will merely initiates a habituated assemblage of movements that operate automatically once put into effect. "Thus," Carpenter explains, "even what we are accustomed to consider our *voluntary* movements are in their immediate and essential nature *automatic*."[44] When Carpenter affirmed that human beings are simply a "bundle of habits" (a phrase William James would eventually adopt), he did not mean that behavior is conditioned and habituated but, rather, that everyday actions such as walking, writing, and speaking amounted to little more than the initiation of automatic processes that could be carried on in the absence of any further voluntary impetus.[45] (For all the forensic dilemmas this conclusion could have presented, it is striking how absent the discussion of medical jurisprudence actually is in Carpenter's writings, given especially that he taught medical jurisprudence at University College London for nearly a decade.[46]) Both Laycock and Carpenter shared the belief that no voluntary conduct took place without the supplementary cooperation of unconscious action. The line of demarcation separating autonomy from automaticity was, at least in practical terms, almost impossible to pinpoint. Carpenter further accentuated the profound ubiquity of automatism by claiming that mental processes, in addition to physical ones, could be automatic as well. He dubbed this mental automatism "unconscious cerebration" to refer to any involuntary thinking that takes place below the threshold of awareness, such as the proverbial tendency to remember the name or location of a forgotten person or item as soon as the conscious attempt to remember it is suspended.[47]

The most striking feature of these British doctrines of automatism, however, was that they could illustrate how pathological states were not deviations from normal physiology but, rather, conditions that emerged out of the normal. What made automatism such a distinctive doctrine was that it could frame both health and disease as part and parcel of the very same set of somatic operations. Carpenter, for instance, emphasized that the complex array of automatisms "performed by us in our ordinary course of life" did not always, or even often, arise from the strict or explicit prompting of the will.[48] It was possible for our automatic processes to be instigated not from within but from without, to thereby transform us into "mere thinking automata, mere puppets to be pulled by suggesting-strings, capable of being played-upon by everyone who shall have made himself master of our springs of action."[49] The most beneficial nature of our neurophysiology—that is, our capacity to instrumentalize our inherent automaticity, to take advantage of the expediency of routinizing behavior and thereby acting in the absence of the constant oversight of the will—was also the very same condition that could explain how we might fall unconsciously under the influence of another's will. Our automatic functions could be easily commandeered through the power of suggestion just as easily as they could be altogether seized and possessed by mental disease. "Nothing in the psychical phenomena of insanity," Carpenter writes, "distinguishes this condition from states that may be temporarily induced in minds otherwise healthy."[50] Laycock espoused an analogous view of the continuity between normal automatic states, suggestibly induced automatisms, and states of dreaming and insanity.[51] For both Laycock and Carpenter, the doctrine of automatism could demonstrate how the physiology of the normal could simultaneously function as the groundwork for the pathological.

The British doctrines of automatism gained increasing international traction and appeal after the 1860s (often in strong association with other scientific exports such as Darwinian evolution). In France, an altogether separate theory of automatism, independently developed by the psychiatrist Jules Baillarger, was revised by the neurologist Jules Luys in light of the work of both Laycock and Carpenter.[52] Like his British counterparts, Luys also proposed that the brain possessed a "spontaneous automatism" (*automatisme spontané*) that represented its normal physiology but, at the same time, acted as "the singular source of all [mental] disorders."[53] During states of health, ordinary mental associations, involuntary emotional reactions, and even daydreams signaled the brain's normal automaticity; but if those excitations were too intense, the brain would have crossed the threshold into a pathological state. Mental disease simply signaled the aggravated exacerbation of the otherwise normal automatisms of the brain.[54] Luys helped popularize Carpenter's notion of "unconscious cerebration" (*cérébration inconsciente*), a principle he asserted best expressed the "rational physiopathology of madness,"[55] and which the philosopher and psychologist Théodule Ribot would eventually come to embrace and promulgate from a position of robust institutional authority.[56] By 1880, a prominent French physician could rightly proclaim, "In recent years, there has been much discussion in the scholarly world [England, Germany, France] concerning automatic and unconscious acts."[57]

Essential Ambiguities

The doctrine of automatism, as it was conceptually developed throughout the nineteenth century, eventually served to accentuate the ambiguities that underlay the supposed boundaries and divisions that long characterized the classical view of human nature. The differences, for example, between volition and involuntary impulses, conscious and nonconscious processes, and even the mind and body were collapsed to the extent that automatism designated the continuity, not the opposition, inherent in those dichotomies. But with particular respect to health and disease, automatism did more than simply denote a porous indistinction. In its most radical articulations, the concept of automatism could reveal that the normal and pathological were grounded on the very same set of physiological principles, such that whatever differences were said to exist between them were ultimately due to the arbitrary imposition of social meaning. If there was one place where the doctrine of automatism was taken to such a conceptual extreme, it was in the writings of the British neurologist John Hughlings Jackson, often hailed as one of the most formative though least universally recognized thinkers of the pathology of mind and brain.

While his professional career at the National Hospital in Queen Square, in central London, as well as at the London Hospital, ran almost parallel to that of his better-known Parisian colleague, Charcot, Jackson's role in the history of the concept of automatism is crucial, though often underappreciated.[58] He was a nexus for the development and international transmission of key concepts in the history of neurology and psychiatry, particularly the most dominant articulations of the notion of the unconscious by the end of the nineteenth century. Jackson cannily and creatively synthesized a patchwork of otherwise disparate writings from both lesser-known physicians and highly venerated physiologists, including Claude Bernard and Laycock, whose writings Jackson prized as an "a priori" clinical expectation.[59] Jackson was considered by his peers and protégés to be as much a philosopher as a physician,[60] not least because of his admiration for the writings of the English polymath Herbert Spencer, whose idiosyncratic theory of nervous "evolution" and "dissolution" Jackson extensively appropriated, though not without modifying and even radicalizing it well beyond anything Spencer might have imagined.[61] Jackson's intellectual influence is even more noteworthy still; Freud and Ribot can be counted as among his eventual adherents, and Jackson's name became a common referent for early twentieth-century neuropsychiatry across Europe and North America (in France, it was called *neo-jacksonisme*).[62] With respect to the history of the concept of automatism—and, for that matter, the unconscious—nearly all roads run through Jackson, at least to some degree.

Jackson fashioned what might best be described as a veritable metapsychology, a term typically reserved for Freud's psychoanalytic writings. In Jackson's case, however, this meant a theoretical and systematic physiopathology of both mind and brain, a conceptual architecture for making sense of normal mental and nervous processes and the deviations therefrom—a *meta-neuro-psychology*, it could be more accurately put, following Jackson himself, who, while often speaking in strictly neurological terms, consistently posited that nervous and mental processes were formally and operationally parallel, declaring at one point, "I shall sometimes, for brevity, use the term Neuropsychology."[63] The proposed architecture of the mind and brain was "evolutionary," in that it was hierarchically organized from "lower" to "higher" states of organizational complexity.[64] "Lower" held numerous meanings for physicians and medical theorists throughout the nineteenth century; for some, like Lombroso, it was defined as a racialized civilizational atavism.[65] Jackson, however, tended to avoid racialized or civilizational analogies. Unlike many of his own British colleagues, Jackson even opposed the theory of the heritability of nervous disease, suggesting that the only thing that could be inherited was a possible disposition to "tissue-change."[66] Adhering instead to an almost off-putting degree of technicality in his writings, "lower" referred simply to the reflex actions that were situated toward the vegetative end of neurological activity, and so to those that were most "automatic" in nature.

The language of evolutionary hierarchies, along with the progression from "lower" to "higher" was a framework that Jackson drew from Spencer, who first formalized his dual, and distinctly non-Darwinian, notions of "evolution" and "dissolution" between 1862 and 1867.[67] Jackson translated this into a structural abstraction of the functions, or operations, of the mind and brain—a functionality that must be understood in terms of an imagined hierarchical configuration that progressed from older, simpler, more engrained, and more automatic nervous and mental processes to newer, more vulnerable, and more voluntary ones. Lower functions were integrated into the higher centers in such a way that the highest functions of the brain were simply more aggregated, more coordinated, and therefore more radically sophisticated articulations of lower nervous processes.[68] The highest neurological and mental functions amounted to the re-embodiment and re-enactment—Jackson called it the "representation," "re-representation," and "re-re-representation"—of the lowest organic sensorimotor processes but at a much greater level of organizational complexity and abstraction.[69] As Jackson somewhat obliquely put it, "A higher level *is* its lower 'over again' raised to a higher power."[70]

Like his physiological predecessors, Jackson also underscored the unbroken continuity linking the "highest" voluntary functions to the "lowest" and most automatic processes, and in the strongest possible terms.[71] Jackson often liked to insist that neurological disorders of any sort were analogous to concentrated "experiments" performed naturally upon the nervous system, thereby revealing its inner workings. What such "experiments" revealed above all was that "even the will is not a separate thing, and that the bulk of the most elaborate actions are automatic, the 'will' being then their leader."[72] The highest intellectual and voluntary processes—the mind and the will, in other words—were simply more refined and complex re-enactments of the body's automatic physiology, raised to a position of organizational supervision; they were not sovereign powers standing apart and beyond those automatic economies.

When patients suffer from mental illnesses or sustain neurological injuries, they experience what Jackson calls a "nervous dissolution," a composite phenomenon made up of two opposing processes, or a "duplex condition."[73] In the first instance, comprising the "negative" side of this double process was the loss of function of the highest nervous and mental processes, which were "temporarily [or permanently] put *hors de combat*," either endogenously, by virtue of disease, or exogenously, as in the case of brain trauma.[74] In states of health, these highest centers were not only "representing" lower automatic activity but *inhibiting* that activity as well. As lower processes were re-embodied into more coordinated and elaborate schemas, a degree of inhibition was normally necessary in order to prevent lower centers from individually breaking out of the new coordinations into which they were being organized.[75] But when higher, inhibiting centers were suspended, these lower centers became disinhibited. The "negative" loss of the higher functions was therefore met by, and "constructively accountable for," the "positive" side of mental pathology— namely, the exhibition and "conservation of all movements more automatic."[76] Mental disorders therefore amounted to a dynamic "reduction" to a now-disinhibited automatic state, a condition dubbed "mental automatism." The entirety of all behavioral, emotional, and intellectual anomalies associated with insanity or brain disease "have one common character," Jackson insisted in one of his earliest and most internationally influential essays. "They are automatic; they are done unconsciously, and the agent is irresponsible. Hence I use the term mental automatism" as a synonym for both insanity and brain disease. The degree or "depth" of the reduction was directly proportionate to the subsequent automaticity. If the reduction was slight, "there is but a slight departure from healthy action," and the patient "will simply do odd things afterwards." But if the reduction was deeper, "his more animal disposition comes out; he may rave and act furiously."[77]

It was here that Jackson was led to make some of his most radical pronouncements concerning the phenomenology of mental disease. For as a mentally ill patient was "reduced to a more automatic condition," the automatism itself, Jackson proposed, underwent a functional augmentation, a "rise in activity" of its own.[78] While nervous dissolutions suspended the activity of the highest neural and mental centers, they did not suspend the process of evolution, or the general increase in organizational complexity. The "mental automatism" of the ill patient—what Jackson described as "the mentation remaining possible to him—his 'pseudo-knowings'"—nevertheless signaled the continued "evolution going on in the lower intact ranges of his highest centers."[79] Even after the onset of pathology, newly disinhibited automatisms were now capable of undergoing their own evolutionary development.[80] This is what accounted for the fact that "elaborate" and "highly compound actions, can be done automatically—i.e., unconsciously."[81] Given that automatisms could undergo an evolutionary development of their own, Jackson's decisive, and often overlooked point, was that mental automatisms were not abnormal processes in and of themselves. "These lower layers are perfectly healthy," Jackson insisted, "although over-acting; that is, they are untouched by any pathological process." The symptoms of insanity were simply "the outcome of over-activity of comparatively slight discharges of perfectly healthy nervous arrangements of the lower layers of the highest level" of the nervous system.[82]

However odd it might sound, the peculiar behaviors characteristic of mental illnesses—from the slightest disorders to the gravest forms of insanity—did not directly signify disease or injury as such; by suspending the highest centers from functioning, disease and injury amount simply to a *loss*, and therefore do not manifest a symptom in any strict sense. Instead, the symptoms associated with insanity, Jackson contended, "signify evolution going on on a lower range than normal of those centers."[83] Pathological symptoms were the expression of lower, though otherwise *perfectly healthy*, entirely normal, and still evolving neural activity. It was the *exposure* of the automatism that was abnormal, not the automatism itself, which otherwise comprised a completely healthy process, entirely "untouched by the pathological processes" that actually exposed it.[84] Even the oddest mental disturbances, Jackson maintained, were still "perfectly normal in the patient, as certainly normal in him as the mentation of the sane is normal in them."[85] The pathological reduction to a state of automatism—the characteristic phenomenon of madness—was simply the manifestation of a hitherto unexposed state of normality. "What we call the insane man's extravagant conduct displays his will," Jackson wrote. "What we call his illusions are his perceptions (memory); what we call his delusions are his beliefs (his reasoning); and what we call his caprice is evidence of his emotional change."[86] To the extent that no degree of mental pathology ever divorces a patient from some state of neurological health, unambiguous determinations of insanity become difficult to achieve, Jackson warned, and they often risk enforcing an overly arbitrary, socially imposed, and potentially artificial standard of well-being.[87]

Jackson would eventually observe that the "double condition" that characterized the dynamism of nervous dissolution—the loss of highest functioning and the upsurge of a lower state—was actually a process that accompanied ordinary and healthy mental life.[88] The mind normally possessed a degree of productive incongruity, a normal "stereoscopy," "mental diplopia," or "double consciousness." The double condition of everyday mental life referred to the ordinary interruptions and superimpositions of ideational associations and memories that flowed into present conscious experience. This "'play' of the mind," as Jackson called it, represented a typical and completely normal and healthy dynamism in which present consciousness would momentarily recede as former and "lower" consciousnesses would "crop-up."[89] Although Jackson used the term "mental automatism" to describe pathological states, we can see that in the final analysis automatism was as much the expression of mental illness as it was of mental health, in the strongest possible sense, since everyday mental activity was comprised by the momentary though recurrent recession of consciousness and the cropping up, or "play," of mental associations, unconscious thoughts, and memories.

It was precisely because "the process of all thought is double"—that is, comprised of a cyclical waning of consciousness and waxing of unconscious states—that incongruous thinking, that is to say, paradoxical reasoning and creative insight, were possible in the first place.[90] To be reduced to a state of automatism could just as easily signal madness as it could the ordinary "play of the mind," and even heightened forms of creativity and intellectual inventiveness. But what automatisms could never *solely* signal was pathology exclusively, for even the most furious ravings of a patient reduced to the severest automatic state still evinced an underlying form of physiological health. In Jackson's pivotal writings we see perhaps the most pronounced realization of what Ey (himself an avowed disciple of Jackson[91]) called the "essential ambiguity" of automatism—a manner of conduct that simultaneously embodied the indistinguishable intersection of mental health, mental disease, and even mental virtuosity.

Unconscious Agents (of Commerce)

By the end of the nineteenth century, the concept of automatism had acquired more than a moderate degree of clinical and theoretical purchase among numerous prominent international thinkers and practitioners of psychological medicine for making sense of the physiopathology and symptomatology of major mental disorders. The concept admittedly introduced something of a moral dilemma to classical conceptions of volition, since taking the ramifications of automatism seriously meant abandoning at least to some degree voluntarist assumptions about human rationality and moral self-governance.[92] But automatism posed more than just a moral challenge to classical conceptions of individualism, for embedded in dominant accounts of the doctrine was the recognition that the gamut of everyday conduct was not always, or at least not fully, one's own, and that automatic processes were so ever-present that they could commandeer the actions of *both the healthy and the ill* alike, in ways that would remain socially imperceptible. Automatism troubled not only the moral norms of self-possessed individuality but also the *social* norms of interpersonal community, the supposition that the truth of another's state of mind could be transparently deciphered on the basis of behaviors and outward comportments.

A young Gertrude Stein, while a student of the philosopher William James and psychologist Hugo Münsterberg at Radcliffe in the mid-1890s, copublished a paper on experimental research she conducted with a fellow student on the normal automatisms of everyday life. It was quite common for perfectly healthy individuals, she explained, "to *act*, without any express desire or conscious volition," insofar as much of our daily conduct, especially those "acts ordinarily called intelligent, can go on quite automatically in ordinary people."[93] The most seemingly purposive and rational performances were often, and often undetectably, *automatic* in nature. Two years later, Stein, at that point a medical student at Johns Hopkins, published a subsequent report in which she noticed that, on some occasions, normal automatisms can be so strong as to amount to an altogether "automatic personality" that "in the ordinary affairs of life . . . obtrudes itself, giving a sense of doubleness, of otherness."[94] Such an "automatic personality," by no means strictly pathological, nevertheless bore a striking affinity, Stein confessed, to one of the most common symptoms of hysterics, what the French psychiatrist Pierre Janet had a decade earlier dubbed their "unconscious personality" (*le personnalité inconsciente*),[95] but which the British psychical researcher Frederic W. H. Myers more generally and, like Stein, less morbidly characterized as ordinary subliminal forces that often "make an irruption into normal life as though for some definite purpose, to fulfill the wishes of some subconscious personality."[96]

Stein's point, in other words, was not simply that the bulk of everyday behavior, even the most seemingly purposive and rational, could be automatic in nature; the more unsettling conclusion was that our most ordinary automatisms displayed a strong formal affinity with pathological conditions, a conclusion similar to the one that Jackson had also reached. If normal automatic behaviors reproduced both the habituated routines of everyday social life and custom and more rarefied forms of intelligent conduct, there was no reason to believe that *pathological* automatisms could not replicate those very same performances—what Jackson called "elaborate" and "highly compound" unconscious acts—with at least enough fidelity to deceive the ordinary observer.

There were few things more clinically astonishing, Jules Luys admitted, than to observe the behaviors of patients in states of elaborate automatisms or what he called "lucid somnambulism," that is, individuals "who speak, answer questions correctly, who write, and give their signature, who, in a word, accomplish a series of apparently conscious acts and who have no idea of what they do!"[97] In what became an internationally known study, the Parisian physician Ernest Mesnet presented the case of a twenty-seven-year-old colonial army sergeant, F., who suffered from a bullet wound in the parietal lobe, an injury that frequently provoked states of automatic behavior. In such conditions, F. exhibited "the habits of waking life continued in sleep," moving and acting "with the appearance of a freedom that he did not have." The former sergeant, Mesnet concluded, was little more than "an automaton blindly obedient to the unconscious activity of his brain," a servant to an "unconscious will" (*une volonté inconsciente*).[98] Although Mesnet confessed, along with many of his French and international colleagues, that contemporary medicine had only "rare examples" of such cases of pathological automatism, even their anecdotal presence forced clinicians to consider the possibility that pathological states could engender conduct that would otherwise appear socially acceptable and indistinguishable from deliberative behavior.

Despite Jackson's insistence that mental automatism was just as much an expression of normal mentation as mental illness, there remained a preponderant conviction, particularly among French practitioners, that strong automatic upsurges were only ever the expression of singularly morbid states of mind. Janet perhaps most influentially argued that automatism was only the symptomatic expression of "suggestible people living in a state of psychological misery."[99] Automatisms, Janet wrote, "must exist in the normal man as they do in the patient; but, instead of existing by itself, as it does for the patient, automatic phenomena are masked and exceeded by more complex phenomena in the normal man."[100] (Claims such as this help explain why one of Janet's own medical students would eventually propose that Janet's theories were actually "an application of [John Hughlings] Jackson's general law" of nervous dissolution.[101]) While Janet for the most part believed that emergent unconscious behavior was mostly the expression of hysterical morbidity, he did propose that nonpathological manifestations of automatic behavior might, in the case of normal subjects, arise as a consequence of a very specific inducement—namely, *hypnosis*.[102]

Some of the foremost European practitioners and theorists of hypnosis similarly contended that hypnotic states were expressions of some form of "cerebral automatism,"[103] "reflex automatism,"[104] or "mental automatism"[105]— that hypnosis was a technique capable of raising unconscious and automatic processes above the threshold of conscious awareness and control, even as there remained strong contention as to whether the very inducibility of automatic behavior signified the underlying presence of mental disease.[106] For prominent clinicians like Charcot, the artificial stimulation of automatic behavior was a profoundly consistent physiological conclusion, since pathological automatisms were nothing more than the overexcitation of normal nervous activity, what Charcot, drawing on British sources, called "an action . . . which has been given the name of cerebral automatism or unconscious cerebration."[107] Indeed, experiments on the electrical excitability of the cerebral cortex had also demonstrated, and further solidified the point, that otherwise ordinary voluntary actions could be initiated artificially in such a way as to bypass the power of the will.[108]

Hypnosis lent itself to other sorts of apprehensive medical conjectures— for example, that certain susceptible populations could be hypnotically induced to commit automatic and unconscious criminal acts, which, as scholars have pointed out, was an especially gloomy though nevertheless tantalizing transatlantic anxiety at the end of the nineteenth century.[109] After all, if dangerous, violent, or criminal behavior could be performed consciously, or even habitually, then it could potentially be replicated unconsciously as well, no matter whether such automatic acts were hypnotically instigated or merely the unintentional effects of mental disease. For most clinicians, however, as Luys himself confessed, the hypnotic inducement to dangerous forms of automatic crime was an idea that for the most part remained "in the realm of fiction" (*le domaine du roman*). Nevertheless, the mere hypothetical prospect that dangerous acts could be perpetrated by an "unconscious agent" (*agent inconscient*), whose seemingly purposive and deliberate behavior was entirely automatic yet artificially induced, meant that clinicians needed to remain vigilant to a potentially "new species of dangerous madman" (*aliéné dangereux*).[110] The perils posed by unconscious agents remained, as one pair of forensic psychiatrists put it, "theoretically not impossible."[111]

Hypnotic inducements aside, a diffuse clinical belief nevertheless persisted that mental pathologies and brain injuries prompted states of often disorderly, if not actually violent, automatism, which were often mistaken for deliberate transgressions and premediated criminal acts.[112] With respect to conduct (or, rather, misconduct) that appears intentional and yet goes on fully unconsciously, wrote the psychiatrist Fulgence Raymond, it is "above all necessary to appreciate, in each case, the danger that it alone constitutes for society."[113] But the social "dangers" these automatisms posed were not necessarily of the violent or disorderly sort. The clinician Adrien Proust (father of Marcel) famously presented the case of thirty-three-year-old Emile X., who often experienced prolonged unconscious automatic states during which he would nevertheless "act according to the habits of everyday life." In late September 1888, after succumbing to an automatic state lasting three weeks, Emile regained consciousness only to discover he had incurred a debt of nearly 500 francs.[114] Luys quickly recognized the extent to which financial activity was undeniably part of the habituated routines of everyday life. The daily management of personal wealth, commerce, wage labor, and the incursion of debt were as much the effects of routine (often more so) as the products of rational volition. It would be reasonable to assume, as Luys did, that patients in states of automatism, whose unconscious behavior "[did] not differ from the waking state," would reproduce, even if only parasitically, certain forms of economic behavior. For Luys, however, what this meant was that such individuals were all the more susceptible to financial deception. "You can force them," Luys warned, "not only to give you a monetary gift, but also to sign a promissory note, a bill of exchange, or to make a commitment of some kind. You can make them write a holographic will, which would be perfectly valid, which they will give you, and the existence of which they will never know."[115]

Despite sympathizing with patients whose latent economic behaviors could be automatically reproduced and unscrupulously taken advantage of, Luys's intention was instead to add to the list of social dangers that automatisms posed. Patients prone to automatic behavior were at risk not only of perpetrating perilous unconscious crimes but also—much more so than a healthy person—of becoming victims of financial wrongdoing, a point the neurologist and hypnosis researcher Albert Moll echoed as well.[116] Their automatisms made them more susceptible to fraud, reluctantly complicit in potential disruptions to the socioeconomic order. It was unsettling, to say the least, to consider the ruptures that pathological automatism might pose to the ostensibly secured patterns of economic life. But an even more unsettling proposition was the inverse possibility: that pathological automatism, far from disrupting the economic order, could actually reinforce it.

"He Seemingly Behaved Like You or Me, Except He Was Unconscious"

Among the many topics that Charcot presented during his so-called Tuesday lectures at the Salpêtrière hospital, his famous unrehearsed weekly patient studies that one English observer described as "[bordering] on the incredible,"[117] were cases of "ambulatory automatism" (*l'automatisme à forme ambulatoire*), or pathologically provoked automatic meanderings. At the end of January 1888 and once again in late February 1889 Charcot presented the case of Men . . . s, a thirty-seven-year-old Parisian deliveryman of fine-art objects.[118] For two years, between March 1887 and April 1889, Men . . . s suffered from seven separate episodes of automatism, lasting anywhere from two hours to eight days, all of which were likely the result of a petit mal epileptic attack wherein automatic behaviors replaced the more clinically recognizable seizures (thus Charcot's eventual use of the more technical term "*comitial* ambulatory automatism" to refer to its epileptic origins). Ian Hacking has called this case Charcot's most paradigmatic study of fugue disorder, a medical entity whose "transient" emergence, popularity, and decline Hacking has explored both historically and philosophically.[119] Although Charcot describes Men . . . s's actions as "fugues,"[120] Charcot nevertheless asserted that automatisms can occur with or without the "impulse to wander."[121] Automatism and fugue were not necessarily identical conditions nor synonymous medical categories, as Hacking implies, for while the reduction to an automatic state could result in unconscious ambulation, fugue disorder, or the morbid impulse to depart and wander, was not in all instances classified as a strict variant of pathological automatism. In fact, some of the most prominent discussions of fugue—for example, Philippe Tissié's *Les aliénés voyageurs*—make virtually no explanatory use of the doctrine of automatism at all.[122]

While nosologically coincident, fugue and automatism differed in one important way. According to Hacking, fugue states were typically characterized as "a sudden departure from the norm," an escapist pathology; but what most defined pathological automatisms, as well as what Charcot most emphasized about them, was that they actually *replicated* the very norms of everyday life, even if they also involved an accompanying impulse to travel.[123] While some cases of automatism (ambulatory or otherwise) could, Charcot warned, take the form of troubling and socially transgressive behaviors, most were quite the opposite, "marked by the unconscious fulfillment of acts of ordinary life," and characterized by only the slightest deviations from conventional behavior. Automatists, Charcot explains, "execute without knowledge or volition, that is, automatically, often highly complicated acts . . . with an often perfect appearance of intelligence and lucidity."[124] This was an observation that Charcot drew from a number of different sources, including Tuke, who was among the few international psychiatrists to visit the Salpêtrière. Tuke was himself an adherent of Jackson's theory of nervous dissolution, making use of it in order to claim that the automatic behaviors that presented during sleepwalking and other hypnotic states were nothing but the expression of normal physiological activity and "at any rate . . . not more pathological than dreaming."[125]

Charcot's patient Men . . . s was, therefore, an exemplary case of pathological automatism far more so than fugue disorder, since what was so unusual about the case was not how much the patient unconsciously departed from his normal routines but, rather, how much he unconsciously reproduced them. "You won't find many stories like this in the medical books," Charcot told his students as he first presented the case.[126] The patient's first bout of automatism lasted for fourteen hours, during which time he traversed the streets of Paris, eventually regaining consciousness at the Place de la Concorde. What was so extraordinary about this particular episode was precisely how ordinary the patient appeared throughout its duration, how undetectable his pathology was to others, and therefore how easily his conduct was mistaken for that of a healthy person. "In the end, he must have appeared acceptable [*sa tenue a dû être correcte*]," Charcot explains, for "otherwise he would have been stopped by the police." Charcot continues, "He must have had his eyes open or otherwise it would have seemed bizarre [*on aurait trouvé cela singulier*], and he would have been taken to the pharmacist. So he seemingly behaved like you and me, except he was unconscious."[127]

Charcot repeatedly emphasized over the course of his two separate pre-

sentations of Men . . . s in 1888 and 1889 the most startling features of the
patient's automatism. For the duration of each episode, Men . . . s maintained
a "tranquil disposition" (*un caractère de tranquillité*), his eyes remained open,
and despite the occasional odd bit of behavior, "he doesn't seem to act any
less reasonably" (*il n'en paraît pas moins agir raisonnablement*) than a normal
individual, and "behaves more or less as if he is conscious."[128] For otherwise,
Charcot regularly pointed out, he would surely have aroused the attention
of passersby. The patient also noted that after he regained consciousness, his
watch still displayed the correct time; to Charcot's pleasant surprise, the patient, in a "mechanical and unconscious fashion," must have continued to
wind his watch. "So here we have a gentleman walking beside you," Charcot
announced. "Are you sure he is awake or not? If there were many such people,
this would not be reassuring."[129]

Some months later the patient experienced a much longer automatic episode, this time lasting two full days. His actions during the course of this
subsequent bout were more than simply ambulatory; they were transactional
as well. The patient purchased some tobacco and a train ticket to travel to
the Bercy area in the twelfth arrondissement. What was so astonishing for
Charcot was not the patient's impulse to travel but that he was capable of
unconsciously performing such elaborate acts, once again without arousing
any suspicions of his pathological state of mind. "He asked the office for a
ticket to Bercy," Charcot noted, which meant that "he had to talk, to take
out his money at the counter; to receive a ticket, and to get on a train," all
while entirely unconscious.[130] "His behavior," Charcot emphasized, "was that
of a normal person."[131] A different sort of transactional form of behavior took
place, Charcot explained, during another automatic episode, which also lasted
over two days. In this case, the patient managed to purchase a meal along with
some coffee. "Did you pay?" Charcot asks the patient in front of his observing
students. "Yes," the patient replies, "I paid 1.15 F. I don't remember drinking
coffee; however, I did find sugar in my pockets."[132] Men . . . s undertook a final
unconscious train excursion in early 1889, this time from Paris to the French
coastal town of Brest. But as Charcot emphasized, what was so important
to note was not simply the sojourn itself, but that during it, "though unconscious or at least subconscious," Men . . . s had "to behave like a man who was
awake, of sound mind, acting deliberately and, in a word, not committing or
presenting anything in his actions, gait, or physiognomy that would allow
him to be deemed a patient [*malade*] or a madman [*aliéné*]."[133]

In addition to replicating routine commercial behavior, Men . . . s also attested to his surprising ability to continue his work as a deliveryman even
while reduced to a state of automatism. During one particular episode, he

explains, and entirely unconsciously, "I carried out my first two errands, on Billancourt street and Clichy boulevard. I then went to Villiers avenue, where I had to pick up a light fixture. I got the fixture, and put it in my coach, without knowing how."[134] During the course of the two Tuesday sessions during which Charcot presented Men ... s's case, the patient was accompanied by a companion, M. X ..., who disclosed to Charcot that he had followed up with the clients that Men ... s had visited during that particular automatic episode in order to see if they had noticed anything strange about his behavior. "He had spoken to them exactly as usual," the companion confirmed, "without exhibiting any bizarre signs."[135] Even when reduced to a state of pathological automatism, Men ... s not only appeared normal but also entirely capable of working and engaging in rudimentary commerce, that is, of reproducing and thereby reinforcing forms of everyday economic conduct.

In seeking to corroborate the medical validity of Men ... s's odd pathological condition, Charcot had a few precedents on which to draw,[136] but one of the earliest and strongest was a paper published by Jackson nearly fifteen years earlier, detailing what at the time he called "the most valuable [case] I possess of post-epileptic mental automatism." It was a paper that Ribot translated and reproduced for the first issue of the *Revue philosophique de la France et de l'étranger*, which effectively introduced Jackson to French philosophy and psychology. Jackson recounted the case of a thirty-one-year-old epileptic man who had been under Jackson's care at the National Hospital and who, as a consequence of his epilepsy, experienced automatic episodes in place of seizures.[137] Jackson reproduced a short note that the patient had written just after regaining consciousness following one particular bout of mental automatism:

20th. Unconscious? for perhaps three-quarters of an hour, remember *ordering* dinner, but not *eating*, or paying for it, but did *both*, and returned to the office, where I *found myself* at my desk feeling rather confused, but not otherwise ill; *was obliged to call at the dining-room to ask if I had been ill, and if I had had any dinner*. The answer was *no* to the former, and *yes* to the latter question.[138]

The patient later relayed a short follow-up to Jackson about a subsequent automatic episode, describing how "three weeks ago I paid for my dinner with half a sovereign ... [the landlady] gave me the change, which I put into my pocket, and very soon afterwards I went to her and tendered a shilling in payment for the dinner, when she told me I had already paid for it, but that I did not appear to remember having done so."[139] What is particularly crucial about this case is not that the landlady was in a position to defraud the patient but, rather, that even in his state of unconsciousness, the patient was able nevertheless to reproduce, like Charcot's patient later would, normative standards of economic behavior. He was unconsciously moved to pay for his dinners, to complete the transactions, to perform the proper financial acts. Doing so, it turned out, did not require the activation of higher mental processes; he was capable of engaging in a relatively correct form of commercial behavior even while in a state of pathological automatism.

While the ostensible misconduct of madmen, reduced to states of automatic uncontrollability, was typically posed as forensic questions concerning legal responsibility and the protection of the social order, the cases outlined by Charcot and Jackson posed a very different sort of conundrum. For at stake in these case studies were forms of pathological behavior that were nevertheless acceptable and appropriate, a clinical paradox of sorts, but one that suggested that social conduct could not always transparently signal the status of an individual's mental health, and that customary behaviors could be normally performed during states of profound psychiatric non-normalcy. Furthermore, the case studies present by Charcot and Jackson exemplified the opposite of a rupture in the economic order. They illustrated, instead, that *successful* daily economic behavior could take place in the absence of higher voluntary and intellectual functions—not only automatically but *pathologically-automatically*. Economic life and pathology, it seemed, were not necessarily incompatible conditions.

At the Boundaries of the Legible

By relying on Jackson's medical precedent, Charcot was able to stress not only how much pathological automatisms replicated ordinary behaviors but also just how ordinary, that is to say, how medically normal automatisms (ambulatory or otherwise) were among the healthy. "It is true to say," Charcot declared, "that in some cases we are all a bit like that."[140] As evidence of such everyday instances of *petit automatisme*, Charcot invoked an anecdote, which he drew from William Carpenter, about John Stuart Mill, who was purported often to be "so immersed in philosophical thought" that he would aimlessly roam the most populated streets of London "automatically unconscious" (*automatiquement inconscient*), yet without standing out as odd, appearing unnatural, or colliding with streetlamps or pedestrians.[141] "Without being a John Stuart Mill," Charcot explained to his students, "how many times has it come to pass that we ourselves, in a moment of great preoccupation, have climbed a staircase without knowing it and arrived at our door very astonished at what we had done unconsciously."[142]

The reference to Mill is notable, as it ultimately takes us back to the topic with which this chapter began, namely, the eccentric, a figure whom Mill championed despite acknowledging psychiatry's far more acrimonious portrayal. While the doctrine of automatism was never formally tied to eccentric behavior, as a medical explication or grounding physiopathology, they nevertheless shared a striking affinity. For if the strongest "nervous dissolutions" to the lowest automatic processes reduced a person to brutish ravings, and if only moderate dissolutions resulted in the automatic replications of ordinary social routines, then perhaps eccentricity could be characterized as the effect of only minimal dissolutions to the most negligible states of mental automatism wherein, as Jackson observed, "there is but a slight departure from healthy action."[143] Describing eccentricity through the lens of automatism, and its essential ambiguities, could not only explain why eccentric conduct often evaded unambiguous determinations of health and illness, but also helps explain why eccentric personalities were sometimes revered as creative innovators (a point I discuss in chapters 3 and 6), since Jackson himself believed that the incongruous and quasi-pathological "double condition" of the mind lent itself to expressions of highly inventive thought.

The eccentric, therefore, might very well have been a kind of automatist, the two figures collectively marking within the turn-of-the-century transatlantic psychiatric imaginary the edges of the "borderland between sanity and insanity," positioned at the boundaries of social legibility in the form of conduct that seemed either decidedly inscrutable (the eccentric) or deceptively clear (the automatist). Borderland typologies did not simply upend classical conceptions of selfhood and individualism, as is often said of the unconscious, but they also unsettled a core supposition of the modern social body: that our conduct is transparent and decipherable and that the social order coheres through the norms of interpersonal legibility. In an effort to hygienically manage social relations, psychiatry instead instituted an irresolvable dilemma at the center of those relations by suggesting through limited though nevertheless emblematic case studies, as well an elaborate degree of trepidation, that with respect to outward behaviors and dispositions, what you see is not necessarily what you get.

If eccentrics, automatists, and other borderland figures were to be classified according to their most dominant attributes, then surely among these would be the fact that they were, for the most part, *men*. This is not merely an empirical claim, that borderland typologies and pathologies were typically or even normatively gendered masculine, or that the borderland was a terrain of specifically masculine embodiments or representative of a highly gendered medical culture—even though this is all undoubtedly the case.[144] Psychiatric encodings of masculine gender norms have not only varied historically and geographically but have also been internally quite incongruous, fragmented according to racial, geopolitical, and class differences. However, one effort that stands out in the numerous and diverse psychiatric efforts to stabilize masculine norms in Europe and the United States has been to anchor them in rationalist conceptions of self-conduct.[145] My contention, however, is that masculinity and rationality were not always normatively united in the transatlantic psychiatric imaginary, and that, instead, we can see in the context of borderland typologies the germination of what will eventually become the *positive* entwinement of masculinity and *irrationality*. In later chapters, I detail how psychiatric thought formalized some expressions of irrationality into highly valued and positively appraised forms of masculine pathologies, valued in specifically economic terms. An emblematic instance, which I detail in chapter 3, concerns the psychiatric discourse on the financial entrepreneur as the instantiation of masculine irrationalism, but an irrationalism that was nevertheless economically venerated. In the context of economic enterprise, in other words, there were right (and profitable) ways to be pathological and, moreover, pathologically masculine.

But how could psychiatric masculinity come to adopt this seemingly contradictory set of attributes, of being both socioeconomically valued and yet deemed medically unsound? I propose that this equivocality emerged in the context of the borderland typology. The forms of conduct characteristic of the borderland, however much they were pathologized, were also expressions of something else: a class of behavioral affordances and performative entitlements to present ambiguous conduct that troubled the boundaries of social legibility—affordances that tended to be framed in relatively masculinist terms. Conduct opacity was a form of behavioral latitude typically affixed to masculine privilege. For while eccentrics were mostly denounced by psychiatric practitioners for their enigmatic behaviors, it was nevertheless a normative presumption within psychiatry that eccentrics were most often and most emblematically men. The ambiguities afforded at the borderland eventually came to function as a *productive precondition* for the apparent contradiction that masculinity could soon come to embody, as a style of performance that could be both medically pathologized and socioeconomically valued. The borderland typology was a psychiatric contrivance that eventually came to serve certain economic ends. Eccentrics, automatists, and other borderland figures represent the beginning of a story concerning the nascent affinity between economic value and pathological states, or the emergence of pathological value.

What Conduct Reveals

"Psychology of Conduct"

The true problem that the borderland typology instituted within psychological medicine was not nosological; it was hermeneutic. The borderland, in other words, was not an isolated disease classification perplexing clinicians and researchers but instead represented an interpretative quandary at the heart of the diagnostic act, one that exceeded the standard clinical query—Is this patient mentally ill?—by foregrounding a more disquieting set of uncertainties: Can mental status be accurately inferred based solely on behaviors and outward dispositions? And does psychiatric medicine really possess the proper hermeneutic tools necessary to make such determinations? While eccentricity and automatism represented the outermost edges of this interpretative dilemma, the borderland between sanity and insanity was not comprised of eccentrics and automatists alone. For it ultimately denoted an expansive continuum of intermediary mental pathologies as well, the "noninsane psychoses" as one prominent American neurologist called them, that is, so-called functional (nonorganic) disorders such as neurasthenia and mild hysteria or what by the end of the nineteenth century would come to be called "neuroses."[1]

The concept of neurosis has been most prominently associated with psychoanalysis to designate pathological dispositions situated on what Sigmund Freud called the "borderline" (*Grenze*) between nervous health and abnormality.[2] While neuroses for Freud included an assortment of semipathological manifestations, from bungled actions and other "parapraxes" to anxieties, compulsions, and obsessive thoughts, neurotic semipathologies were not simply the hallmarks of psychoanalytic thought alone. In the mid-1890s, Daniel Hack Tuke, who had years earlier warned of the interpretative frustrations of borderland behaviors, synthesized an assortment of psychiatric theories that had circulated for decades throughout Europe, on compulsive behaviors, obsessive thoughts, and nonsensical phobias, anxieties, and doubts into a single medical concept, which he dubbed "imperative ideas."[3] In contrast to what would eventually become the overwhelmingly psychoanalytic approach to neurotic behavior, Tuke's attitude toward imperative ideas was instead *Jacksonian*; he entreated his readers as John Hughlings Jackson had once done to direct the clinical focus not on the severest mental disorders but on only the "the slightest departures from the standard of mental health."[4] For, in doing so, what the clinician discovers is that all imperative ideas are united according to a single feature: "their *automatism*, the overwhelming and recurring tendency to be haunted by a certain idea, to perform certain acts, or to use certain expressions against the will."[5] Far from being an acute form of insanity, imperative ideas, whether in the form of inexplicable impulses, compulsively repeated thoughts, senseless emotions, or even so-called earworms, were actually quite common occurrences even among otherwise healthy individuals.

It was, as Tuke wrote, "the absolute innocuousness and sanity of a large number of imperative ideas and acts" that remained their most defining feature; they are the psychological irregularities that "haunt" everyday healthy mental life but nevertheless seem "consistent with sanity" in part because we are "quite able to perceive their groundlessness and their ridiculous character."[6] For Tuke, imperative ideas must be understood "in Jacksonese," as an exemplary instance of the mind's normal "double condition," when the slightest nervous dissolutions are met with emergent mental automatisms.[7] "Imperative ideas are very common and . . . nearly everyone has some," wrote the psychiatrist George Savage in response to Tuke.[8] They are merely "the trifling forms of obsession to which we are all liable."[9] The French psychologist Alfred Binet agreed that imperative ideas and other common "automatic manifestations" comprised a branch of mental science that "merits closer study."[10]

Imperative ideas revealed a subtle though significant psychiatric paradox concerning neuroses and other intermediary or borderland pathologies: that the mind could act anomalously, even abnormally, yet without necessarily compromising a person's sanity in any strict medical sense. If too overwhelming, imperative ideas could certainly be detrimental. But mental health could never be defined through the total absence of these semipathologies. For Tuke and others, it was precisely how a person *behaved*, not in the absence of imperative ideas but despite their ubiquity, that most defined what it meant to be sane. These intermediary mental pathologies, therefore, not only comprised the bulk of what defined the borderland between sanity and insanity, but they also incited many clinicians to redirect the psychiatric gaze, to turn away from efforts to diagnose the mind alone and to look instead to the status of *conduct* as the true measure of mental health.

The turn toward conduct was by no means a unique or anomalous psychiatric reorientation; it characterized a transformation in the clinical attitude of some of the most prominent and influential turn-of-the-century psychiatric practitioners. Pierre Janet recounted late in life having undertaken just such a transformation in his own clinical work, away from a medical focus on the mind and toward conduct as the primary object of psychiatric concern. Beginning initially with his appointment at the Sorbonne in 1898, but much more substantially when he succeeded Théodule Ribot as chair of experimental psychology at the Collège de France in 1902, Janet increasingly found himself drawn away from the confines of narrow clinical analysis toward a more holistic "practical psychology." The problem with conventional psychiatry, he wrote, was that it had fundamentally "evolved from Cartesianism," and thereby "regarded thought as the most primitive phenomenon and action as a consequence or secondary expression." Janet realized, however, that the subordination of action to thought would need to be reversed if psychology were to have any sort of purchase beyond the narrow confines of clinical diagnostics. "We are obliged to formulate a psychology," he argued, "in which externally observable action is the fundamental phenomenon, and in which inner thought is only a reproduction, a combination of these outward actions in a reduced and specialized form."[11]

The inversion of the primacy of the mind was undoubtedly a strike against the implicit Cartesianism that had pervaded not only French psychiatric medicine but virtually all psychological doctrines that espoused the mind's medical and scientific preeminence. It was, after all, in a famous passage from the Second Meditation that René Descartes remarked that upon glancing out his window and observing men crossing the square, he was always inclined to say that it was real men that he perceived. "Yet do I see more than hats and coats," Descartes asked, "which could conceal automatons?" No, he replied, since "I *judge* that they are men." For Descartes, the mind and its rational faculty of judgment always accompanied all sensory perceptions, effectively overriding the vulgar simplicity of physiological apprehension. That men *appeared* to be automata was a perceptual postulate that was inevitably superseded by the rational realization that they certainly could not be.[12] Things, however, were quite different for Janet as well as his predecessors and international colleagues. As I discussed in chapter 1, in glancing at a busy street in Paris or London sometime near the end of the nineteenth century, one might very well have failed to notice the ambulating automatists moving through the crowd, invisible amid the clamor of modern life, raising no apparent alarms, since even while reduced to a state of pathological automatism these unconscious agents would nevertheless have managed to reproduce the routines of everyday life. In much of the collective imaginary of turn-of-the-century transatlantic psychiatry, a hat and coat could very well conceal an automaton; a psychological Cartesianism would be just as impractical in such a circumstance as it would in trying to adjudicate the behavioral aberrancies of eccentrics and even the everyday neuroses among the healthy.

What Janet had in mind for his so-called practical psychology was an approach capable of transposing interior experiences into the language of actions, what he dubbed a "psychology of conduct." The strange terminology was nevertheless important to retain, for through it Janet could signal "a broader and higher form than behaviorism."[13] Behaviorisms, like the one proposed by the American psychologist John Watson, might suitably be applied to animals, but their application to humans would always be thwarted by the fact of consciousness. Janet refused to follow Watson's behaviorist prescription to "dispense with consciousness in a psychological sense."[14] But this did not mean that interior life had to be sanctified as inaccessible and unobjectifiable. Instead, "one must regard the phenomenon of consciousness as specialized conduct," wrote Janet, "a complication of the act which is superimposed on the elementary conduct." Mental phenomena could be reimagined as "higher forms of conduct," gradationally linked to more automatic physiological activity.[15] For Janet, the matters of fact that defined human life and experience could be expressed as conduct through and through.

If the borderland typology instituted a novel dilemma with respect to the norms of interpersonal social legibility, then this chapter explores the hermeneutic strategy—the style of psychiatric reasoning—that arose as a consequence of the interpretative frustrations that the borderland introduced. That strategic form of reasoning was to turn to conduct alone, to evaluate social performances on their own terms without the added burden of trying to infer an underlying mental state. Unlike the mind, which continued to be medically assessed according to its conformity to the bounds of reason, the operative criterion for behaviors and social performances was not rationality as is often believed but *propriety*, the determination as to whether an act was appropriate or acceptable. Conduct, in other words, is a socio-moral category, not an epistemic one. Much of the opacity of borderland behaviors could be resolved, or at least avoided, if psychiatric clinicians simply asked whether the conduct in question was tolerable or customary according to *some* social benchmark.

The turn to conduct, then, represented a subtle hermeneutic shift for psychiatric diagnosis; the goal was not to *interpret* the underlying mental truth of a particular performance but merely to *appraise* it according to its outward social merit, to ask whether the performance was valuable or costly with re-spect to the social order. In that respect, mental status was not always, or did not always need to be, the primary concern for psychiatric evaluation, and whatever mental abnormalities might have existed for an individual could be offset by behaviors and actions that typified the norms of social propriety. The central claim that I propose toward the end of this chapter is that in order to appraise the social merits of otherwise illegible behavior, turn-of-the-century transatlantic psychiatry tacitly turned to a specific social benchmark that fell outside the remit of psychological medicine—namely, economic norms of conduct.

By relying on economic benchmarks of behavioral propriety, psychiatrists were able to appraise the social value of the gamut of questionable behaviors, from the plainly inscrutable to the deceptively plain. But in addition to the evaluative advantages that such benchmarks provided, the economic herme-neutic enabled clinicians to adopt a seemingly paradoxical conviction that a person could act entirely appropriately, perhaps even successfully, with re-spect to the management of wealth, finances, and business endeavors while at the same time possessing a mind that may very well have been unsound. Eco-nomic conduct could serve not only to counterbalance the possibility of mad-ness but even, in some cases, to *redeem* it—that is, to assign a degree of social worth to behaviors that would otherwise have signaled the presence of men-tal disease. The psychiatric turn to conduct, which this chapter examines, represents a style of reasoning that eventually came to embody an unstated economic logic and a conviction that certain economic behaviors could com-mandeer psychiatric determinations of health and disease and therefore serve as the very best indicator of psychological health and morbidity.

Conduct and the Human Sciences

The psychiatric turn to the norms and bounds of conduct was therefore not only a way to obliquely assess a person's mental state but also functioned as a means by which to diagnostically offset the medical significance of the in-escapable abnormalities of the mind. But the turn to conduct was also repre-sentative of the role psychological medicine played in the larger history of the human sciences. The classification and enumeration of forms of right conduct were practices that dated back at least to the so-called courtesy literatures of the Middle Ages, which described techniques for cultivating proper manners and developing social virtues through instruction in personal, domestic, and reli-gious hygiene.[16] These literatures bore the traces of an Aristotelian tradition of ethical "habituation" (*ethismos*), the presumption that people have some part to play in modifying the intensity and degree of their feelings and the nature of their actions, and that one's character, like an instrument, can be regulated for the purpose of developing virtue.[17] After the eighteenth century, however, with the emergence of the human sciences, the inquiry into human behavior took on a very different tone. Within the broad array of knowledge fields that emerged throughout the nineteenth century, and that included an assortment of disciplines—psychology, linguistics, sociology, anthropology, economics, and, to some degree, the humanities as we know them today—conduct was not simply framed as it once had been, as a set of normative practices and techniques for moral self-cultivation. Instead, human conduct took on the quality of an epistemic *problem*—an unknown, an object of inquiry, anxiety, and contestation that demanded resolution, interpretation, and certitude.[18]

When Wilhelm Dilthey first distinguished the human sciences (*Geisteswissenschaften*) from the natural sciences in 1883, he was quite categorical that to whatever degree human activity might be free from the axiomatic strictures of the natural world, the human was still an undeniably natural entity.[19] At the same time, however, the naturalness of human nature did not imply a reduction to physical determinism. One of the first and most crucial discoveries in nineteenth-century biology was that the vital order itself could not always be faithfully mapped onto the rules of the physical universe. While the physical world could answer to mathematical principles, the world of living beings was instead organized according to what Georges Canguilhem dubbed an "order of properties," perennially subject to anomaly and variation.[20] The human sciences began to grapple with the meaning of human nature at a moment when a univocal conception of the natural world was coming undone.[21]

Hitherto dominant ontological precepts of human nature were subsequently revised throughout the nineteenth century—no longer was human essence defined solely by recourse to a substance or principle, whether it be the soul or an immutable doctrine of matter. To know what the human *is* became instead the task of knowing what the human *does*, and it was precisely the human sciences that undertook such a charge. Michel Foucault proposed just this point at the end of *The Order of Things*—that it was the *function*, not the substance, of human nature that was at stake for the human sciences. To live, to speak, to work, to act—whichever functions were prioritized, the fact remained that for the human sciences, the meaning of human nature was to be found in "conducts, behaviors [*des comportements*], attitudes, accomplished gestures, and uttered or written phrases."[22] And so it was that the human sciences undertook to stabilize the truth of conduct through the most compelling technique available to them: the countless developed "theories of all that action is," that is, "praxeologies," as the historian Paul Veyne has proposed.[23]

The historical recognition that conduct was indeed a *problem* was in the first place tied to the systematic appreciation of how capacious and elusive the concept of conduct even was, how prone it was (and continues to be) to persistent "colloquial instability," as the philosopher Helen Longino has observed.[24] A palpable definitional slipperiness in part explains the relative scarcity of scholarly analyses on the topic of conduct itself. "It is astonishing," writes Arnold Davidson, "and of profound significance, that the autonomous sphere of conduct has been more or less invisible in the history of modern (as opposed to ancient) moral and political philosophy."[25] Explorations in the "autonomous sphere of conduct" have not proven to be particularly evident or forthcoming in the histories of the human sciences either, which is more astonishing still given how vital the problem of conduct has been for the modern study of human nature.

Much of the difficulty revolves around the challenge of isolating precisely what sorts of expressions and activities properly fall within the rubric of human conduct. Among turn-of-the-century psychologists and progenitors of modern social theory, it had become alarmingly clear that the domain of conduct was limited neither to doctrines of rational moral behavior and deliberative action nor to protocols for manners or other social codes of "strategic self-enactment."[26] Max Weber admitted that it was not possible empirically to draw an absolute line between meaningful action and "merely reactive [*reaktives*] behavior," that the border between the two was "absolutely fluid" (*durchaus flüssig*).[27] Other prominent figures wrestled with the question as to what precisely the range of human conduct was a property or expression of. To what degree, Émile Durkheim asked, were ordinary actions and behaviors a manifestation of the latent collective impingements of social institutions?[28] According to other social and psychological theorists of the late nineteenth and early twentieth centuries, the most premeditated undertakings could be imperceptibly warped or commandeered by the influences of crowds, by the rhythms of metropolitan life, by the inculcation of bodily routines and formalized practices, or simply by the profound interactional dynamism of society itself (to say nothing of biological propensities).[29] Was it ever possible decisively to determine, Freud wondered, whether an ordinary performance or disposition was the expression of agency or merely the externalization of an unconscious impulse or a "counter-will" (*Gegenwille*)?[30] Could not the very belief in one's own autonomy be nothing more than the expression of a historically fashioned "habitus," a situated moral economy, or a social "field"?[31]

By the turn of the twentieth century, at the moment of the historical ascendancy of the human sciences, human conduct designated a domain of

behavioral opacity, a reservoir of often incompatible performative possibilities, and a problem site for a modern and scientifically inflected theory of
human action where the social and the individual, the conscious and the unconscious, the deliberative and the unreasoned, and the voluntary and the
automatic intersected, overlapped, and occasionally fused. An emerging and
uneasy socio-medical conclusion held that conventional behavior and the
execution of everyday socio-moral norms were as much expressions of self-
direction as they were, simultaneously, confirmation that people were often
acted upon by forces "beyond" themselves. We are inclined to believe that the
ramifications of these intellectual developments fell squarely on the status
of consciousness, interiority, and subjectivity. It was psychological medicine,
perhaps above all, that played the most influential role in framing conduct as
a problem by the end of the nineteenth century. In 1884, under the banner of
the "New Psychology," John Dewey posited that novel theories of the mind
emerging from the data of physiological psychology, developmental studies,
and clinical research into insanities and mental disorders revealed how much
our mental life remained "suppressed," entering consciousness only in vague
and transitory ways.[32] The so-called new psychologies soon branched out into
an array of research fields, from abnormal psychology, mental physiology,
and neuropsychiatry to the various "experimental psychologies," which had
vastly different meanings in France, Germany, and the United States.[33] But the
hand-wringing these psychological discoveries occasioned with regard to the
challenges they posed to classical conceptions of subjectivity was secondary
to the apprehensions such discoveries presented to the domain of *behavior*, to
bearings and comportments externalized as actions in the social world. The
idea of unconscious, repressed, or otherwise vague and transitory thoughts
was certainly an unsettling theoretical proposition; but unconscious or otherwise indeterminate *conduct* was more than just a medico-philosophical curiosity. It was a potential threat to the social order.

Medicine of the Moral Order

One of the earliest and most salient ways psychiatric practitioners demonstrated that their medical remit encompassed the adjudication and management of right conduct was how they formalized the medical equivalence
between mental health and moral action. During the first several decades of
the nineteenth century, clinicians throughout Europe and the United States
introduced a novel array of diagnostic categories that increasingly emphasized
the possibility that madness was not strictly dependent on intellectual impairments alone. This spectrum of diagnoses, including the French "reasoning

madness" (*folie raisonnante*) and the English "moral insanity," comprised a diffuse and often nebulous set of umbrella categories that were linked by an emergent medical conviction that morbidities with respect to moral character and proper social conduct could be manifest without any discernible debilitation to the intellect or intelligence.[34] It had become possible, in other words, for someone to be behaviorally, and thus morally, deranged while at the same time entirely rational.

In their earliest introduction, these new psychiatric categories marked a self-conscious transition away from an eighteenth-century tendency to conclude that the essence of madness resided in a debilitation of cognition alone, in the form of either false judgment or morbid sensibility.[35] In 1801, the French physician Philippe Pinel introduced some of the first clinical accounts of mental disorders that seemed to arise without marked intellectual impairment.[36] Pinel's newly cataloged cases of *manie sans délire* (mania without delirium) and *folie raisonnante* (reasoning madness) were rare clinical instances of nondelusional insanities during which the intellectual capacities seemed not only to be unaffected but actually quite acutely expanded. These conditions, Pinel contended, "would have appeared completely enigmatic if we were to follow the ideas that Locke and Condillac have given about the insane"—namely, that madness was nothing but unreason.[37] Analogous and independent clinical observations of the pathological derangement of conduct alone, without apparent impairments of the mental faculties, were recorded and described within the next decade across the North Atlantic.[38]

This is not to suggest that a strictly mental view of insanity was being abandoned; many prominent early nineteenth-century clinicians continued to hold onto the belief that insanity could refer to nothing but the unsoundness of mind.[39] Nevertheless, the most indelible mark that psychiatry first made within the developing human sciences, what defined its earliest medical contribution as distinct from prior and more philosophically oriented theorizations of human nature, was that *reason and madness could in fact coincide*, that a patient's character and conduct could destabilize even as the mind's rational capacities remained intact. "Now-a-days," wrote the German physician and phrenologist Johann Gaspar Spurzheim in 1817, "it is well ascertained that, in insanity, the power of judging is not always deranged. Many insane persons . . . reason with perfect consistency; nay, in many that power is increased."[40]

For Spurzheim, the very idea of a *rational madman* meant that something more was at stake in insanity than a threat to the dominance of reason alone. It was not as intellectual disability but, rather, as moral incapacity that insanity posed the most troubling threat to the social order.[41] Madness represented a loss to moral freedom, and thus to the possibility of human agency and responsibility upon which social, legal, and economic institutions were most dependent. Early psychiatry seemed to embrace the controversial proposition that Friedrich Nietzsche would not make until 1887, in the second essay of *On the Genealogy of Morality*: that the idealization of the concept of the "sovereign individual" (*souveraine Individuum*) was not predicated on human reason but on the moral capacity for responsibility (*Verantwortlichkeit*), the ability to make promises, and to formalize contractual obligations.[42] Concerned with how people governed themselves, how they ought to act, and the costs associated with anomalous behavior, psychiatry was more than just a behavioral medicine—it was a tacit form of practical reason, a moral discourse.

It was in the context of German thought that the relationship between madness and moral freedom was first most systematically explored. For German Romantic psychiatrists and medical philosophers in the first decades of the nineteenth century, including Johann Christian Reil—who in 1808 first coined the term "psychiatry" (*Psychiatrie*)[43]—and Johann Christian Heinroth, madness represented moral alienation in the form of absolute unfreedom.[44] For others, like Johann Christoph Hoffbauer, madness was a disruption in the ability to conduct oneself, an implication that society would benefit more from a public capable of self-control and self-accountability—of governability, which reason alone did not necessarily guarantee.[45] These German Romantic medical philosophies provided an additional authoritative validation for the existence of behavioral pathologies for European clinicians in the ensuing decades. The prominent French psychiatrist Étienne Esquirol saw in Hoffbauer's discussion of intermittent and transitory mental disorders confirmation of his own concept of "monomania," a broad disease category that referred to a diffuse genre of partial and affective insanities that had enraptured the French medical and legal imaginary since the 1810s.[46] While Esquirol's earliest formulations of monomania included impairments to the intellect in the form of marked delirium,[47] he discovered under the rubric of a particular subcategory of monomania, which he dubbed "homicidal monomania," an isolatable derangement of moral volition that seemed to present "no appreciable impairment of intelligence or emotion."[48] Homicidal monomania was the violent pathological deprivation of moral freedom, the loss of control over emergently erratic and disturbing behavior.[49] As Esquirol's student Étienne Georget would contend, mental illness was the deprivation of liberty, the loss of voluntary self-control, and thus the privation of moral autonomy and legal agency.[50]

The British analogue to *folie raisonnante* and monomania was dubbed "moral insanity" by the physician James Cowles Prichard, who like his international colleagues felt that a strictly mental approach to madness was "far too limited an account" and far too deferential to a philosophical rather than a medical view of human nature.[51] Moral insanity designated a disorder of emotions, inclinations, impulses, habits, and moral character but "without any remarkable disorder or defect of the intellect or knowing and reasoning faculties."[52] Not only could the morally insane reason, but they even exhibited "great ingenuity in giving reasons for the eccentricities of their conduct."[53] Moral insanity often signaled nothing more than the inability to "[conduct oneself] with decency and propriety in the business of life."[54] By the 1840s, despite the nuanced differences that ultimately distinguished them, *folie raisonnante*, monomania, and moral insanity were effectively perceived as variations of the same medical possibility and collectively enjoyed a substantial degree of transnational circulation, as clinicians sought to incorporate and modify the formulations of their international colleagues.[55] The somewhat unified perception of these various disorders had much to do with the extent to which they shared several diagnostic features. For example, they were collectively viewed as relatively isolatable disease states, unconnected to some degree from the patient's overall psychical makeup. They represented a somewhat discrete abnormality that could remain latent and unnoticed, not immediately revealed by any other feature of the patient's state of mind. As such, the pathologies appeared to manifest themselves in unpredictable and often inexplicably irruptive ways.

Alexandre Brière de Boismont, another of Esquirol's students, described the case of a man who, "enjoying perfect health at the moment he leaves his family," and exhibiting no particular "derangement of the intellect," was suddenly "seized with a fit of furious madness" in the middle of the road and murdered three bystanders.[56] In the United States, the psychiatrist Isaac Ray found much in the concept of moral insanity to explain how and why mental disease could "[change] the peaceable and retiring individual into a demon of fury," converting the tranquility of a person's "lawful and innocent occupations" into "shameless dissipation and debauchery, while the intellectual perceptions seem to have lost none of their ordinary soundness and vigor." Moral insanity, Ray professed, had become "too well established, to be questioned by those who have any scientific reputation to lose."[57]

Early advocacy for the professional relevance of psychiatry owed much to the promulgation of the belief that moral pathologies were latent, unpredictable, and irruptive.[58] Such pathologies effectively dislodged an entrenched and otherwise commonsensical relationship between reason and madness. Not only could the apparent rationality of morally insane patients provide no warning for the emergence of their erratic behavior, but reason was to some degree powerless in the face of moral insanity itself; it was powerless as a precautionary gauge of the pathology that could arise, and powerless also as a counterforce to stop it. Ray, who continued to promote his conviction in the reality of moral insanity for decades, described how patients "with intellect unwarped by the slightest excitement or delusion, and with many moral traits" could be nevertheless "irresistibly impelled to some particular form of crime," an impulsion that was "as much a mystery to themselves as to others."[59] In many cases, these individuals appeared to others and also believed themselves to be acting "with perfect freedom of will, unconscious of any irresistible bias."[60]

What numerous international clinical studies appeared somewhat systematically to reveal during the first half of the nineteenth century was that the morbid incitement to deranged behavior could persist along with the continued integrity of some intellectual and rational capacities. Clinicians even pointed out that many patients could be entirely aware of their criminal transgressions or moral improprieties even as they were otherwise compelled to commit them. What mattered was not cognizance of right and wrong but a patient's "power of control *over* his actions."[61] As one British psychiatrist put it, the experience of moral disorders could be summed up as "Yes I knew perfectly well what I was doing, and that I was doing wrong, but I had not the *power to control it*."[62]

Dramatic and scandalous cases of the momentary loss of self-restraint at the behest of an inner drive or morbid impulse soon came to represent in Europe and the United States a developing clinical truism about social and economic class disparities, of poor or marginal populations perceived as pathologically ungovernable and who needed to "be restrained . . . for the good of the public."[63] According to the American psychiatrist John Gray, the problem of self-control and self-governance was especially true for so-called pauper populations, who Gray believed were incapable of self-preservation and "unable to govern themselves, or direct their energies."[64] Moral disorders became a signifier for populations situated at the socioeconomic periphery of European and American society. Thomas Laycock wrote of a "large number of criminals, termed in France the 'classes dangereuses,' and in English phrase, 'known to the police,' and another still more numerous body, not exactly of this class, but incorrigible vagabonds, drunkards, mendicants," who numbered in the tens of thousands and who possessed "no self-control."[65]

The morally disordered, those whose freedoms were curtailed not by the abrogation of reason but by the loss of self-control, were defined through the entangled encoding of pathological and socioeconomic class markers; the poor, those disinclined to work, and men (in particular) who posed a

threat to the social order were all potential members of a so-called homicidal class, whose insanity was "transitory," passing inexplicably into unno-
ticeable normalcy just as quickly as it erupted.[66] While insanity never lost
its primary meaning as a disease of the mind, grounded ultimately on some
disease of the brain, still from the standpoint of clinical observation, normative adjudications of mental health by midcentury came to rely heavily on
the propriety of a patient's conduct. Appropriate comportment and suitable
behavior often functioned as the benchmarks of normality, moral freedom,
self-accountability, and socioeconomic suitability.

By the 1860s, however, two increasingly dominant trends began to emerge
in relation to the various classifications of moral pathology that had circulated for decades throughout Europe and the United States. The first was the
increasing resistance to the singularity of the diagnosis of moral insanity—
that is, to the belief that moral pathologies were isolatable and discrete disease
states, unconnected to the total symptomatology of the patient. Moral insanity came to be increasingly viewed as part of a patient's global psychiatric
makeup, either in the form of a prodromal manifestation of mental derangement, or as part of a patient's underlying degenerative constitution.[67] Second,
and particularly through the popularization of degeneracy theories, clinicians
were increasingly likely to view moral pathologies as congenitally inherent in
certain population types.

According to the psychiatrist Bénédict Augustin Morel, one of the central
architects of French degeneracy theory, what had formerly been judged as
moral insanity was merely a variation of "hereditary madness" (*folie hérédi-
taire*).[68] Patients suffering from this sort of disorder exhibited a congenitally
inherited nervous temperament that often resulted in transitory eccentric and
transgressive behaviors.[69] What differentiated Morel's morally insane patient
from earlier iterations of moral pathology was the presumption of hereditary
defect, one that underwent continued generational decline. Valentin Magnan,
France's premier degeneracy advocate after Morel, would write later that it
was precisely Morel who inaugurated "a new era by attaching *folie raisonnante*
to hereditary insanities and by highlighting the close bond that unites it, in
generational successions, to imbecility and idiotism, that is to say, to degeneration." Magnan concluded that "reasoning maniacs" (*maniaques raisonnants*)
were ultimately "hereditary degenerates in whom we notice as always the lack
of mental balance [*d'équilibre psychique*]."[70]

Henry Maudsley, perhaps the most prominent English advocate of the degenerative theory of insanity, was equally convinced that "moral peculiarities
are constitutional,"[71] the expression of an underlying "hereditary taint."[72] Many
such "tainted" individuals, it was feared, lived undetected among an otherwise

healthy social body. "A great number of insane people live among us," wrote the French psychiatrist Ulysse Trélat, "blending into our activities, our interests, and our emotions." This shrouded population of the congenitally ill, the "lucidly insane," as Trélat called them, bore the distinctive features of moral insanity. They "were delirious in their actions but not in their speech," and their deceptive "air of reason" often enabled them to acquire a degree of social authority and to sow discord among friends and family.[73]

After the 1860s, then, while moral pathologies continued to exhibit some of the same crucial diagnostic symptomatologies of the earlier moral insanity and *folie raisonnante,* they differed from their pathological predecessors not only by being tied to so-called degenerative populations but also by the extent to which they were characterized as prodromal precursors to total mental derangement, rather than purely isolatable states of disease.[74] Prominent psychiatric practitioners eventually transformed the underlying nosology of moral disorders; they did not doubt the clinical reality of the *symptoms* described by earlier categories such as moral insanity or *folie raisonnante.* Instead, they simply opposed the divisibility of pathological behaviors from intellectual derangements and, therefore, the very possibility of a partial or isolatable disease state.[75] Moral disorders were simply expressions of other more comprehensive pathologies that soon came to dispense fully with the otherwise important distinction between acquired disease and constitutional attribute. Moral pathology was the primary attribute of the worst kinds of "dangerous madmen" (*aliénés dangereux*).[76] The reason, wrote Richard von Krafft-Ebing, was because the moral madman was ultimately a congenitally "moral idiot," someone who lacked the innate *intellectual* capacity to comprehend moral judgments.[77] Moral insanity was no longer a strictly behavioral disorder but an intellectual disability, a form of pejoratively labeled "moral stupidity and idiocy" (*moralischen Blöd- und Schwachsinn*).[78] It was precisely because moral deficiencies could manifest during a "prodromal period" (*Prodromalperiode*) often lasting for years that they could easily be mistaken for discrete disruptions to conduct alone.[79]

By the last decades of the nineteenth century, moral pathology had all but finalized its transition away from an irruptive and singular disorder of conduct to a congenital and developmental deficiency of intellect and personality. The Swiss psychiatrist Eugen Bleuler associated "moral depravation" (*moralische Depravation*) with the degenerative "inferiority" (*Inferiorität*) of neurological organization, a form of psychosis that biologically undergirded criminality through the underdevelopment of moral capacities.[80] Among German and Austrian clinicians in particular, the language of personality inferiority came to pervade the analyses of moral pathology.[81] Emil Kraepelin explained in an early edition of his psychiatric textbook that both intellectual and moral disturbances arose from the "organic burden" (*Belastung*) of accumulated neuropathic traits. For Kraepelin, *moralisches Irresein* or "moral madness"—what he in parentheses additionally labeled "*folie morale*, moral insanity"—was one of the main forms of "congenital mental inferiority" (*angeboren Schwachsinns*), a disorder of the temperament and a "deficiency or weakness in the counter forces that restrain social men from the reckless satisfaction of their egoistical inclinations."[82] Other clinicians would go on to propose that the very term "moral insanity" "should be complete struck out" and replaced with a broader category of psychopathy that could encompass both imbecility and degenerative insanity;[83] that moral deviance was simply rooted in "mental inferiorities" (*psychische Minderwertigkeiten*); and that "moral insanity [*Moral Insanity*], what the French called *Manie raisonnante* [is] the next closest pathogenetic disorder to congenital feeble-mindedness [*angeborenen Schwachsinne*]."[84]

Madness Unbound from Reason

During the first several decades of its emergence, psychiatry introduced a radical possibility that it eventually tempered before largely abandoning: that madness could be severed from the bounds of reason, that there could be a *madness of conduct* alone. While presentist hindsight might chalk up the recognition of the madness of conduct to simple diagnostic error, what we must recognize is that at the time of its promulgation, the possibility of a madness unmoored from reason amounted to a *discovery*, a revelation that transformed a core supposition about human nature: that the precondition for entry into the social order, and our attendant sense of moral autonomy, were not predicated on reason as an Enlightenment sensibility might have had us believe. It was not rationality that held the social fabric together but *propriety*, and as prominent clinicians observed, the adherence to social norms, conventions, and everyday practices need not have been prompted by reason alone. Propriety, it turns out, was a performance that could be executed by the rational and irrational alike, for pathologies could disrupt socio-behavioral norms just as readily as they could reproduce them, as we saw in the previous chapter.

The madness of conduct was, furthermore, a psychiatric discovery predicated on a style of reasoning capable of supposing that the psychological and behavioral dimensions of a person could exhibit divergent morbidities and, consequently, that the disorders of the mind did not necessarily presuppose the disorders of conduct. But even as the prominence of the moral insanities waned, the style of psychiatric reasoning that they inaugurated did not.

For just as the moral insanities of the early nineteenth century appeared to illustrate that an otherwise reasonable mind could be betrayed by deranged behavior, so did the neuroses of the turn to the twentieth century seem inversely to confirm that an unsound or at least semi-pathological mind could be allayed by proper social action. If psychiatrists were indeed the medical stewards of the social order as they claimed, then the dangers of moral pathologies and the mitigations of psychoneuroses depended at least in part on the medical capacity to diagnostically partition conduct as a discrete domain of medical analysis. For Tuke, moral insanities and neuroses shared an underlying etiological affinity. Drawing once again on Jackson, Tuke argued that the aberrations of moral behavior, temperament, and character were—like imperative ideas—nothing but the upsurges of automatic processes that "have fuller play than is normal" and that "are themselves healthy [but] are only out of proportion."[85] Moral insanity, Tuke maintained, had become far too synonymous with depravity, heinousness, and the performance of "absolute crime,"[86] when in reality, moral disorders were for the most part only slight deviations from conventional behavior.[87] In that sense, even eccentricity was a kind of moral pathology, a breach in the normative customs and conventions of conduct, typologically consanguineous not only with the automatist but also with the neurotic.

To the extent that the borderland between sanity and insanity was always liable to frustrate behavioral conventions as well as the customary norms of interpersonal social legibility, in order to answer the question of when and under what circumstances a social performance could be acceptably pathologized, psychiatry needed to transform its tacit style of reasoning about human conduct into an explicit medical program. And there was no better embodiment of such a program than the writings of Tuke's colleague the English clinical psychiatrist and self-styled philosophical anthropologist Charles Arthur Mercier. Mercier was an ardent disciple of Jackson, an advocate of the belief that insanity was only ever comprehensible through the "law of dissolution,"[88] even admitting at one point that it was "not always easy to separate the views which have been derived from [Jackson's] direct inspiration from those which I have arrived at by independent pursuit of the methods of thought which he teaches."[89]

Unlike his mentor, however, Mercier's writings eventually adopted a much wider perspective on the social, political, and philosophical ramifications of mental illness.[90] Throughout his career, Mercier was singularly focused on one task: to establish conduct, and conduct alone, as the primary object of psychiatric judgment, to remake the medicine of the mind into a medicine of social practice. Mercier latched onto and revitalized what in the history of psychiatry was only an implicit style of reasoning, thereby transforming conduct not only into an explicit cornerstone of psychiatric medicine but also into the organizing principle of a philosophy of human nature. (In order to bolster the legitimacy of his distinctive style of medical reasoning, Mercier went so far as to publish a four-hundred-page tome on propositional logic and analytic reasoning, a so-called new logic.[91]) The stakes of Mercier's program are evident in the opening lines of a monograph he published in 1911, titled *Conduct and Its Disorders*, which professed to be one among a lamentably few thoroughgoing studies on the topic of human conduct itself—its nature, permutations, and classifications:

> It would appear, *prima facie*, that few studies are more important than that of Conduct; for man is engaged in conduct during the whole of his waking life, and even the times, places, and occasions of sleep are parts of conduct. Conduct is what we are all engaged in, from birth to death.[92]

Conduct and Its Disorders represented a modernized gloss on what had once been the aim of premodern conduct literatures, a theory of human action in relation to philosophical ethics, newly elucidated in light of conclusions drawn from the mind and brain science, thus comparable to John Dewey's later *Human Nature and Conduct*.

For Mercier, conduct was the single most enduring expression of human nature. One could speak of the conduct of children, even infants, as readily as one might the conduct of the elderly; health, class, religion, gender, and ethnonationality were all expressions and subdivisions of the human condition that could be described according to the parameters of conduct alone—a subject matter, moreover, that could be investigated and cataloged without recourse to murky propositions concerning the nature of interiority or the metaphysics of the mind. Mercier believed his study of conduct was more than a medical inquiry; it was a systematic science of human action—a "praxiology," as he dubbed it.[93]

Mercier's "praxiology" should not be misconstrued as a simple behaviorism. Far from eschewing the experiential data of psychology, Mercier's study of conduct was more akin to the psychological functionalism of his American contemporary the philosopher and psychologist George Herbert Mead, for whom psychology was behavioristic only to the extent that "the approach to experience is made through conduct."[94] In slight contrast, and by adopting a sweeping *biological* functionalism, Mercier cast a wide classificatory net when he defined conduct as any purposive action oriented toward an end, even if the purposiveness was only evolutionary in nature. A deliberative act, even mental reasoning, were as much expressions of conduct as were automatic

behaviors, habits, and instinctive impulses, the latter of which were undoubtedly "reasoned" and goal-driven from an adaptational point of view.[95] Purposiveness and meaning were not limited to the realm of psychology alone
nor restricted simply to actions to which conscious subjective value could be
attributed; involuntary behaviors, which made up such a large portion of ordinary life, were profoundly suffused with purpose and sense.[96] It was not, in
other words, a simple behaviorism that Mercier espoused but a radical theory
of action, profoundly informed by the data of biological and psychological
medicine.

Psychiatric Praxeology

By the end of his career, Mercier had become particularly disparaging with
respect to psychiatric theories that blindly endorsed a purely mental view
of madness, which ultimately garnered the former president of the British
Medico-Psychological Association a reputation as an especially pugnacious
colleague and interlocutor.[97] The mental view of insanity, Mercier wrote at
one point, "which is sanctioned by universal consent and unanimous practice, has for some years past been challenged by me." He continued: "I define
insanity not in terms of mind, but in terms of conduct and action. . . . He is
insane, not whose mind is unsound, but whose conduct is unsound."[98] While
few of his colleagues accepted this proposition in principle, Mercier felt that
the profession nevertheless confirmed it in practice; since direct insight into
the state of the mind was impossible, psychiatrists had no choice but to look
to conduct as the only arbiter of mental health. As one English physician
recalled, when Mercier first presented his theory of conduct to the British
psychiatric community, it was initially judged to be specious and inexact, only
later to be so ubiquitously adopted that "it was argued . . . his views were common property and required no assistance to recommend them."[99]

Mercier first formally presented his position on insanity as the derangement of conduct in 1881, a view for which he advocated on both methodological and conceptual grounds (and which he continued to defend until
his death in 1919). Mercier never denied that insanity was certainly a mental
disorder. However, access to a patient's mental state was only ever possible
through inference, speculation, and trust in self-reporting. Conduct, on the
other hand, amounted to incontrovertibly observable data.[100] But there was
an even more substantial rationale for why conduct should have been the
"primary element in the concept of insanity," not "merely a symbol of the
mental condition."[101] Mercier's commitment to a Jacksonian view of nervous
physiopathology meant that he too adopted the belief that insanity referred to

the nervous dissolution of higher brain processes and the consequent overaction of disinhibited lower functions. Since the highest levels were merely the most complex coordinations of sensorimotor processes, it would be unlikely for true insanity *not* to manifest in the form of behavioral abnormalities. Nervous dissolution could not, Mercier believed, be limited to intellectual impairments alone because the highest nervous centers governed the most developed functions related to intellect, character, *and* behavior. Disorders of conduct, then, were not incidental phenomena but, rather, the direct manifestations of insanity.[102] "The alteration of conduct," Mercier explained in a distinctly Jacksonian register, "is seen in the relinquishment of work and in the 'automatic' movements which these patients perform."[103]

Approaching insanity through the standpoint of conduct yielded important forensic advantages as well. As to whether a patient should, for instance, be deemed capable of "managing himself and his affairs," there was, according to Mercier, only one avenue to making a proper medical adjudication—through "inference from observation of whether he *does* manage himself and his affairs capably."[104] Mercier never denied that some mental disorder was always present in insanity; but it was only present in the same way that "salt is always present in sea-water," that is, as an integral component but not as its essential identity.[105] For Mercier the disorders of the mind could designate a broad array of psychological irregularities, from perceptual errors to emotional instabilities and deficiencies in reasoning. Such states could be described as mental abnormalities, but they did not necessarily warrant the label *insanity*.[106] A patient's insanity should be judged "not on what he thinks or feels but on what he *does*; not upon inferred disorder of mind, but on perceived disorder of conduct."[107] Mercier's point was not to dissociate insanity from a disorder of the mind but simply to point out that fully tethering insanity to the mind alone would present significant diagnostic disadvantages; if we were to label every mental abnormality as insanity then, given that almost everyone experiences some degree of neurotic irregularity, we would all by necessity be deemed insane.[108]

Mercier was certainly aware of the counterintuitive novelty of what he proposed, given that the purview of psychiatric medicine had traditionally been restricted to the health of the mind alone. To extend that purview to a holistic appraisal of behavior, as well as to the questions of context, legibility, and propriety that would inevitably ensue, would necessarily alter the boundaries of professional inquiry. For Mercier, conduct was more than a simple diagnostic convenience; it was an expression of the vital essence of human life itself. In his earliest writings, and (like Jackson) drawing heavily on Herbert Spencer, Mercier adopted a conception of normative vitality, or health, which he defined as an organism's successful "adjustment" to its environment.[109] In the broadest sense, conduct referred to how any organism sought to modify itself in the face of environmental fluctuations; conduct was, in this sense, an organism's adaptational self-tuning to the vicissitudes of its world. "The dynamic adjustment of the organism to its environmental conditions," Mercier wrote, "which we call conduct, is the outward manifestation of intelligence. The only criterion of intelligence is conduct, and by conduct alone can we judge of the amount or even the existence of intelligence in others."[110] Conduct was "intelligent" insofar as an organism could successfully adjust itself to the demands of its environment; intelligent conduct was an expression of vital health. An unsuccessful modification in relation to an environment— a maladjustment, we could say—would be the occasion for pathology.

By representing a person's inability to adapt and acclimate, disorders of conduct could on their own, and without any reference to the mind, signal the presence of insanity. Mercier's theory of conduct as a vital praxeology had one particular benefit for psychiatric medicine. It could demonstrate that the discovery of moral pathologies, of a madness of conduct alone, was not an incidental finding in the history of psychiatry but that it represented the truest expression of the psychiatric enterprise. Psychiatry had erred, both practically and theoretically, when it abandoned its earliest approach to the moral insanities, when it sought to recouple madness to reason and the intellect. "The battle as to the existence of moral insanity is not over," Mercier wrote, in an article he coauthored with George Savage. "We, on our part, wish to re-state our belief in moral insanity, and to go one step further and show that breaches of the conventional as well as the moral laws of society may be but symptoms of disorder or disease of the higher nervous system."[111] The disavowal of moral insanity amounted to a professional liability because without the ability to distinguish mental pathologies from behavioral pathologies, psychiatrists would be incapable of making sense of rarer incongruencies between conduct and the intellect.

In order to make the case for the hermeneutic advantages that the category of moral insanity provided, Savage presented his own case of an intellectually impaired young woman who had inherited a sizable property (he does not specify the nature of that impairment, save that she was incapable "of doing the simplest sum or of making the most ordinary money transaction" or of "reading or taking in from books any simple abstract notion"). What was striking about this case, however, was that the woman represented something of an antipode to the typical structure of moral insanity, of a rational mind betrayed by deranged behavior. For in this case, the patient presented an allegedly impaired mind that was, however, mitigated with proper and normatively acceptable behavior, especially regarding the management of her estate, precisely because she "understood her incapacity, and allowed others to act for her."[112] Against the standard portrayal of moral insanity, Savage's patient represented the inverse case of a purportedly impaired mind that was, however, offset by *appropriate* action. By surrendering the management of her financial affairs, the patient demonstrated her psychiatric health because while her mind might have been deemed medically diminished, her conduct certainly was not.

The clinical divisibility of mind from behavior was an affordance that only the style of reasoning about moral insanity and disorders of conduct made possible; thus, its clinical repudiation, as Savage saw it, ultimately hindered psychiatry's hermeneutic precision. There was tremendous value in carving out conduct as an autonomous sphere of medical analysis, which was why Savage and Mercier in particular continued to deploy the category of moral insanity for decades.[113] Savage's patient was not in the end morally insane, but it was nevertheless the structure of moral insanity, and the style of reasoning on which it was predicated, that allowed him to recognize her health and, subsequently, her moral autonomy. But, as we must not fail to note, her health and autonomy were ultimately redeemable on *economic* grounds, an effect of what was perceived to be her proper financial conduct. It was not just *any* behavior that could offset her supposed impairment and preserve her from the stamp of madness and psychiatric morbidity.

Economizing Madness

By turning to conduct as the crucial metric for psychiatric appraisal, it was almost inevitable that embedded norms of social propriety would surreptitiously, if not overtly, function as surrogate benchmarks for clinical judgments. Economic behavior was undoubtedly the most readily available and methodologically expedient proxy for behavioral propriety because it was a form of conduct that already possessed an internal set of value metrics. The medical and moral virtue of any social performance could be transposed into the simple arithmetic calculation of economic utility: remunerative forms of behavior, actions that either protected or accumulated wealth, could be assigned the status of medical normalcy and social propriety, while forms of conduct deemed costly or perceived to be depreciating or financially ruinous could be assigned the status of irregularity, disorder, and disease. It mattered less whether the economic conduct was truly prompted by reason or incited by pathological impulses; to that end, madness was not, as we will see, merely incompatible with proper economic action. At stake was ultimately the question of whether value was perceived to be gained or lost, for it was around this appraisal that normalcy and propriety were determined.

As Mercier observed, the disorders of conduct posed a threat to the social order less because they brought dangerous patients before the law but more frequently because they "[brought] scores of spendthrifts to ruin every year."[114] A distinctly economic style of appraising conduct is evident in Mercier's writings above all in large part because Mercier, more than any of his colleagues and medical contemporaries, took the style of reasoning about conduct disorders and moral insanities to its furthest logical conclusions. If conduct truly was, as Mercier insisted, "the most important of all the factors that the alienist has to consider, for by conduct and by conduct alone, can we judge of sanity or insanity," then how ought the clinician evaluate a patient who presents what would otherwise count as the most pronounced illustration of genuine madness—namely, a delusional state of mind, wherein a patient "sees a spectral cat where no cat is"? For Mercier, the answer was simple; hallucinations alone were not sufficient grounds for a clinical diagnosis of insanity. "If the appearance of the cat does not evoke from him any corresponding phase of conduct," Mercier explained, "then although we admit that his mind is disordered, we do not regard him as insane."[115]

Hallucinations were simply the strongest expression of a fact that held true for all people—that no mind was ever fully unencumbered by some degree of unsoundness. What this implied is that if mental abnormalities alone were the sufficient basis of an insanity diagnosis, then anyone could be counted among the clinically insane. How was it possible to admit to the ubiquity of mental abnormalities while still managing the boundary between health and pathology? The answer relied upon the diagnostic capacity to partition the mind from conduct in order to distinguish the healthy from the insane on the grounds of conduct alone, even if the conduct in question might have mitigated something as seemingly grave as a hallucinatory episode. "If [a man's] mind be subject to delusion," Mercier continued,

> so that he fancies, for instance, that he is a man of unlimited wealth, or that he is in dire poverty, yet if, in spite of this delusion, he manages his financial and other affairs with prudence and success, it would be impossible to regard him as insane. But the moment that he begins, in consequence with his delusion, either to draw cheques from impossible amounts, or to refuse himself and his family the necessary means of subsistence, from that moment he is regarded as insane.[116]

Few clinicians were willing to follow Mercier to such ends (even Savage was skeptical).[117] For Mercier, however, as I elaborate in more detail in chapter 3, an economic metric proved to be the most effective, efficient, and robust means of assessing the medical status of conduct. This had much to do with the fact that psychiatry and economics shared a commensurate professional commitment to the study of human behavior. While Mercier remained convinced that psychiatrists were society's most capable praxeological experts, he nevertheless observed in the opening pages of *Conduct and Its Disorders* that praxeological expertise was something psychiatrists shared with another "department of conduct," namely, political economists, who had been pursuing the "systematic investigation" of conduct "for generations."[118]

Mercier acknowledged that, in different ways than psychiatry, economics was as committed to a praxeological theory of action as it was to setting the terms of behavioral propriety and the normative adjudication of right conduct—a point the Austrian economist Ludwig von Mises would confirm decades later in *Human Action*, his "praxeological" study of free market capitalism (von Mises relied upon French sociologist Alfred Espinas's notion of "praxeology" rather than Mercier's, which he does not mention and likely never heard of).[119] The groundwork for economic analyses of right conduct emerged with the rise of marginalist economic theories in the 1870s. Marginalism represented a narrowing of the field of economic inquiry, a shift from questions of production and labor to more "microeconomic" analyses into the nature of consumption and the so-called subjective theory of value in relation to scarcity—a development in which the economic problem of utility maximization intersected with questions concerning psychology and social behavior.[120] Just seven years after Mercier published *Conduct and Its Disorders*, the American economist John M. Clark—son of John Bates Clark, an early adherent of marginalism in the United States—asserted that economics was ultimately "a science of human behavior."[121] (Marginalism would lay the groundwork for what would eventually become the veritable synonymy between economic action and rational behavior *as such*.[122])

Given their respective praxeological orientations, then, it was possible to view psychiatry and economics collectively not just as subbranches of the human sciences but as complementary expressions of novel moral reasoning, newly adapted into medical and economic terms.[123] Mercier for his part was adamant that it was the duty of the psychiatrist to "trench as alienists so often have to trench, upon the provinces of the jurist and ethical philosopher."[124] And it was precisely an economic style of psychiatric reasoning that facilitated just such an encroachment into the moral domain. Even Freud observed the kinds of medical and moral revelations that economic behaviors afforded. In a brief digression on the "ailments that afflict a greater number of my healthy acquaintances as well as myself," Freud described what he took to be the frequent act of forgetting to pay a bill or to pay off a debt, a neurotic breach of "conventional duty" (*konventionellen Pflicht*). "Among the majority even of what are called 'respectable' people," Freud explained, "traces of divided behavior can easily be observed where money and property are concerned." There appeared to be in commercial society a morbid "counter-will [*Gegenwille*] against paying out money."[125]

What many psychiatric practitioners, Mercier being most singular among them, seem to have implicitly recognized by the turn to the twentieth century was that human conduct had come to possess a form of social value that was at once, and indistinguishably, moral, medical, and economic in nature. It was simply the relative unambiguousness of economic forms of valuation that allowed them to function as the most felicitous gauge of medical normalcy and moral propriety. The psychiatric turn to conduct in general, and economic conduct in particular, could therefore be viewed as an effort to resolve the hermeneutic challenge concerning the legibility of social performances—a hermeneutic challenge that psychiatry itself instituted. Economic behaviors were among the most socially embedded forms of everyday action, and the most readily transposable to a simple, and *profoundly legible*, arithmetic of valuation; for that reason, focusing on a person's economic conduct could help bypass the diagnostic indecipherability that the borderland typology introduced.

Whether eccentrics or neurotics were truly healthy or whether routine social performances actually masked an underlying pathology were each medical quandaries that could be readily adjudicated by simply looking at how wealth, money, and financial affairs were managed. Through such observations, pathological suspicions could be mitigated, and in some cases absolved, by proper economic behavior. And from this possibility emerged the startling clinical supposition that pathology and economic aptitude could in fact coexist, that a person could be mentally disturbed and yet economically quite normal and sane. The coincidence of madness and economic competence, furthermore, takes on another appearance, which I discuss in the next chapter—that madness was more than simply consistent with proper economic conduct; it could, as we will see, *benefit* certain kinds of economic ventures as well, and it may have been not only valuable but quite lucrative to allow a degree of madness to circulate throughout the economic order. Madness could very well bestow upon normal economic behavior a novel and potentially profitable form of pathological value.

3

From Disorders of Enterprise to Entrepreneurial Madness

The Psychiatry of Enterprise

When nineteenth-century psychiatrists discussed economic behavior, they often did so by turning to the alleged hygienic and therapeutic effects of occupational toil and employment. The moral and, after the eighteenth century, medical value of personal industry and labor over idleness was an imperative that preceded the formalization of psychiatry, and it became an obligation that psychiatrists adopted and refashioned. English private asylum owners of the early nineteenth century described daily labor as both a prophylactic against mental disease and a means by which "incurable maniacs . . . may still acquire the habit of rendering themselves useful to society."[1] In the United States, Isaac Ray suggested that what he perceived to be the great frequency of insanity among women was potentially due to their exclusion from the workforce, a point the American neurologist Silas Weir Mitchell echoed some years later.[2] Since labor practices within asylums were normally disconnected from market economies, the value of therapeutic labor was not monetary so much as it was about forging and reinforcing the presumption that health and productivity were linked, that mental soundness signaled a propensity toward productive work.[3]

As the association between psychiatric well-being and occupational productivity grew and normalized throughout the nineteenth century, so too did class-based assessments of mental illness and, along with such evaluations, the formalization of population typologies such as the "pauper insane" and institutional disparities in available treatment, from public asylums to private clinics and "middle-class" hospitals.[4] Although entire genres of mental pathology were developed in specific relation to industrial labor (the topic of chapter 5), the psychiatric preoccupation with class difference did not simply amount to the wholesale pathological condemnation of the poor and working classes. Daniel Hack Tuke described the pathogenic propensities among the wealthy, writing, "Certain mental causes of lunacy, as over-study and business worry, produce more insanity among the upper than the lower classes."[5] The English neurologist William Gowers claimed that slight moral abnormalities had to be measured by their deviations from socioeconomic standards of conduct. "Many actions would be more distinctly pathological in a man of refinement," he wrote, "than in an ill-tempered man of the lower classes."[6] While disorderly and potentially criminal pathologies were more likely to be medically attributed to the poor and working classes, eccentricities and moral lapses tended to be more apparent among wealthier patients. An indisputable market ideology saturated the psychiatric imaginary, as a means by which contingent socioeconomic structures and social ideals related to work, wealth accumulation, and class difference could be infused with medical validity and thereby normalized.[7]

An economic style of psychiatric reasoning, which this chapter will begin to explore in earnest (and the diversity of which will be delineated in subsequent chapters), was undoubtably the expression of just such an economic ideology. At the same time, however, such a style of thought was not simply reducible to the category of *ideology* alone, if that term is meant strictly to denote a dominant and obfuscatory system of ideas, attitudes, and representations whose function it was to mystify material modes of production and the realities of class struggle. In the previous two chapters, I sought to explain why an economic style of reasoning would become crucial for psychiatry, as a stratagem that could ensure behavioral intelligibility and social coherence, and as a technique capable of recasting the constitutive components of human nature according to novel benchmarks of propriety and acceptability. In the first instance, an economic style of psychiatric reasoning signaled the formation of a new regime of truth with respect to conduct, a reappraisal of the very ontology of social performances according to forms of value that had become indistinguishably moral, medical, and economic. Such a style of reasoning, and the new ontologies of human nature that it inaugurated, while not a vulgarly simplistic expression of capitalism, nevertheless served undeniably ideological ends, in part because of how propitiously it broadened the types of behaviors that could now be perceived as profitable. For through this style of reasoning, health alone no longer occupied the sole medical condition conducive to economic prosperity; madness too, as we will see below, could be just as lucrative, if not more so.

As a point of entry, this chapter begins with a particular figure that frequently circulated throughout nineteenth-century transatlantic psychiatric discourse—namely, the agent of commerce, who was not solely defined through labor imperatives or class affiliations but through a behavioral aptitude with respect to business, a capacity to engage in commercial activities, to partake in trade and market transactions, and to accrue and manage wealth. For the agent of commerce, the benchmark for mental health had long revolved around success in enterprise and prosperity in commercial conduct, while mental illness was believed to be precipitated by, and thus coincident with, shortfalls and financial losses. When, in 1829, Étienne Esquirol listed the admission numbers to his asylum at Charenton according to profession, landlords (*propriétaires*) and shopkeepers (*commerçants*) ranked just under military professionals.[8] For the French forensic psychiatrist François-Emmanuel Fodéré, the order was reversed; "shopkeepers and merchants who had made ruinous speculations [*des speculations ruineuses*]" were the most common professions to be represented in the asylums, followed by members of the military.[9] What was clear for medical experts was that the agent of commerce took on a substantive degree of both financial *and psychological* risk. The source of such risks were the ever-present "reversals of fortune" (*revers de fortune*) that accompanied commercial enterprises, and were viewed as among the most potent sources of insanity. For some clinicians, such reversals were likened to a blow to the head or even a fall, given the degree to which they could induce pathological modifications to one's personality.[10] According to Henry Maudsley, the terror of poverty and the longing to become rich were among the most harmful attributes of modern civilization. Particularly in the context of trade and commerce, "speculations of all sorts are eagerly entered on, and . . . many people are kept in a continued state of excitement and anxiety by the fluctuations of the money market."[11] George Savage reported that the British Lunacy Commission in 1911 listed "business anxieties and pecuniary difficulties" as among the top causes of mental disease.[12]

Financial despair was long viewed within transatlantic psychiatry as an indelible pathogenic danger, and one to which a particular economic cross section was most susceptible—namely, the North Atlantic middle class. Psychiatric clinicians were rarely explicit about class identity, though it is clear that the agent of commerce could represent a relatively broad economic spectrum, from traders, shopkeepers, and independent proprietors of the lower middle merchant classes to so-called white collar professionals, which included not only physicians, lawyers, and engineers but also bankers, stockbrokers, and clerks, that is, the purveyors of financial capital, who were among the fastest-growing workforce across the nineteenth century.[13] In that sense, the agent of commerce, who was as much a dutiful *homo economicus* as an enterprising entrepreneur, did not represent a strict socioeconomic category so much as a "class situation" in Max Weber's sense—that is, a style of life characteristic of how one's "life chances" and economic interests were determined with respect to market conditions.[14] (And for that reason, the agent of commerce undeniably reflected the class bias of psychiatrists themselves, an inadvertent commentary on their own economic position as members of a rising professional group.) The agent of commerce was therefore situated somewhere between financial elites whose wealth was likely inherited and did not rely on salaried income, and who would not likely feel financial successes or losses quite so keenly, and the laboring classes who might not have had the resources to partake in financial enterprise in the first place; this is what made the agent of commerce most susceptible to the assortment of injuries—financial and psychological—that often arose from business ventures.

On the other hand, the long-held belief that financial enterprise was pathogenic was not simply due to the assumption that economic ruin alone was a source of distress; shocks associated with sudden financial gains could be comparably traumatic. At the same time, the agent of commerce was not portrayed solely through the morbidities of business life. For some clinicians, financial enterprise could embody the most authentic expression of mental health and biological vigor, indeed to such an extent that the successes brought about through economic endeavors could potentially temper medical suspicions of insanity, even when all other signs pointed to mental disease. The dichotomy of economic success and failure, as well as that of health and pathology, made up a dynamic interplay for the agent of commerce, the so-called (and unambiguously gendered) businessman, who by the turn of the twentieth century was as much a model for neurotic vulnerability with respect to the vicissitudes of economic modernity as he was a paragon of adaptational prowess, fitness, and financial expertise. I refer, in this second sense, to a crucial thematic that will be discussed below: the formalization of the concept of the entrepreneur, the so-called man of action, who amplified and even distorted many of the attributes of the ordinary financial enterpriser. In many respects, as we will see, the entrepreneur must be understood as having been fashioned at the intersection of economic *and psychiatric* thought. To the extent that the entrepreneur was claimed to be in possession of a mind that deviated from the ordinary reasonings of *homo economicus*, a deep and underlying affinity linked entrepreneurial virtuosity to conceptions of mental pathology—a pathological disposition, however, that was deemed vital for economic success.

This chapter, therefore, explores one particular iteration of an economic style of psychiatric reasoning, in this case organized around how the portrayals of business enterprise and financial conduct—including commerce, speculative ventures, and wealth management—were linked to diverse psychiatric appraisals in such a way as to function as robust benchmarks with which to evaluate health and pathology. To that end, this chapter examines three interwoven psychiatric suppositions that increasingly gained traction by the start of the twentieth century. I present them here sequentially only for didactic purposes (and not necessarily in the order that I discuss them below), for they must otherwise be understood as concurrent and interlocked components of a single and comprehensive constellation of psychiatric thought.

First, it was commonly held that business enterprise could be, and often was, pathogenic, in terms of the anxieties and mental turmoil it caused, the losses it could produce, and the sudden and excessive windfalls that it could bring about. However, and slightly mitigating the first supposition, is the second—that many clinicians came to the conclusion that financial conduct did not actually demand that an economic actor be mentally healthy, that sanity was not in fact a prerequisite for *homo economicus* at all. In this respect, madness and financial competency could coincide in a number of ways, all having to do with what I described in the preceding chapter as the partitioning of conduct from the intellect. A medically impaired mind did not guarantee that conduct would be equally diminished, and many observations seemed to demonstrate that insane patients were perfectly adept with respect to money.

Third, and entangled with the other two suppositions, was a combined psychiatric-economic hypothesis that posited that sometimes a bit of madness was quite integral to economic success. This latter point concerns the developing formulation of the entrepreneur as a financial virtuoso—a genius whose proximity to madness was predicated on an ability to perceive things that were not actually in the world, what both economists and psychiatrists claimed was the ability to observe "new combinations" in the data of ordinary experience. Entrepreneurial virtuosity was more than simply compatible with madness; it relied upon it, often at great risk, for "virtuosity" was the term used to describe the entrepreneur who succeeded in instrumentalizing irrationality for the sake of profit, whereas "madman" was the term used to describe the enterpriser who failed. Therefore, with regard to financial endeavors, we can contrast *pernicious* pathologies, which were costly and ruinous both fiscally and psychologically, with *lucrative* pathologies, which not only signaled an (advantageously) abnormal intellect but also the likelihood of turning a profit. We can see in this dichotomy how the boundary separating mental health and pathology was recast according to the benchmarks of economic value, and in such a way that madness was not merely something to be discredited, for, as it turned out, pathologically derived forms of value could be far more rewarding than any wealth derived from sound and rational minds alone.

At the heart of this chapter, however, lies an essential conflict, which I propose an important though often overlooked attribute of entrepreneurialism could properly resolve. How was it possible to be *profitably* mad without, however, being driven insane by the volatilities of capitalism and market economies? What assurance was there that the right form of madness could operate in the right way during the course of business undertakings, and in such a way as not to yield to the pernicious morbidity ordinarily associated with commercial life? That assurance, I will argue below, hinged on the fact that entrepreneurialism was not simply a pathological form but also a form of what I call *pathological privilege*, that is, a performative entitlement largely organized around a very specific set of gender and racial boundaries. Entrepreneurialism functioned in large part as a kind of inoculant against the virulence of capitalism, for like an inoculant it possessed something akin to a viral constituent—namely, a degree of madness that could potentially immunize an agent of commerce against the pathogenic ravages of commercial life. For if the entrepreneur were already a little mad, then the pathologies associated with capitalism could be co-opted as an asset—a tool for success—rather than as a liability to which an economic actor might otherwise succumb. Entrepreneurial pathology was the protective bulwark that enabled an agent of commerce to be profitably mad without falling victim to the pathogenic perniciousness of economic life.

And yet, the privilege of successfully adopting and enacting the profitable *and protective* madness of entrepreneurial virtuosity was not available to just anyone, and it is here where the gendered and racialized attributes of the early twentieth-century North Atlantic entrepreneur come sharply into focus. As this chapter goes on to argue, entrepreneurialism was a form of lucrative pathological behavior that doubled as a particularly inflected performance of both masculinity and whiteness. It was not simply that the entrepreneur implicitly signified a North Atlantic white man, which was undeniably the case. But more than that, entrepreneurial pathology and masculine whiteness came to possess an interdependent and mutually enabling relationship with one another. On the one hand, it was by way of masculine whiteness, which already possessed a substantial degree of economic privilege, that entrepreneurialism could circulate through the social body as an irrational prerogative, an entitlement to pathology. Conversely, by being partially restyled through the norms of entrepreneurial conduct, even the most virulent and morbid characteristics of masculinity and whiteness could be recoded with social and economic value and thereby perversely redeemed. To the extent that entrepreneurial madness was a privilege of masculine whiteness, lucrative economic pathologies were not only forms of conduct structurally foreclosed to nonwhite enterprisers but, in fact, contingent on the fact that the only madness nonwhite enterprisers could experience were the traumatisms of racialized economic violence. As we will see, the entrepreneur's entrenchment within the logic of racial capitalism was precisely what assured both its profitability and its immunity against capitalism's pathological perils.

The Diagnostic Value of Economic Conduct

As we observed in the previous chapter, it was Charles Mercier's strenuous belief that a patient's conduct alone could reveal the truth of their psychiatric health. But when he proposed that even in delusional states, the appropriate management of financial affairs mitigated and even rebuffed the presumption of insanity, Mercier inverted the causality of a standard and somewhat commonsensical *juridical* approach to the relationship between mental illness and financial self-management that had long prevailed throughout much of Europe and the United States. Isaac Ray summed up that approach best when he wrote, "The business of the jury . . . is, to ascertain whether the individual is mentally capable of managing his affairs," an assessment grounded on a principle of protectionism for the sake of both the individual and "those who deal with them, unacquainted with their mental condition."[15] This common juridical style of reasoning proposed that the fact of insanity voided in advance and without qualification the possibility of sound economic conduct. For this reason, it was necessary to assess the mental status of a person *before* allowing them to engage in economic activity, for an insane person was believed in principle to be incapable of properly entering the economic domain.

In contrast to the juridical style of reasoning, which mirrored legal codes concerning the limitation of economic rights for those dubbed non compos mentis, an economic style held that pathology and economic fitness could actually coincide, and that psychiatric assessments concerning financial competency should not precede but, rather, follow actual economic activity, for only then could the clinician truly appraise a person's state of health. This was not the first time that idiosyncratic styles of thought within psychiatry opposed commonsensical juridical conclusions; the very category of moral insanity effectively defied a long-standing legal supposition that madness was an intellectual impairment alone, functioning in that regard as the proxy for a professional boundary dispute between psychiatric and legal expertise.[16]

Mercier was not the only clinician to promulgate this economic style of psychiatric reasoning; he was simply the most emphatic. Decades earlier, Ray had come to some very similar conclusions in the United States. "We see some [moral maniacs]," he wrote, "managing their affairs with their ordinary shrewdness and discretion, evincing no extraordinary exaltation of feeling or fancy, and on all but one or a few points, in their perfect enjoyment of reason." For Ray, an economic style of psychiatric reasoning underwrote a broader commitment to unqualified economic liberty and the belief that the sanctity of property and wealth accumulation was too fundamental a right to be interfered with by physicians or courts over the mere suspicion of unhealth. Preventing purportedly mentally ill individuals from managing their affairs, or invalidating their contracts under the guise of protectionism, amounted to a form of civil injury that far outweighed whatever potential financial damages might have otherwise putatively arisen.[17] Although the injunction against the financial control of property and wealth was invoked in almost all serious cases of mental pathology, Ray felt that such invocations often lacked merit or, at the very least, precision. It was almost impossible to identify the exact "degree of imbecility which is incompatible with the management of property."[18]

But beneath this noninterventionist commitment to economic freedom lay the belief that madness and economic fitness could coincide. Even the mentally ill continued to possess the "natural right" to manage and dispose of their own fortunes, especially when their "mania does not involve [their] notions of property." To prevent a mentally ill person from continuing to enjoy economic freedoms would be "as unjust and irrational, as it would be to inflict upon a felon convicted of theft, the penalties attached to the violation of every article in the criminal code." For even in the profoundest states of insanity, economic competency was not in principle diminished. An individual might, Ray explained, suffer from severe hallucinations, even with respect to their wealth and property; such a person might, for example, espouse the idea that they are in possession of "immense wealth, or that every ship which enters the harbor is his and freighted with his goods." But even here, "we are not too hastily to strip him of what is really his own, for he might, nevertheless, in the management of it, evince the most commendable prudence and economy." Ray reminds his readers of a "remarkable, but not an uncommon fact" that "monomaniacs often make no practical application of their insane notions to their own conduct or concerns, but continue to manage both, as if no such delusion existed."[19] If the severest forms of mental illness did not in principle abrogate economic competency, then even the presence of insanity itself was not sufficient grounds to exclude the mentally ill from the economic sphere.

The compatibility of madness and economic capacity was, however, hardly a foregone conclusion within transatlantic psychiatry but, rather, subject to much contestation. In one of the first commentaries on the newly established French Civil Code of the early nineteenth century, the jurist Charles Toullier examined the topic of the legal interdiction of the mentally ill, that is, the certification—at that moment still performed only by judges—of suspected madmen, rendering them unable to manage their affairs.[20] While some undeniably severe states of insanity demanded interdiction, Toullier argued that mental illness did not, in principle, abolish a person's right to manage wealth. "To justify interdiction," Toullier wrote,

> the absence of reason must be relative to the ordinary affairs of civil law, that is, to an individual's administration of his person and property. He who goes astray in palpably false speculations, a delusional man [*un homme à visions*], should moreover not be restrained [*interdit*] if he governs his affairs well and the public has nothing to fear from his madness [*déraison*].[21]

Despite writing from the standpoint of a jurist, Toullier nevertheless posited that some forms of conduct could be analytically partitioned from mental states, thereby espousing what I have been calling an economic—rather than juridical—style of psychiatric reasoning. The more staunchly juridical appraisals of the economic competency of the mentally ill were, perhaps counterintuitively, found among the *psychiatric* reactions to Toullier's conclusions. "I do not believe [Toullier's] sound opinion," wrote Étienne Georget. For Georget, the behaviors of the mentally ill were by definition uncertain and unforeseeable. "One can never trust a madman."[22] Despite being one of the earlier promulgators of monomania and the genre of moral pathology more broadly, when it came to money, Georget adopted a far more intellectualist view of economic capacity. To the extent that insanity signified a loss of moral freedom in principle, a mentally ill person was inherently incapable of enacting the economic agency necessary for financial self-determination; in the long run, the patient's madness would eventually overwhelm any semblance of economic aptitude. "Left to their own devices, [madmen] give into excesses," Georget's mentor, Esquirol, would write some years later. "They dread the duties and the subjugations of the world and the anxiety of business [*la tracas des affaires*]."[23] The insane, Esquirol concluded, have de jure no head "to conduct commerce, to run their business affairs, to fulfill their jobs or positions."[24]

In England, and echoing his French colleagues, James Cowles Prichard likewise claimed that madness *in principle* precluded any possibility of managing one's own affairs. The mentally ill, Prichard warned, inevitably squandered their wealth, often to the detriment of their families, and therefore warranted the immediate "interference in the exercise of personal rights."[25] Affording an insane patient the freedom to conduct business would yield nothing but financial ruin, no matter how apparently competent such a patient might initially appear. For Prichard, Esquirol, and other early architects

of the moral pathologies, a juridical style of reasoning most convincingly elucidated the economic possibilities of the mad. These early clinicians did not,
in other words, take their own conclusions about the divisibility of conduct
from the intellect nearly as far as subsequent adherents (such as Mercier)
eventually did. Psychiatric conclusions concerning the compatibility of insanity with economic aptitude and financial self-management, therefore, could
vary quite categorically, and while an economic style of psychiatric reasoning
did not replace or oppose but rather coincided with a juridical style, there was
one way that both forms of reasoning were ultimately compatible—in their
shared conviction that economic conduct could always be looked upon as a
reliable indicator of the status of a person's psychiatric health.

The Parisian clinician Alexandre Brière de Boismont described the case
of a merchant who was brought to his private *maison de santé* for "acts of
licentiousness." In the earliest stages of his mental disorder, the patient's behavior did not seem out of the ordinary. "He went to the Stock Exchange
every day," Brière de Boismont explained, "had a lot of contact with people
in his profession, and no one noticed or reported anything about his mental state." Even after he was brought to the sanatorium, the patient still appeared reasonable, particularly with regard to his business affairs. Over time,
however, it was the patient's economic behaviors that first and most palpably
signaled the presence of mental disease. "His commercial position quickly
took a grave turn," wrote Brière de Boismont. Eventually, "the commercial
courts charged him with fraudulent bankruptcy; an arrest warrant was issued,
and a bailiff came to my establishment to execute the warrant."[26] Joseph Williams, a London physician, also observed that changes in economic behaviors
functioned as the clearest indication of the onset of mental illness. Certainly,
dramatic shifts in character and temperament could raise the suspicion of
possible illness. But what best revealed that the mind "has been 'turned'"
in a pathological manner were the emergent bizarreries in commercial and
financial activity; they functioned as a behavioral gauge—particularly in light
of losses incurred through unreasonable expenditures or extreme financial
speculations—through which latent psychiatric morbidity could be most singularly exposed.[27]

Economic conduct, therefore, possessed a polysemous indexical value.
Depending on the nature of financial activity, typically perceived through a
simple arithmetic of monetary gains or losses, behaviors in the economic domain could redeem the disavowed social status of the mentally ill and alleviate suspicions of madness among borderland patients just as saliently as they
could signal the likely presence of insanity among the otherwise seemingly
normal. Although Williams warned that erratic and disastrous financial activ-

ity likely augured psychiatric decline, he also insisted that "so long as a man manages his property with discretion, and neither injures nor threatens to injure himself or others, however eccentric in other respects he may be, yet he is not a fit object for control or for confinement." Eccentrics who were "in other respects very ridiculous," were often quite capable, Williams maintained, of managing their affairs and even acquiring large fortunes.[28]

This was the reasoning that in part underwrote a legal appeal by the German appellate judge Daniel Paul Schreber, who was removed from the bench in 1893 on the basis of mental incompetence after suffering from a serious episode of paranoid and delusional schizophrenia (Schreber recounted his experiences in his *Memoirs of My Nervous Illness*, which Freud famously analyzed in 1911[29]). In his efforts to have his position reinstated, an ultimately failed legal pursuit that lasted several years, Schreber insisted that "the only question" for determining competence was "whether I possess the capacity for reasonable action in practical life," that is to say, the capacity simply to "manage one's affairs." Even the severest mental abnormalities were immaterial, according to Schreber, with respect to the most pivotal matter of concern: whether he was able to *conduct* himself appropriately in everyday socioeconomic life.[30] In that sense, Schreber too deployed an economic, rather than juridical, style of psychiatric reasoning.

But economic conduct did not always function to extenuate psychiatric suspicions of mental illness. Some clinicians took a relatively equivocal tone, particularly when apparently sound financial activity coexisted with symptoms that they felt were simply too aberrant to ignore. The English physician Daniel Noble remarked upon his own frequent clinical encounters with patients who appeared "thoroughly sensible in the affairs of business, and in all the ordinary events of life," while nevertheless believing themselves to be "a prophet or an emperor," in possession of a "second spirit," or that their "arms are made of earthenware."[31] In such cases, psychiatrists were sometimes left at what they believed was a diagnostic impasse, finding themselves making seemingly paradoxical claims about the equivalence and compatibility between psychiatric health, economic competence, and social inclusion. Maudsley illustrated this impasse when he concluded that while "an insane person is sometimes competent to manage his affairs who is not fit to be entirely at large," there were also some who, "not being competent to manage their affairs, might very well be permitted to be at large after fitting legal provision had been made for the proper management of their property."[32]

The compatibility of madness and financial aptitude was, therefore, not a univocal psychiatric conjecture. It was a medical supposition that could assume numerous permutations and degrees of acceptance. The number of

clinicians willing to assert that an individual could be absolutely and unqual-
ifiedly insane while also entirely proficient with respect to financial affairs
were few, even if they were, like Mercier, quite adamant and outspoken in
their views. Others were less inclined to partition mental health from economic capacity quite so drastically. That being said, there was one conceptual
work-around through which clinicians could tacitly admit to the compatibility between mental illness and economic capacity without having to make
explicit declarations in that regard. Maudsley touched upon it when he observed that a great deal of the activities of everyday professional life become
profoundly routinized, often to the detriment of the continued hygienic cultivation of the intellect. Once a person has become acclimated to "an important
business or profession," the requisite conduct necessary to successfully carry
out that business could be performed "without much real mental activity—
almost automatically, indeed." Professional and economic undertakings, upon
habituation, could be carried out "unconsciously . . . [as] a sort of acquired
instinct."[33]

Here, we return to the notion of unconscious automatism, which, as I
detailed in chapter 1, was rooted in a widespread assumption within transatlantic psychiatry that everyday activities, including those of the economic
sort, could be habitually and unconsciously performed. But given its "essential ambiguity," it was never clear whether unconsciously automatic behaviors
expressed the normal routinization of everyday habits or whether they were
the manifestations of mental pathology. The acceptance of the *normativity*
of everyday automatisms introduced a somewhat radical possibility: that
pathological automatisms too could possess at least some degree of social acceptability so long as the conduct in question continued to adhere to social
customs and conventions. And as we saw with the case studies introduced by
Jean-Martin Charcot and John Hughlings Jackson, they often did.

Mercier, in particular, seized upon precisely such a possibility, that proper
forms of automatic conduct could be pathologically instigated. Following
Jackson and Charcot, Mercier likewise confirmed that many of his patients
in postepileptic automatic states "so behave as to pass muster as normal persons." Their pathologically induced unconscious behaviors so profoundly
mimicked ordinary conduct that for most observers, there was no reason to
suppose "that they were other than normal sane persons." What did it matter, Mercier provocatively proposed, whether they were mindless "automata"?
"Supposing that they went about their business," Mercier wrote,

> bought and sold, married and were given in marriage, performed all their
> duties to themselves, their families and society, in a normal manner, what is it

to a bystander whether they have minds or not? It would never occur to any onlooker to raise the question. They would go to their graves with the reputation of having been sane all their lives.[34]

For Mercier, the difference between healthy automatic conduct, which was merely the expression of routine, and pathological automatisms, which were morbidly induced, ceased to become a difference that mattered once psychiatrists admitted that behavioral propriety was being observed in either case. So long as a person's conduct, economic or otherwise, remained proper and acceptable, so long as customs and social conventions were adhered to, it would be unreasonable to conclude that such an individual posed a threat to the social order.

What ultimately mattered to the social body, Mercier effectively argued, was not whether people were rational, or even conscious for that matter, so long as they conducted themselves appropriately—an aspiration that depended far more on habituation than on rationality. When actions and behaviors conformed to social norms, the status of psychiatric health was not absolved or nullified so much as it was subordinated to a secondary matter of concern. Pathological automatisms that abided by social norms could be clinically disregarded, which is to say medically permitted, under the presumption that much of our healthy conduct was customarily performed unconsciously as well. The *source* of our behavior—reason or pathology—was ancillary to the more pressing concern in adjudicating mental health and illness: whether a patient's conduct fell sufficiently within the boundaries of social correctness. Consequently, while many clinicians might not have been willing to conclude, as Mercier so strongly did, that a delusional madman could be economically competent, the broader psychiatric concession that much of our conduct was unconscious and automatic opened a space for imagining precisely how mental pathology and economic competency could notionally coincide at least to some measure.

The supposition that some economic activities could be performed unconsciously and automatically did not, however, *blur* the boundary between the normal and the pathological so much as it rendered that boundary inconsequential, since the standard by which economic behavior was measured and appraised ultimately came down to whether it was properly performed, whether the correct procedures were followed, or whether the right rules were adhered to in the right ways. Consequently, and somewhat implicit in Mercier's economic style of psychiatric reasoning was a conclusion concerning the nature of *homo economicus* as such. When we speak of automatisms with respect to economic behavior, there is a tendency to turn to practices of industrial labor and the effect of treating workers as machines.[35] Here, however, the idea of automatism is ultimately a metaphor, a figurative analogue for the monotonies and mechanical doldrums, to say nothing of the dehumanizing degradations, of industrial wage labor.

In contrast, what Mercier's presumptions about the social normativity of automatic conduct reveal is that the most authentic economic automatist was, in fact, neoclassical *homo economicus* all along, whose most singular attribute was an unstated compliance with and conformity to what Thorstein Veblen somewhat reprovingly called the "hedonistic theory of conduct," that is, to rules of action (or "laws of human enjoyment," as William Jevons called them) that could be deciphered and predicted, rules that in the first instance were likely fixed through the repetitions of ordinary life.[36] The conduct of *homo economicus*, as Mercier likely saw it, was ultimately the expression of rule-boundedness, obedience to a proceduralism that arose more from custom and routine than from sustained calculations. If, as Mercier implies, the function of *homo economicus* could just as well be performed unconsciously and automatically, then it was a genre of conduct that could be embodied by the healthy just as readily as it could be reenacted by the mentally ill. No manner of economic conduct, it seemed, was in principle opposed to the fact of madness.

Economic Vitality and the "Man of Action"

Mercier's position that conduct alone was the key to determining psychiatric health and that even delusional individuals could, through their (largely economic) behaviors, evince a sufficient degree of sanity never reached the degree of endorsement and embrace that Mercier had hoped for from his British and international colleagues. Unfortunately, Mercier's skeptics mistook for an overly forensic approach to insanity what was instead a profoundly holistic theory of health, human nature, and social inclusion. At issue for Mercier was not simply that competency in the management of one's own affairs could function as a clinical stopgap against diagnostic suspicions of madness. Something more was at stake, for conduct was not only an indicator of mental disease but also a gauge of vitality, a benchmark that could signal medical vigor and well-being. It was a person's actions, nothing more or less, and their retroactive coherence and justifiability that above all defined organic success and social viability. In his systematic account of conduct, Mercier developed a praxeological theory of health, which was epitomized by a particular anthropological model of vital success—the agent of commerce, better known as the "businessman."

As I detailed in chapter 2, conduct for Mercier denoted the successful adjustment of an organism to its own specific environment by way of activity that could be described as "purposive." Purposiveness for Mercier did not necessarily entail psychological intent but simply the appearance of teleological behavior. Proper environmental adjustments were never contingent or accidental; they were ends-oriented processes that responded to specific external triggers and exigencies.[37] The criterion of success was retroactive, based simply on whether an organism could maintain its vitality as before. Pathologies such as insanity were simply "disorders of the adjustment of the individual to his surroundings, or of the organism to its environment."[38]

But adjustments could vary on the basis of their success, that is, on the quality of adaptation itself, which Mercier judged on the novelty and the complexity of the adjustment. In this way, some forms of conduct could be considered more successful, more intelligent, and thereby healthier than others. The most exemplary forms of conduct among human actors, Mercier explained, were to be found among certain select social types, namely, scientific intellectuals, artistic virtuosos, and above all the "wealthy parvenu—the man who has forced his way to the front over heads of numberless rivals, the man whose business capacity has enabled him to make a fortune early in life." For Mercier, the enterprising prodigy, "said to exhibit superior ability," represented an exemplary case of vital conduct and adaptational success. "He is a man of unusual intelligence," Mercier wrote. "Hundreds of his fellow-citizens entrust their money to the companies of which he is director, on the strength and in the belief that he is better able to carry out adjustments to circumstances than they are themselves." The business*man*—gendered exclusively male—represented, in his conduct, archetypal adaptational success and intelligence, which he displayed "solely in dealing with circumstances in such a way as to extract from them the maximum benefit."[39]

Conduct, then, was not simply defined according to its disorders and deviations from normalcy but also in terms of its level of health, robustness, and intelligence. The healthiest and most intelligent kinds of conduct displayed a variety of crucial characteristics. They were novel but also elaborate, capable of adjusting to "circumstances more and more distant in space; more and more separated in time; more special; more comprehensive; more complex."[40] Mercier illustrated the point through an allegorical comparison of professions, hierarchically depicted according to what he believed was the requisite intelligence of conduct needed to successfully carry out the tasks of each occupation. Laborers, for instance, whose occupation Mercier characterized as beholden to a mechanistic regimen, demanded the least elaborate and novel conduct. Their environmental circumstances, he felt, likely did not

demand new combinations of behavior. Laborers were, therefore, ranked at
the bottom of the professional hierarchy, followed in ascending order by arti-
sans, managers, manufacturers, and finally statesmen.

Another indicator of healthy conduct was the conservational strength embodied in the adaptation—that is, the degree to which any single adjustment
could sustain and promote the general preservation of the organism in the
face of future environmental changes. Mercier, again, deployed a financial
illustration. "It is not merely ability to conduct business successfully that constitutes intelligence of [conservation]," Mercier explained, "for if a man conducts his business capably and successfully, but squanders the income that he
makes, or lives beyond it, he is considered less sensible, that is less intelligent,
than if he were to live within his means and lay up something against a rainy
day."[41] The man who saved, who acted prudently in the present according
to potential and unknown risks of the future, was most effectively "[secur-
ing] the full benefit from his circumstances." And the greater the amount of
"benefit" that could be extracted, the more intelligent could the conduct be
said to be. It was precisely this "shrewdness or common sense" that Mercier
proclaimed to be "the distinguishing characteristic of the 'practical' man."[42]

"Shrewdness" was a crucial notion for Mercier, a quintessential attribute
of praxeological intelligence. Shrewdness was idealized on the grounds that
the condition most diametrically and morbidly opposed to it was insanity
in the most definitive sense. "However deficient a man may be in either of
the other forms of intelligence," Mercier wrote, "he is not considered insane
unless this last form [shrewdness] is defective; and when there is defect in
this, he is allowed to be insane in however high a degree he may exhibit those
other forms."[43] In no other context was shrewdness better demonstrated than
in business and financial dealings. Consider, Mercier insisted, how readily the
evidence of shrewd financial behavior could offset misgivings about otherwise
eccentric personalities. Take, for example, occultists, spiritualists, fanatics—
"no one would dream of calling such people insane, so long as their business was successfully conducted and their homes properly maintained—so
long, that is to say, as they contrive to extract from their circumstances a due
amount of benefit."[44]

The question of whether people could successfully manage their affairs
was not associated with healthy conduct in a merely contingent fashion. Successful economic adaptation, or shrewdness, was emblematic of health in the
fullest sense, and its failure was the very epitome of pathology. And the difference between the two was surprisingly severe. Mercier described the hypothetical case of a man whose mind and money were thoroughly occupied in
some "chimerical object," to the point of appearing eccentric to those around

him. If, Mercier warned, such a man were to lose himself in this pursuit—to sink his fortune entirely into it and to neglect his family, friends, and self— then "he is no longer considered of sound mind."[45] And yet, Mercier proposed, let us imagine that after having sacrificed everything, his misfortunes take an auspicious turn, and his chimerical object becomes a success. What happens then?

> He takes out his patent; his device is adopted. Manufactures compete for his services, wealth returns, friends return, wife and children are reinstated in their position. He becomes chairman of companies, enters Parliament, he is knighted—ennobled. Where are now those who said he was mad? They acknowledge their mistake. They were blind, foolish, misled.[46]

The fact of this hypothetical individual's earlier bizarreries had not altered; instead they were retroactively reinterpreted as an enigmatic expression of virtuosity rather than mental disease. But, Mercier cautioned, if this very same man had prematurely died before his chimerical object had been completed, such that its success could never have been observed and confirmed, what would the onlookers have thought then? "He would still have been mad," Mercier concluded. "Ought we then to blame the onlookers? By no means. Praise and blame would be alike inappropriate."[47]

In this sense, and in the determination of the shrewdness of conduct, there was no such thing as action viewed in principle, that is, in isolation from its context, which would determine on its own the medical, moral, *and* financial value that ought be ascribed to it. The very same action performed in two separate circumstances might generate profit in one context and give way to loss in the other. "We must recognize," wrote Mercier, "that their criterion of sanity was the success of the man in obtaining benefit from his circumstances. So long as the balance was in his favor, the question of his sanity did not arise. When he began to decline in wealth, people began to whisper that he was 'not right.'"[48] Even the strangest, most apparently destructive behaviors could be retroactively justified through the demonstration of adaptational success—in this case, the shrewdness of invention and business acumen.

For Mercier, the quality of behavioral adjustments, that is, the intelligence and health of conduct, corresponded to a tripartite social pyramid. Most people adjusted to their circumstances in typical, habituated, and uninventive ways, according to what Mercier somewhat contemptuously viewed as the conduct of the masses. Above them, however, stood a smaller population of the ordinarily to moderately ingenious, those whom Mercier dubbed the clever and reasonably shrewd and whose adjustments displayed some degree of elaborateness, novelty, and conservational strength. At the top of this pyramid, however, towering above the rest, was the *genius*, a "rare" social type "who originates extremely novel ways of dealing with circumstances, and who adjusts these novel operations to the end in view with such accuracy as to ensure its attainment."[49] Enumerating social types eventually became something of an intellectual diversion for Mercier. Toward the end of his life, he anonymously published over the course of six months a series of short character sketches, organized loosely around the notion of "human temperament," in the pages of a weekly medical journal.

Mixing an incipient personality psychology with an unmistakable interest in philosophical anthropology, these sketches enumerated broad dispositional traits within the social body. Some traits, like the "artistic temperament" or the "suspicious temperament," were little more than loose taxonomies, while others, such as the "business man" or the "philosopher," were more formalized classifications.[50] Mercier soon republished the sketches collectively into a single volume, *Human Temperaments*, making no changes to the individual articles save altering the order in which they appeared. The "suspicious temperament," for example, was originally published as the last entry, but in the first edition of *Human Temperaments*, it was the "philosopher" that appeared as the final entry—a surprising decision for Mercier, fixated as he was on the concept of conduct, for as he concluded, "With action the philosopher as such has nothing to do."[51]

In 1918, however, just a year before his death, Mercier revised his study of human temperaments a final time, adding four new types, and changing their order once again so as to conclude the volume with entries for the "man of business," the "practical man," and the "man of action"—typologies that not only shared an underlying affinity with each other, but that collectively embodied the substance of Mercier's medical philosophy. While the "man of business" was characterized as the one "who extracts the maximum benefit from his circumstances," a business temperament was not an essentially economic disposition, for it was defined by a shrewdness that could just as easily appear in distinctly noneconomic circumstances. According to Mercier, Napoleon Bonaparte, on account of his political and military virtuosity, best personified "the supreme example of the business temperament."[52] The "practical man" was similarly defined through the lens of the health and intelligence of conduct, as a personality trait that was in this case energetically and impulsively moved to action—sometimes detrimentally, but never without purpose.

It was, however, the "man of action," who took up the position of the final and culminating human temperament, the one who most quintessentially consummated what Mercier had all along sought to express through his vital praxeology. The "man of action" displayed a strong "proclivity to action," but a proclivity that was met with self-control. "The man of action is the born leader of men," Mercier explained, someone who possessed the shrewdness of the "man of business" and the vibrancy of the "practical man."[53] "Men of action" were not quite the revolutionizers that their counterpart "men of thought" were—such as the philosophers whose ideas could well come to shape the contours of civilizations. Unfortunately, Mercier observed, while "men of thought" were often capable of leaving a mark on *history*, their actual lived lives were rarely defined through the rubric of adaptational success, which is why "the man of thought does not enter into his kingdom till after he is dead." The "men of action," on the other hand, did nothing but exhibit a profound health and intelligence of conduct. Their contribution was neither posthumous nor historical but defined through a remunerative value—we can call it profit—that could be imminently realized and expeditiously procured. "The man of action reaps his reward during his lifetime," Mercier concluded, and while history may not remember these exemplary archetypes of human conduct—these prodigious praxeological virtuosos, in Mercier's mind—their role and function in the ongoing development of society as a matrix of customary actions and behaviors could not go unremarked.[54]

Toward an Entrepreneurial Pathology

Mercier was hardly alone in valorizing the "man of action." Far too striking to ignore are the set of affinities that linked turn-of-the-century psychiatric depictions of prodigious agents of commerce to comparable portrayals by early twentieth-century economic theorists. Shortly before Mercier sang the praises of the business temperament and the men of action who supposedly stood atop the social hierarchy, the Austrian economist Joseph Schumpeter inaugurated his own conception of the "man of action" (*Der Mann der Tat*), a leader among the masses, a dynamic and energetic figure who enacted new possibilities and who also stood at "the top of the pyramid of society."[55] In the sphere of economic activity, this "man of action" was dubbed the "entrepreneur." While abstract conceptions of businessmen and enterprising actors were long present in modern political economy, economic theories of the early twentieth century reframed these figures as singular and extraordinary forces, indispensably valuable to the entire business process.[56] Veblen described how the businessman, the entrepreneur (or "undertaker"), had become the central coordinator of industrial processes and transactions performed for the sake of business ends and financial gains.[57]

Schumpeter's earliest comprehensive account of the entrepreneur ap-

peared in the first edition of *Theory of Economic Development* (1911).[58] The
description itself was built on a dichotomy central to Schumpeter's thinking—
a dichotomy he was not the first to introduce, and had already broached in
his earlier *The Nature and Essence of Economic Theory*—that economic conduct could be differentiated into two classes of behavior: static and dynamic.[59]
Static economic conduct was defined by its habituated and automatic passivity. "Statically disposed individuals," Schumpeter explained, "are characterized by essentially doing what they have learnt, by moving within the received
boundaries and by having in a determining way their opinion, dispositions,
and behavior influenced by the given data of their sector."[60] This kind of ac-
climatized and reflexive behavior was itself directed by the predictability and,
from an economic standpoint, the viability of hedonistic motives, namely,
pleasure and pain. "In reality," Schumpeter explained, "static behavior and
hedonic motives de facto coincide."[61]

What this implied was that static hedonism was the crucial characteristic
trait of the rational economic actor, for whom desire to accumulate wealth
and aversion to loss were the primary doctrines of action. "The principle of
rational behavior [*das Prinzip des rationellen Verhaltens*]," wrote Schumpeter,
"is indeed the key to the processes we observe in the world of statics."[62] The
"rationality" of *homo economicus*, then, was not defined through attributes
such as prudence and calculation. Rational economic actors in the ordinary
sense were ultimately defined by their conformity, adherence, and regularity.
As the American economist Frank Knight would later observe, "Economic
man is the individual who obeys economic laws."[63] Not only did Schumpeter
share with Mercier an inclination to valorize the "man of action," but the
two shared a comparable view of *homo economicus* as the paradigmatic social
automatist, the one whose conduct was secured through routinization, proceduralism, and an adherence to hedonistic rule-boundedness. While Schumpeter's static actor was a category capacious enough to designate the conduct
of the ordinary economic masses (the lowest form of fixed and uninspired
conduct, according to Mercier), the static actor also personified the behaviors
of the pathologically unconscious—the actions performed by the sorts of patients that so intrigued Mercier's clinical forebears (as I discussed in chapter
1). Admittedly, for Charcot and Jackson, even the most normal, unsuspecting,
and "tranquil" automatist was undoubtedly suffering from some form of insanity. But such an assumption, as we have seen, never sat well with Mercier,
for how can anyone who so adequately and indistinguishably performs what
even an economist would call perfectly rational economic behavior ever be
deemed insane?

In opposition to static, hedonic behavior, which seemed to some measure to be compatible (at least from Mercier's standpoint) with ostensibly pathological states, Schumpeter identified a second, dynamic and energetic type, with which, as we will see, madness was not only compatible but actually quite requisite. In its earliest formulation, the static/dynamic dichotomy did not simply correspond to forms of action but to types of individuals as well.[64] Schumpeter understood the difficulty in imagining "that there should be a type of economic behavior that is different from hedonic behavior," that is to say, of the rational economic actor. "It seems there cannot be any goal besides satisfying wants."[65] Nevertheless, such a type did in fact exist, he argued, "a new and independent agent [Agens] in the economy" whose task was "construction in the economic sphere."[66] This figure, whom Schumpeter called the "man of action" (Mann der Tat), was the inverse of the static economic subject.[67] Schumpeter's man of action was a leader (Führer), "characterized by perceiving new things, by changing the received boundaries to their behavior, and by changing the given data of their sector."[68] Schumpeter explained that while "a static economic subject" (ein statisches Wirtschaftssubjekt) drew passively from available data in the world,

> our man of action shapes them. He puts them in new forms and puts them in new contexts, like the great, creative artist does with the traditional elements of his art. He changes the economic behavior the "static" actors could only carry out, year in year out. His action . . . is substantially different from static behavior. It also generates substantially different phenomena.[69]

The man of action was an agent and enactor of change and novelty as well as a destroyer of the old.[70] In the economic realm, this agent, the "entrepreneur" (Unternehmer), had a very specific function: to extract the component elements of a static economy from their prior static uses and to proceed "to use them in a different manner" in order to generate "new combinations" (neuer Kombinationen) in the economic sphere.[71] Such "new combinations"—a key phrase for Schumpeter, to which I will return below—could refer to the creation of new financial objects and commodities, new business methods, or even new markets.[72] And it was precisely this task of "pushing through new combinations" that Schumpeter called "economic enterprise" (wirtschaftliche Unternehmung).[73]

In his earliest writings, Schumpeter proposed that the entrepreneurial drive resided ultimately in the psyche (der Psyche) of only a select group of economic actors. "Most people do not see the new combinations," he wrote. "They do not exist in them. . . . They do not have the disposition to experiment with something new. . . . This is the masses." The authentic entrepreneur was not only able to perceive these new combinations but was also able to act (*handelt*) on them as well. "It is this type that scorns the hedonic equilibrium and faces risk without timidity," Schumpeter wrote.[74] To be an entrepreneur, then, was precisely to abandon the hedonic performances that centrally defined the rational economic actor; the entrepreneur had sometimes to transgress the very rules that economic man was obliged to obey. This is why, from the standpoint of the ordinary economic actor, "behavior that differs from [static hedonism] . . . seems to be irrational [*vernunftwidrig*], as it were. The so-called higher motivations do not seem to exist here." When encountering the entrepreneur's "ahedonic" conduct, "the only possible explanation seems to be an error [*Irrtum*]."[75] While Schumpeter stressed that dynamic conduct belonged ultimately to a "healthy mental life" (*gesunden psychischen Leben*), and, later, that the carrying out of new economic combinations demanded a robust intellectual aptitude, nevertheless he concluded, "His conduct and his motive are 'rational' in no other sense. . . . Hedonistically . . . the conduct which we usually observe in individuals of our type would be irrational."[76]

In subsequent editions of the *Theory of Economic Development*, and in Schumpeter's later writings, the division between static economic subjects and dynamic entrepreneurs was described less in terms of human typologies and more as an entrepreneurial "function,"[77] one specifically oriented toward "the creation of enterprise."[78] Schumpeter consistently juxtaposed entrepreneurial conduct to the actions of rational economic actors in an effort to formalize a type of economic agency distinct from capitalists and laborers who, though opposed from the standpoint of class, were allied from the standpoint of the static hedonism that defined their behaviors. Entrepreneurs were in practice neither capitalists nor laborers, even if capitalists and laborers could occasionally enact an entrepreneurial dynamism (a supposition that Schumpeter did not originate so much as promulgate).[79] And although Schumpeter increasingly defined the entrepreneur as a function, capable of being operationalized by an assortment of economic subjects, he never relinquished the idea that there were entrepreneurial personalities or "aptitudes . . . present in only a small fraction of the population" that "define *the entrepreneurial type* as well as the entrepreneurial function."[80]

It is not difficult to understand why Schumpeter's entrepreneur, his "man of action" who possessed an irrational ability to create, innovate, and also destroy, a capacity that placed him at the apex of the social order, and whose "activity impresses the stamp of its mentality upon the social organism," would call to mind the idea of the creative "daimonic" genius.[81] Schumpeter clearly sought to install the entrepreneur as a preeminent twentieth-century figure of prodigious virtuosity, a genius among artists and other creative mavericks, and he was not alone in this regard. Frank Knight too noted that the "business

man has the same fundamental psychology as the artist, inventor, or statesman."[82] And even John Dewey ambivalently observed that "captains of industry are creative artists of a sort, and that industry absorbs an undue share of
the creative activity of the present time."[83]

At the same time, the entrepreneur's increasing promulgation as a novel
typology of social genius was undeniably entrenched within perniciously
medicalized conceptions of social hierarchies and, with it, efforts to exercise
biopolitical control over marginalized populations. In a letter to his colleague,
the Yale economist Irving Fisher, Schumpeter extolled the virtues of *How to
Live*, a small but very popular volume on social hygiene that Fisher cowrote
with the physician Eugene Lyman Fisk.[84] *How to Live* advanced a number
of techniques and models for hygienically maintaining the well-being of the
social body, the most alarming of which was the book's defense of eugenic
practices including segregating and sterilizing select populations in order to
preclude the dissemination of socially "undesirable" traits.[85] Drawing upon
many sources, not the least of which was Francis Galton's *Hereditary Genius*,
Fisher and Fisk sought to partition the social body according to "one standard
of judgment," a single "line of division," namely, "fitness," according to which
the authors believed valuable heritable traits could be distinguished from
those they felt were costly, thereby demanding elimination. Below the demar-
cating line of fitness were positioned the "feeble-minded, paupers, criminals,
insane, weak and sick, who are a burden, economically and socially." Above
it rose an ascending order of the socially valuable, an order that culminated
with the luminary of the biopolitical social hierarchy—the "genius."[86]

But in addition to the defenses of social eugenics to which it was tied, the
conception of the entrepreneur as a kind of genius embedded another tacit
(I might even say "unconscious") medical legacy—in this case, a *psychiatric*
inheritance. The notion of the entrepreneur-as-genius, Schumpeter's in particular, could not entirely depart from a lineage of psychiatric appraisals of
the genius as consanguineously related to the madman, a pathological disposition, however, that was not only integrated into economic depictions of entrepreneurialism but was recast as a lucrative virtue. In many late nineteenth-
century psychiatric studies, genius was not always portrayed as the epitome
of mental health—as Mercier might have otherwise argued—but, rather, as
an anomalous mental condition akin to insanity.[87] Maudsley described the
category of genius according to its deep proximity to the "insane temperaments,"[88] because of the tendency among so-called geniuses to impulsively
flaunt conventions and "act independently . . . in the social system." Invoking
John Stuart Mill's paean to eccentric individualism, Maudsley conceded that
geniuses tended to introduce a "novel mode of looking at things" that typi-

cally "[transcend] the established routine of thought and conduct."[89] The genius appeared "strange [and] not quite right" to most, and their actions were
often quite "startling . . . to those who work with automatic regularity in the
social organization."[90]

In a passage uncannily presaging the Schumpeterian language of entrepreneurial dynamism (and the "new combinations" creatively generated therefrom) and its opposition to static automatism, Maudsley explained that the
creative inspiration of the genius is "the exact opposite in this regard to habit
or custom—that 'tyrant of custom' to which the minds and behaviors of most
people have been yoked." Maudsley continues:

> In the inspiration of a great thought or deed there is the sudden starting forth
> into consciousness of *a new combination* of elements unconsciously present in
> the mind; these having been steadily fashioned and matured through previous
> experience.[91]

While Maudsley was careful not to entirely conflate genius with madness,
still, these two conditions were nevertheless united to the extent that they
stood out quite analogously from the population of normal individuals who
were otherwise obedient to social habits and customs.[92]

What Maudsley cautiously described as merely the "kinship of genius to
insanity" was, for other medical thinkers, a far more thoroughgoing equivalence.[93] For Cesare Lombroso, the boundary separating genius from insanity
was impossible to fix; an "actual continuity" linked the physiology of the former to the pathology of the latter.[94] But it was a continuity that need not be
denounced, Lombroso insisted, for what the continuity itself revealed is that
"the destiny of nations has often been in the hands of the insane," and that
the mad, in their proximity to genius, "have actually been able to contribute
so much to the progress of mankind."[95] The popular science writer John Fer-
guson Nisbet also noted how frequently insanity "treads upon the heels of
genius,"[96] to such a degree that the two conditions dovetailed into one another
not only diagnostically but also physiologically.[97]

In another Schumpeterian anticipation, Nisbet noted that geniuses and
madmen shared a penchant for novelty and mutually experienced "a spontaneous morbid activity of the nerve-cells and fibers of the brain whereby
new combinations of sensory impressions and memories are constantly being
formed."[98] As an example of the proximity of genius to states of pathology,
Nisbet focused on the "minor geniuses" who possess a prodigious capacity
to make money. "Men who amass wealth or who initiate and successfully
carry out great commercial schemes," he wrote, "are not usually accounted
geniuses, but there is no doubt that the qualities of mind which bring them

to the front are in many cases identical in their origin and nature with the literary or artistic gifts." Like their artistic counterparts, financial geniuses often exhibited a predisposition toward severe neuropathic traits, in this case avarice and extravagance.[99]

Not every clinician or medical writer, however, would go so far as to frame genius as a pathological form. Some presented far more tempered views on the link between insanity and genius,[100] while others entirely opposed the equivalence.[101] This is simply to say that the psychiatric status of genius was a contested matter, part of what William James called the "genius controversy" that pervaded psychological medicine at end of the nineteenth century. James himself fell on the side of extolling the neuropathic dispositions of those who supposedly possessed heightened intellectual and creative aptitudes, and of insisting that traits normally framed as afflictions should instead be revalued as assets, that "we should broaden our notion of health instead of narrowing it."[102] It would likely be a gross understatement, James years later cautioned, to unequivocally declare that a "neuropath" can never reveal new truths.[103]

Even the notion of innovation, taken alone, was anything but value-free from a psychiatric standpoint; it contained its own discrete pathological inheritance, not exclusively tied to medical discussions of genius. While Schumpeter sought to distinguish his concept of innovation, or the creative act of getting things done, from a more general notion of *invention* as a mere production of ideas, later commentators, such as the economist Adrien C. Taymans, proposed that innovation in Schumpeter's sense likely owed a conceptual debt to the idea of "invention" as outlined by the French sociologist Gabriel Tarde.[104] For Tarde, "imitation" and "invention" were the two primary and intermingled engines of social development and change, the "elementary social acts."[105] Taymans saw in Tarde's concept of invention (which was, for Tarde, "the first cause of wealth"[106]) an incipient formulation of economic innovation, an interpretation to which Schumpeter was, for his part, quite receptive.[107]

The difference between Tarde's "inventor" and "imitator" was itself quite analogous to Schumpeter's dichotomy separating the dynamically creative man of action from his static counterpart. According to Tarde, one of the most elementary facts of society was the "personal ascendancy of one man over another," a structure, he maintained, that effectively divided the social body into leaders and followers, according to a relationship that was more or less analogous to that between a "hypnotizer" (*suggestionneur*) and the "hypnotized" (*suggestionné*). The masses were effectively characterized by their imitative submission to customs and conventions according to a deep "passivity, credulity, and docility—as incorrigible as it is unconscious." As a result, "the crowd of imitators is a kind of somnambulist," Tarde explained, social automatists who are induced to obey the biddings of "the inventor [*l'inventeur*], the initiator of all kinds," who by virtue of "his strangeness, his monomania [*monomanie*], [and] his imperturbable and singular faith in himself" is "a kind of madman" (*une sorte de fou*). The elementary struc- ture of society, Tarde contended, was of creative mesmerizers leading a flock of passive automatists, "fools guiding somnambulists." Yet this relationship nevertheless fulfilled society's "logical ideal," for while the "sheepishness" (*moutonnerie*) of imitators functioned to preserve society, the audacity of the inventors served to elevate and expand it.[108]

The assertion that the inventor was a kind of "madman," a "monomaniac" no less, was not a casual remark, for Tarde would surely have been familiar with the legacy of the "inventing monomaniac" in early French psychiatry. "Some [monomaniacs]," Esquirol maintained decades earlier, "believe them- selves to be savants noted for their discoveries and inventions."[109] Ulysse Trélat eventually provided an extended discussion of "inventive monomania" (*la monomanie des inventions*).[110] He recounted, among several other examples, the case of a skilled optician who earned a not insignificant 8 to 10 francs a day. "After having for many years tasted, along with his family, the benefits of his situation," Trélat wrote, "he suddenly had the unfortunate idea of making a marvelous invention [*une invention sublime*], and from that moment on, not only did he stop working, but he devoted everything he had ever acquired, all of his resources, to that discovery." The optician eventually sold his furniture, ruining his family in the process, all for the sake of bringing an illusory con- trivance to fruition. "Inventors," Trélat concluded, "are incurable."[111]

For some clinicians, financial innovations, whether in the form of new commodities, projects, or markets, seemed to be grounded on a capacity for speculative foresight that was itself not devoid of pathological underpinnings. According to Tuke, the proximity of pathology and speculation was ultimately a feature of the modern financial era.[112] As early as 1816, the English physi- cian John Reid observed that, in the "grand mart of trade," psychiatric health vacillated in relation to the "vicissitudes of commercial speculation." Reid recounted the case of a young man who fell victim to a disastrous "variety of mercantile adventures," which resulted in the loss of not only his finances but his higher faculties as well. "The demon of speculation," Reid explained, "which had before misled his mind, now possessed it entirely." The patient be- gan to devise a number of audacious, even bizarre, financial schemes, some of which reminded Reid of "another madman that I had heard of, who planned, after draining the Mediterranean, to plant it with apple trees, and establish a cyder manufactory on the coast."[113] By the start of the twentieth century, the American neurologist Moses Allen Starr continued to affirm the not uncommon tendency among the mentally ill to "[undertake] the most extravagant and impossible schemes," the most "colossal combinations and extensive business organizations."[114]

But this brief digression only raises a more pressing question: What was actually at stake in attributing the rank of social genius to the entrepreneur? Certainly, to suggest that a financial enterpriser could display a degree of virtuosity tantamount to that of an artistic master could assign cultural status to a type of expertise that might otherwise have been perceived as nothing more than the crass skill of moneymaking, a vocation perhaps only slightly more reputable than professional thievery. Furthermore, to suggest that entrepreneurs were among the "fittest" members of the social body would go a long way to justifying that financial professionals ought to stand at the summit of the *political* order, just as they were purported to stand at the apex of the social pyramid.

Yet there was more to the conception of entrepreneurial genius than merely a vaunting of professional supremacy and a defense of an egregious social hierarchy. In fact, a psychiatric legacy—the indelible trace of pathology—was crucial to underscoring the key function of entrepreneurial exceptionalism, namely, the capacity to apprehend and act upon "new combinations," to do "new things," to innovate.[115] Instituting novelty in the social order, bringing about change and innovation amid the otherwise customary flow of social convention, demanded a degree of "abnormality" that only a conception of genius firmly attached to madness could best express. Weber made note of this very point when he observed, on ethnological grounds, that "the most important source of innovation [*Neuordnung*] has been the influence of individuals who have experienced certain 'abnormal' states (which are frequently, but not always, regarded by present-day psychiatry as pathological) and hence have been capable of exercising a special influence on others." A certain quantity of force was necessary to "overcome the inertia of the customary," and with respect to social conventions, such a force would inevitably be marked as "abnormal."[116] It is precisely in its abnormality—what Schumpeter called the entrepreneur's apparent "irrationality"—that entrepreneurial genius shared a fundamental affinity with madness.

This pathologically inflected notion of entrepreneurial genius effectively encapsulated a theory of social change, but one almost entirely recast in economic terms. To bring about an abrupt social transformation, such a theory proposed, demanded the capacity to break from the customs and conventions of the social order—to stand apart from them in order to introduce an eruptive shift. Such a capacity, however, belonged to a typological continuum of so-called abnormals, bookended on one side by madmen and on the other by prodigies and virtuosos. But whether an abrupt change would come to be admired or reproached—revered as a masterly creation or harshly pathologized as folly—came down to whether the emergent novelties yielded remuneration. This was the line that separated "innovation" from *aberration*, a distinction analogous to Mercier's hypothetical case study that I enumerated above concerning the inventor and his "chimerical object." The value of social change could never be appraised in principle, this line of reasoning proposed—that is, outside of an *economic* context.

Whether the particular "abnormal" who institutes transformation should be labeled a genius or madman, and whether the change itself should be called innovative or simply aberrant, depended almost exclusively on whether the "new combinations" generated a fiscal return. While the prospects of lucrative innovation ultimately relied upon an abnormal impetus, still, the perceived benefits that could arise often justified the perils of pure speculative madness. The madman/genius/entrepreneur was unquestionably the inaugurator of great social perils; the only question was whether the perilousness could be mitigated, revalued, and redeemed through healthy returns.

American Disorders

Thus far I have explored how accounts of business enterprise and financial conduct were multifariously interwoven with psychiatric appraisals of health and illness, part of what I have characterized as just one iteration of the economic style of reasoning that suffused turn-of-the-century transatlantic psychiatry. Not only could performances of economic propriety temper suspicions and even redeem the undeniable presence of insanity, but, conversely, some genres of madness appeared to demonstrate an underlying compatibility with rationally static economic action. At the same time, the drives necessary for engendering "new combinations" in the economic domain appeared not only consistent with pathological states but actually quite dependent upon them; the entrepreneur, in other words, had to be a little bit mad. If the automatic proceduralism of *homo economicus* could be pathologically replicated, and if economic innovations called upon an entrepreneurial creativity and irrationalism inextricable from madness, then with respect to both maintaining and augmenting the economic sphere, it was not necessary for psychiatry to excise insanity but simply to make sure that the right sorts of pathologies were operating in the right sorts of ways. Confronting these entwinements of madness and enterprise, however, was a long-standing observation, a rather entrenched hypothesis within the history of psychological medicine, that commerce and the vocations of moneymaking were in themselves deeply pathogenic practices.

Clinicians and medical practitioners had long held the conviction that business and commerce could precipitate mental disorders, though not always because of the monetary losses and "reversals of fortune" they occasioned. The eighteenth-century physician Thomas Arnold contended that, given England's industrial successes at the end of the eighteenth century, the forms of madness that arose as a consequence of commercial enterprise were more prevalent in England than in any other country because British ventures were simply the most lucrative. "I see no other way of accounting for this vast increase of the disorder [of insanity]," Arnold wrote, "than by attributing it to the present universal diffusion of wealth and luxury through almost every part of the island."[117] Reid corroborated the point when he described the psychological effects of the South Sea Bubble of 1720, claiming that while a "few lost their reason in consequence of the loss of their property . . . many were stimulated to madness by the too abrupt accumulation of enormous wealth."[118]

By the mid- to late nineteenth century, however, it was the United States that came to hold the unhappy status in the psychiatric imaginary as the most economically pathogenic nation. Americans were the population most consumed with making or advancing personal fortunes, Ray wrote, whether by skill, luck, or "daring speculation." Consequently, the erratic unsteadiness and agitating unpredictability of the "money-market," combined with the overall "mercantile spirit" of American life, produced far too much mental "wear and tear."[119] Mitchell observed that in the two decades between 1870 and 1890, major commercial cities such as Chicago had experienced such immense economic growth that they had become "a place today to excite wonder, and pity, and fear." Everyday life had adopted the very tempo—as well as the anxieties and psychological turmoil—of commercial markets. The trouble with "making money fast," Mitchell wrote, is that it often means "accumulating a physiological debt of which that bitter creditor, the future, will one day demand payment."[120]

Richard von Krafft-Ebing noted that nervous disorders most associated with the freneticism of commercial life were, by the end of the nineteenth century, on the dramatic rise throughout much of Europe (the German industrialist Walther Rathenau would allegedly refer to Berlin as "Chicago on the Spree"[121]); nevertheless, the most severe cases of business disorders were still to "be found in America, where the haste of life is almost proverbial and time means money [*Zeit Geld bedeutet*]."[122] It is not at all clear whether Krafft-Ebing knew that the phrase "Time is money" was the very lesson that Ben-

jamin Franklin taught in his *Advice to a Young Tradesman*, and that Weber
would invoke (*Zeit Geld ist*) in *The Protestant Ethic*, twenty years *after* Krafft-
Ebing.[123] According to Krafft-Ebing, the disorder most associated with the
freneticism of commerce was nervous exhaustion or neurasthenia, and the
clinician who most advanced and explicated that disorder was the American
neurologist George Miller Beard, who had already contended that the "Americanization of Europe," in terms of industry *and insanity* was well underway.[124]
The Americanizing process, furthermore, extended well beyond the bounds
of major European cities; in 1897 a provincial French psychiatrist recorded
numerous cases of delusional disorders of great wealth and grandeur among
itinerant French workers and Algerian migrants whose economic ambitions
drew them toward major cities and toward the prospect (failed in most cases)
of great fortune.[125]

What made American "nervousness" so distinct from other transatlantic
and colonial contexts was not only the intensity of commerce, financial capital,
and economic industrialization evident in the United States but also the sheer
potency of nervous illnesses related to "modern civilization."[126] American disorders, like neurasthenia, were in a class of their own, even as they were being
exported across the Atlantic.[127] One prominent American neurologist would
refer to neurasthenia as "our one great national malady," so affixed as a disease
form to the nation-state that it could even function as "an important stimulus
to patriotism and racial solidarity!"[128] In looking to the specific causes of
these American disorders, neurasthenia in particular, Beard proposed that
the most common were the demands of wage labor, which often committed
an "overdraft" of the nervous reserves.[129] Cases of "nervous bankruptcy" were
especially evident in the specific context of industrial labor, which, given its
physical hardships, manual repetitiousness, and poor hygienic conditions,
was one of the root causes of "the increase of insanity and other diseases of
the nervous system among the laboring and poorer classes."[130] But laborious
bankruptcies could arise in professional forms of labor as well, even (and
often especially) among financial professionals. "The proper care of the boss
is as essential as the proper conduct of the business," wrote another notable
clinical neurologist. "The modern millionaire business man is a new factor in
the problem of [mental] sanitation."[131] For different reasons, though all related
to workplace overexertions, psychiatric thought was aligned in the belief that
the "mercantile spirit" of American market capitalism took a toll on the entire
spectrum of the social workforce.

Beard enumerated several other pathogenic sources for these so-called
American nervous disorders. One particularly inauspicious cause was economic liberty itself. "A factor in producing American nervousness," Beard

wrote, "is, beyond dispute, the liberty allowed, and the stimulus given, to Americans to rise out of the position in which they were born, whatever that may be, and aspire to the highest possibilities of fortune and glory."[132] Clinicians had suspected for some time that the excesses of economic liberty could prompt mental disease.[133] But Beard enumerated the virulence of freedom far more candidly than most. The freedom of commerce was in itself hardly an assurance of financial security; it was often nothing more than an imposing incitement to trade, invest, and take financial risks, which for prospective agents of commerce translated into the perpetual dread of potential loss and, with it, "the stress of poverty, the urgency of finding and holding means of living, the scarcity of inherited wealth, and the just desire of making and maintaining fortunes." For the poorer and working classes, however, who did not possess the financial privilege to take such monetary risks, economic liberty amounted merely to the freedom to choose simply how to suffer. "Not 'How shall we live?,'" Beard explained, "but 'Can we live at all?' is the problem almost every American is all his life compelled to face."[134]

The Privilege of an Enterprising Madness

Psychiatric accounts of the pathological perils of business and commercial activities raise a conceptual dilemma with respect to the relationship thus far enumerated between insanity and enterprise. This is what in the introduction to this chapter I described as an essential conflict, for how was it possible that financial ventures could be simultaneously deemed pathogenic while also compatible with, and even dependent on, forms of madness? How, within the terms of psychiatric thought, could an agent of commerce take advantage of the pathological affordances necessary for financial success—that is, to be profitably mad—without, however, surrendering to the virulent distress of commercial life? What could ensure that the right forms of madness were operating in the right ways in the economic domain, that *lucrative* economic pathologies could be properly balanced against their pernicious counterparts?

This balancing act, I propose, was organized around what I have called the pathological privilege that tacitly underwrote normative portrayals of the North Atlantic entrepreneur—a privilege hinging on the fact that entrepreneurial pathology was fundamentally gendered and racialized as masculine and white. If economic innovations demanded the seemingly irrational transgression of financial conventions, this meant that enterprisers had to confront the inevitable hazards of economic risk head-on. However, the ideal enterpriser, the lauded entrepreneurial "genius," was able to occupy the liminal space where the pathologies *caused* by enterprise could be transformed into the innovative madness *requisite* for it. But occupying such a position within the tumult of financial life meant that the entrepreneur had to be structurally sheltered from the worst forms of economic injury. It was precisely through the behavioral encodement of masculine whiteness that entrepreneurial irrationalism could be formalized as an entitlement that insulated it from the deepest economic perils. It did so not only by restricting the privilege of an enterprising madness to white economic actors and agents of commerce but, even more fundamentally, by structurally ensuring that nonwhite and, particularly in the United States, Black enterprisers experienced only the most pernicious forms of economic injury. Through a white masculine encodement, entrepreneurial conduct circulated through the social body and the economic sphere as an acceptable and even idealized pathological form, lauded precisely for its remunerative unconventionality. This lucrative advantage, however, was built atop the assurance that, across the North Atlantic and in the United States especially, Black enterprisers, perhaps most acutely, were all but destined to suffer in the financial domain. So rather than achieving the status of creative entrepreneurial irrationalism, Black enterprisers were forced instead to endure a different and far more destructive "madness," in Frank B. Wilderson III's sense, which a universalist entreaty to financial and commercial equity was unlikely ever to cure.[135]

In the first place, and in its early twentieth-century depictions, entrepreneurialism was firmly affixed to a broadly heteronormative masculinist appeal. The purported "passivity" of the static economic actor in Schumpeter's rendering marked it as acutely feminized in contrast to the "*man* of action," who not only worked hard but speculated perhaps too intensely, taking the occasionally detrimental and sometimes pathogenic risk. The entrepreneur was merely one articulation of a diffuse turn-of-the-century discourse on North Atlantic masculinity, the boundaries of which were organized around heterosexual dichotomies and evolutionarily inflected ideals related to virility, aggression, and social dominance. While undeniably variable as a gender performance and style of conduct, masculinity was nevertheless profoundly embedded within the language of class situations and a broad political-economic rationality.[136] As transnational anxieties over the rise of social pathologies, cultural decadence, and a looming national decline generated broad "crises" with respect to the status of a bourgeois masculine identity, such crises were tempered through invocations of ideal and remedial economic behaviors.[137] Robust masculine norms hinged on benchmarks of economic conduct, financial success, and occupational status.[138] Achieving a socially acceptable and even admired masculine standing—becoming a "self-made" man, for example—was deeply bound up with the right forms of conduct with respect to the market economy.[139]

At the same time, masculine propriety was equally transposed into a language of medical health, through which salubrious forms of masculinity could be differentiated from "a pathological lack of male behavior." But even medicalized appraisals were also attached to conceptions of economic productivity, since the truly healthy and vigorous man was he who not only demonstrated a willingness to work but could also bring about financial successes in the workplace.[140] And yet economic productivity, as we have seen, always brushed up against the pathogenic perils of overwork; healthy economic masculinity was situated always on the edge of a capitulation to the so-called American disorders. In such cases, pathological breakdowns of masculine economic actors were costly affairs, often linked to the rise of unstable markets and financial crises. When working- and businessmen went mad, in other words, everyone seemed to pay the price.[141]

But the real problem with economic pathologies was that they were simply too prevalent to be entirely avoided. Businessmen, as we saw above, could likely never evade the virulence of economic life, particularly since an ideal economic masculinity demanded not just laborious exertion but, often, a costly overexertion. Instead of striving to shelter an idealized conception of masculinity from the pathologies arising from economic behavior, what if there were a way to articulate masculinity and economic madness in relation to one another, but in a manner that was productive, positive, and remunerative? This is precisely what the concept of entrepreneurialism fulfilled, by providing an acceptable and even rewarding convergence of madness and masculinity. In contrast to, say, the diagnostic category of psychopathic personalities—one of the central means by which early twentieth-century psychiatrists defined a distinctly *masculine* pathological form[142]—the entrepreneur could co-opt madness into a socially appropriate model of masculine abnormality and irrationalism, one that could circulate throughout everyday market society as a style of performance suffused with social and economic value. It was not that the entrepreneur was strictly opposed to the psychopath, since entrepreneurs, as we have seen, could very well and were to some extent expected to exhibit certain psychopathic characteristics. What differentiated the psychopath from the entrepreneur was what correspondingly distinguished the madman in general from the entrepreneur: that the former was perilously detrimental to society whereas the latter generated economic value.

What entrepreneurialism in all its creative and lucrative madness demonstrated was that pathology could be gendered as a desirable masculine trait, and for that reason, it was not that masculinity was pathologized but, instead, that *pathology was masculinized*, that is, transmuted into a profoundly value-laden possibility and behavioral encodement. While initially and perhaps most emphatically anchored in male bodies, entrepreneurialism could be detached from those bodies in order to circulate as a normative gender performance capable of being adopted by other bodies as well (a "female masculinity," for instance, could undoubtedly appropriate an entrepreneurial style[143]). But there was an intrinsic limit, I would argue, as to how expansively an entrepreneurial style of pathological conduct could be appropriated, and this had to do with the fact that the entrepreneur was not only gendered masculine but also racialized as white. Gender norms and performances possess a mutability that racial identities do not. The entrepreneur's masculinity could be appropriated by other (nonmale) bodies in a way that the entrepreneur's whiteness could not, at least not so unconditionally or without substantive costs.

As Cheryl Harris has argued, whiteness is best understood as a form of property, the possession of which grants the privilege of economic access. *Being* white means possessing the benefits of whiteness.[144] So if masculinity is what coded certain pathological performances as socially acceptable, then it was really whiteness that ultimately bestowed upon those performances the inferred privilege of social and economic value. To that end, the relationship between entrepreneurial pathology and masculine whiteness was mutually enabling, even exploitatively interdependent, in some measure. For it was through masculine whiteness that entrepreneurialism could circulate through the social body not only as an acceptable and even valuable abnormality but also as a discretely masculine and white *prerogative*, an entitlement to be irrational, at least in the economic sphere. On the other hand, that entitlement to economic madness also meant that masculine whiteness could be coded as value-generative through and through, such that even the most unreasonable, abnormal, and even morbid characteristics of masculine whiteness nevertheless enjoyed the potential of being economically redeemed (a point I illustrate in detail in chapter 6).

The context in which this supposition is most effectively corroborated, however, is in the counterexample—that is, in the *thwarted* entrepreneurial aspirations of nonwhite economic actors who, in their restricted access to pathological privilege, experienced the madness of capitalism as a pernicious form of psychical trauma rather than as a lucrative mechanism for financial success. An emblematic case study concerns the enterprising ambitions of middle-class Black men in the early twentieth-century United States, whose plight was so succinctly captured by the sociologist E. Franklin Frazier.[145] Frazier's views on Black enterprise shifted dramatically in his writings beginning in the 1920s, where his early advocacy for economic success as the key to political equality transformed into the jarring realization that in the economic domain the potential profitability of entrepreneurial irrationalism was unfailingly frustrated by the pathogenic economic violence of anti-Black racism. Frazier's observations substantiate the assertion that entrepreneurial immunity against the virulence of capitalism was conditional on the pathological privilege of masculine whiteness.

Frazier first focused his sociological attention on financial developments taking place in Durham, North Carolina—his "city on a hill," according to the historian Leslie Brown, a city that boasted its own Black Wall Street where Black men could articulate a "masculinist perspective on black freedom" through a "masculinist vision" of Black economic achievement.[146] It was there where Frazier argued that Black middle-class men could assume "the role of the modern business man," having "mastered the technique of modern business and . . . the spirit of modern enterprise."[147] The youngest generation embodied the strongest entrepreneurial spirit, Frazier wrote, having been "saturated with the psychology of the capitalist class" in an effort to "expand their businesses and invade new fields." Black enterprise entailed the endeavor to innovate and create, which was itself (as we have seen) an arduous undertaking, one for which these young agents of commerce came eventually to adopt the disposition of the "tired businessman who does not know how to enjoy life."[148]

The real revelation in Frazier's study of Black enterprise, however, appeared several decades later when, upon returning to the plight of his young entrepreneurs, he came to adopt a much grimmer appraisal of the state of the Black middle class. As he observed, Black businessmen had initially cultivated themselves, socially and morally, around the task of making money because they believed that "money will bring them justice and equality in American life." In the intervening decades, however, the harsh realities of anti-Black racism demonstrated that money alone could never alleviate the injuries of institutional racial violence. Instead of retreating from the false promise of economic equality, Frazier explained, the Black middle class sought instead to unhealthily entrench themselves even further in the inevitably exclusionary ideology of capitalism, to "escape in the delusion of wealth" and "make a fetish of material things or physical possessions." Their reactions, however injurious, were entirely understandable; the "quasi-pathological character" of American society and its immanent logic of racial violence was enough to drive anyone mad.[149] Frazier's observations correspond to W. E. B. Du Bois's portrayal of the period of post-Reconstruction in the United States, during which a developing Black bourgeoisie sought to develop an elite class identity through economic advancement, only to have that dream collapse by way of the devastating realization that capitalism, moored as it continued to be to white supremacy, could never function as an avenue to Black liberation.[150] As Cedric Robinson writes, "The violence and terror that descended upon Blacks during the fifty years that followed Reconstruction left the Black elite shaken and pared down to its opportunists."[151]

At the same time, Frazier strongly believed that American anti-Black racism was itself an expression of a deep-seated mental pathology. "The behavior motivated by race prejudice," he wrote some decades earlier, "shows precisely the same characteristics as that ascribed to insanity."[152] White supremacy was a "dissociation of consciousness," Frazier claimed, behavior to be expected among hysterics or somnambulists—that is, a form of *automatism* or, in this case, an unconscious automatic reaction specific only to the topic of Blackness among white North Americans. "Southern white people write and talk about the majesty of law," Frazier observed, "the sacredness of human rights, and the advantages of democracy—and the next moment defend mob violence, disfranchisement, and Jim Crow treatment of the Negro."[153] Racial violence was not the expression of delusional unreason so much as it was a form of moral insanity, a madness of conduct. When performed as an exalted and comparative racial identity, whiteness was both property *and pathology*. But what Frazier does not acknowledge is that the madness of white supremacy may very well have been the very same pathology that underwrote the behavioral abnormality of the (masculine white) entrepreneur, an ultimately lucrative irrationalism glaringly inaccessible to Black populations. For when Black businessmen sought to emulate it, as Frazier observed, the only thing they experienced were the maddening effects of exclusionary racial violence.

This is not to suggest that, in the United States or elsewhere across the North Atlantic, financial freedoms were not available to Black and nonwhite populations. Instead of speaking of *the* entrepreneur, Frazier reminds us that we should speak of an assortment of entrepreneurial aspirations, whose varying degrees of financial success were ultimately dependent on the access any given enterpriser had to the *privilege* of pathology—of transmuting the virulence of commercial life into a lucrative resource. In that sense, entrepreneurialism embedded a logic with respect to the difference between the lucrativeness and perniciousness of economic madness. For some to access the beneficial pathological wellspring of entrepreneurial creativity, it was structurally necessary, indeed conditional, that others disproportionately bear the brunt of the destructive virulence of economic trauma; and masculine whiteness chiefly demarcated the boundaries between these two possibilities. As Frazier observed, Black businessmen were in large part denied access to the privilege of entrepreneurial madness, and when they sought to inhabit the entrepreneurial form, what they instead experienced was not the madness that made money but the disorders of a racialized marketplace and its attendant forms of economic violence, exclusion, and erasure. Masculine whiteness made it possible for some entrepreneurs to take advantage of the profitability of madness according, however, to an entitlement to pathology denied to Black and other nonwhite enterprisers.

To put it another way, the very concept of the entrepreneur was, and likely remains, firmly entrenched within the structure of racial capitalism.[154] If racial capitalism is at least in part defined as racialized expropriation, then in the context of entrepreneurialism, expropriation plays out around the binary of lucrative versus pernicious pathologies. In order for some to overwhelmingly benefit from the remunerative advantages of economic madness, it was incumbent, almost in principle, that others, typically racialized others, significantly bear the violent burden of pernicious economic traumatisms. Entrepreneurial success, or the pathological privilege of white masculine economic irrationalism, was contingent on the fact that Black and other nonwhite enterprisers were essentially driven insane by the structural racisms constitutive of the history of capitalism. Only in this way could the white agent of commerce become profitably mad without surrendering to the virulent distress of commercial life, because it was ultimately his nonwhite counterpart who was forced to submit to that virulence.

Over the course of the twentieth century, entrepreneurialism became entrenched within an increasingly global cultural vernacular as a quasi-democratic reservoir of performances, a discernible and discrete form of behavior and style of conduct directed toward creative novelty and the engendering of value, and a supposed vessel of economic freedom available to everyone, the detriments of which arise only when entrepreneurial behaviors are deployed unscrupulously. Against this diffuse entrepreneurial optimism, however, scholars have posited far more critical appraisals of the entrepreneurialization of the self, the reconfiguration of "the human as an ensemble of entrepreneurial and investment capital," just one effect of the multifarious neoliberal transformations that have globally emerged since the end of World War II.[155] As I have argued, the concept of the entrepreneur, from its earliest twentieth-century articulations (so, in other words, prior to a discernible neoliberal moment), was not only fundamentally anchored in a psychiatric legacy but also, and precisely through that legacy, *irremediably affixed* to the logic of racial capitalism and the pathological privilege of masculine whiteness endemic to it. Even before it was recoded as a diffuse neoliberal style of self, the concept of the entrepreneur was already heir to a genealogy of madness, an inheritance through which the entrepreneur's consanguinity with the structure and history of racial capitalism was most acutely on display.

In their multifarious deployments of an economic style of reasoning, psychiatric researchers and clinicians articulated an assortment of ways that business enterprise could function as a gauge for assessing mental health. Admittedly, whether financial conduct could encapsulate an individual's psychiatric condition was a matter of some debate, and what economic behaviors actually measured was not always immediately clear. But what the economic style of psychiatric reasoning conceptually facilitated—what it allowed prominent clinicians to propose—was that against the frenetic and often pathogenic background of modern economic life, the relationship between capitalism and psychopathology was not as straightforward as it seemed. For it was not simply the case that modern capitalism drove people insane. In some cases, madness was perfectly compatible with economic competency, and in other cases, madness was a desideratum for an entrepreneur's economic success.

Availing oneself of the creativity and profitability of entrepreneurial madness, however, demanded balancing the most lucrative forms of economic pathology against the pernicious morbidity of commercial life. And such a balancing act depended on an entitlement to pathological conduct intrinsic to the entrepreneurial form, an inoculative bulwark against the psychical tumult of business and commercial undertakings organized around the boundary of masculine whiteness and therefore structurally foreclosed to nonwhite enterprisers. The pathological privilege underwriting entrepreneurial success reveals a troubling incongruity about entrepreneurialism as a social form and style of life conduct: that in the final analysis entrepreneurialism remains ethically irredeemable while nevertheless being *economically* redeemed time and time again to such a degree that its unrelenting amassment of financial value has come to stand in for its moral worth. Here we begin to observe a central conclusion of this book, which I will continue to elaborate in the pages ahead, which is that the specter of sociopolitical irrationality that haunts the modern liberal social order will never be exorcised from the social body so long as these irrationalisms can continue to be imagined and perceived as generators of wealth. Reason alone will never upend the menace of sociopolitical unreason, for the latter can always continue to validate itself according to an economic rationality.

The Aphasic's Will,
or Dispensations for the Propertied

Can an Aphasic Make a Will?

One of the first systematic accounts of aphasia, or the general neuropathology of language disturbances and speech loss, was presented in 1865 by the Parisian physician Armand Trousseau in the second edition of his lectures on clinical medicine. Among his numerous clinical observations, which were recounted and reinterpreted by clinicians and aphasia researchers for decades to come, was the case of a fifty-seven-year-old patient, Mr. X., whom Trousseau had encountered only two years earlier and who, Trousseau stressed, "possesses great wealth." Mr. X. had suffered from a major neurological episode, likely a stroke, that left him not only with hemiplegic paralysis on his right side but subsequently deprived of the faculty of speech. By the time Trousseau saw him, however, Mr. X. had regained his strength and, so it appeared, much of his mental integrity. Although he was unable to utter anything other than the monosyllable "yes," Mr. X. seemed able to understand and make his thoughts known to others, so much so that he was able to continue to play a role in the administration of his wealth and estate. While his son managed his affairs, Mr. X. continued to be consulted on matters concerning leases and contracts. "The son tells me," Trousseau reported, "that his father can indicate perfectly well, by way of gestures that become intelligible for those who are habitually around him, what elements of the contracts do not suit him, and he is only satisfied when modifications, which are typically useful and reasonable, are made."[1]

It is not unlikely that the case of Mr. X. was on the mind of the British physician and neurologist William Henry Broadbent when he recounted, less than a decade later, a remarkably similar clinical observation of his own. Broadbent's case concerned a seventy-year-old widow, Mrs. W. B., who had some years earlier suffered the effects of a likely stroke that, as in the case of

Mr. X., left her hemiplegic on her right side and without the ability to either write or verbally express anything more than a few monosyllabic utterances. "Her relatives," Broadbent pointed out, "treated her as if she was utterly incapable of looking after her own affairs," which, in the period immediately following the onset of her aphasia, grieved her immensely. Eventually, however, after some convincing that she was in fact quite cognizant of her financial circumstances, which were indeed comfortable, Mrs. W. B. managed once again to take possession of the management of her wealth. "During the same time," Broadbent wrote, "she had been making a new will," which she did with her sister-in-law's help. "It often took a long time to do it to her satisfaction," Broadbent explained, "but she never rested till every particular was exactly as she wished it; when the right guess was made she showed her satisfaction quite unmistakably by her gestures, and by saying 'Yes, yes, that's it.'"[2]

Among the growing and increasingly shared international compilation of clinical observations concerning the neuropathic loss of language, these two case studies stood out for many years because of how clearly they demonstrated an aphasic's capacity both to manage wealth and, perhaps even more importantly, to exhibit the testamentary capacities sufficient to write a will.[3] They were, however, far from the only examples in the growing compendium of known aphasics to achieve this particular renown. By the 1870s, barely a decade into the formal study of aphasia, among the questions that clinicians and neurological researchers posed—What are the symptoms and taxonomies of language disorders? What are their underlying neurological impairments? How can these disorders be treated?—was another uncertainty that circulated with equal urgency: Can an aphasic make a will?

This was precisely the question that John Hughlings Jackson posed as part of a three-article series devoted to "affections of speech," which first appeared in the inaugural volume of the journal *Brain* (of which Jackson was a founding editor). There was, Jackson argued, "no single well-defined 'entity'—loss of speech or aphasia." The reason for this is that far too much variability existed among patients, in terms of which neurological processes were disrupted, to what degree, and how quickly. Aphasia could never be diagnosed *in principle*, Jackson insisted; its fundamental contingency and variability were not simply incidental features but, rather, central attributes of the disease. "Thus, to state the matter for a particular practical purpose," Jackson concluded, "such a question as, 'Can an aphasic make a will?' cannot be answered any more than the question, 'Will a piece of string reach across this room?' can be answered. The question should be, 'Can this or that aphasic person make a will?'"[4]

The inquiry into an aphasic's ability to write a will might seem like a somewhat narrow medico-legal concern were it not that a prevalent though transient feature of late nineteenth-century aphasiology was to consider the degree to which aphasics could be said to retain and demonstrate testamentary capacities and thereby continue to function as economic agents capable of owning and distributing property and wealth.[5] Jackson's succinct phrasing of the question continued to be cited and invoked as a veritable shorthand for this testamentary quandary for forty subsequent years. The aphasic's testamentary capacity was, in other words, an inquiry that ran parallel to the first formative decades of the study of the pathology of speech loss.

A month before the final installment of Jackson's three-article series was to appear, a reviewer responded to the first two articles, rather optimistically: "We confess that we have hopes that, before Dr. Hughlings Jackson has finished with the subject, he will leave us a few suggestions as to the medico-legal importance of the most prominent symptoms detailed above." Given that nothing close to a uniform assessment of an aphasic's testamentary capacities yet existed, the reviewer had hoped that Jackson could fill that void by providing the international medical community of aphasiologists with a technical resolution to this conundrum. "A few simple tests of the patient's capabilities," the reviewer wrote, "would be most welcome from the pen of so experienced a practitioner"—tests that, in the able hands of someone of Jackson's repute, could be "fashioned into weapons sufficiently keen and powerful to enable us to determine exactly the medico-legal position of any given case of aphasia."[6]

Sadly, however, Jackson never fulfilled the reviewer's hopes—not in the third article nor in any other of his subsequent writings. Nor, for that matter, did any other clinician or neurologist, at least not in any comprehensive sense. This is not to say, however, that there were not extensive efforts to construct the sort of "weapons" for which the reviewer pined. In the ensuing decades, aphasiology researchers generated not only much more expansive taxonomies of the different modalities and varieties of neuropathological speech loss—apraxia, agraphia, aphemia, ataxia, alexia, asemia, aphrasia—but also a variety of administrative protocols by which the various aphasias might be successfully applied in the civil domain. Compare, for instance, British neurologist Henry Charlton Bastian's diagrammatic classification of aphasia from 1869 to the much more expansive "Table of the Aphasias" produced by the American physician Frank Langdon almost thirty years later—a taxonomy that was prepared precisely with an eye for how pathological subdivisions might bear out practically and administratively (figs. 4.1 and 4.2).

But even these expansive taxonomies could never fully resolve the enigmatic nature of aphasia, which, above all, unsettled a commonsense belief

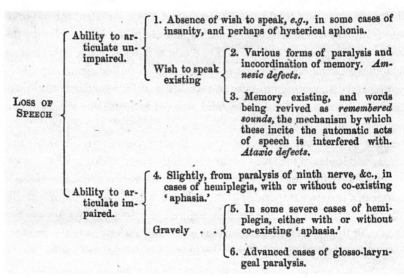

LOSS OF SPEECH

Ability to articulate unimpaired.
1. Absence of wish to speak, *e.g.*, in some cases of insanity, and perhaps of hysterical aphonia.

Wish to speak existing
2. Various forms of paralysis and incoordination of memory. *Amnesic defects.*

3. Memory existing, and words being revived as *remembered sounds*, the mechanism by which these incite the automatic acts of speech is interfered with. *Ataxic defects.*

Ability to articulate impaired.
4. Slightly, from paralysis of ninth nerve, &c., in cases of hemiplegia, with or without co-existing 'aphasia.'

Gravely . .
5. In some severe cases of hemiplegia, either with or without co-existing 'aphasia.'

6. Advanced cases of glosso-laryngeal paralysis.

FIGURE 4.1. Henry Charlton Bastian's 1869 arrangement of speech loss in "tabular form." From "On the Various Forms of Loss of Speech in Cerebral Disease," *British and Foreign Medical Chirurgical Review* 43, no. 85 (1869): 209–36

that civil freedoms and economic aptitudes were constructed upon an unadulterated continuity between competency, volition, and transparent self-expression. Aphasia not only disrupted normative appraisals of civil agency, but did so in ways that were unexpected and almost consistently anomalous. Langdon, himself, described the case of Jackson Butler, a fifty-five-year-old Black man who worked as a tobacco warehouse porter in Cincinnati, Ohio. Butler was admitted to the Cincinnati Hospital in 1896 with aphasic symptoms, suffering from what was likely a series of strokes. On testing his word-perception and word-utterance capacities, Langdon reported that "words pronounced to [Butler] could be repeated immediately in the case of some words. A notable exception to this last statement, was the word 'no.' On being asked to say 'no,' [Butler] seemed to make a severe mental effort and finally said 'I can't say no.'"[7]

In the history of aphasiology, anomaly was the norm. Aphasia represented a disease state very different from the sorts of mental pathologies that might have ordinarily been classified under the heading of insanity. Indeed, for many clinicians, it was vital to distinguish aphasics from the insane, for whom they were often mistaken by both the public and some medical experts. Not only was aphasia a pathological category onto itself, but it was also a category that ultimately failed to sufficiently generalize every given case of language disorder. Aphasiologists were inclined to admit that there was no such thing

FIGURE 4.2. Frank Warren Langdon's 1898 "Table of the Aphasias." From *The Aphasias and Their Medico-Legal Relations* (Norwalk, OH: Laning, 1898), 16.

as aphasia *as such* but rather *aphasias*, the degree and variability of which varied widely. And it was frequently in discussions concerning testamentary capacities that aphasiology researchers emphasized the nongeneralizability of language disorders. As the New York neurologist Joseph Collins explained, determining which specific variety of aphasia a patient suffered from was "the most important thing in determining the patient's capacity to make contracts, wills, checks, and indulge in other civil matters."[8]

Psychiatric clinicians and researchers had long been interested in the testamentary capacities of insane patients, a forensic apprehension that well preceded the emergence of the discourse of language disorders.[9] Aphasia, however, represented something quite different, for it did not map analogously onto insanity, even as researchers would eventually insist that aphasia entailed more than simply the loss of language capacities but also a degree of global cognitive decline. This chapter explores how, through an economically inflected style of diagnostic reasoning, aphasiologists ultimately concluded that, in contradistinction to the classically insane patient, an aphasic's purported testamentary capacities could never be abrogated in principle, no matter how severe the state of language loss might have been and no matter to what degree clinicians came to view aphasia as commensurate with an inevitable enfeeblement of cognitive powers. As we will see, aphasiologists almost always defended an aphasic's capacity to bequeath wealth.

This dominant tendency to make sense of and ultimately substantiate the testamentary capacity of aphasics was not, however, organized around a concern with patients' civil rights. The discourse was instead anchored on an underlying recognition that a degree of economic value could be extracted from this neuropathological condition—that aphasia was a disorder from which value could be derived. It might seem unusual at first to consider that aphasia could be value-generative in any respect. Aphasics were, after all, hardly comparable to entrepreneurs whose purported pathological creativity was regarded as the source of potential profit (as I discussed in the previous chapter). Few aphasic patients rigorously engaged in ordinary acts of business enterprise and financial speculations. But what we must instead appreciate is that writing a will was not just a legal right or civil freedom but a form of economic conduct that presupposed the possession of transferable wealth. An aphasic whose testamentary capacity was in question was a patient who ultimately enjoyed an auspicious financial status, and so the question "Can an aphasic make a will?" was a concern most germane for individuals whose economic activity would have been entirely organized around acts of inheritance. The magnitude of this otherwise prosaic observation comes into focus when we acknowledge that the most prevailing method of wealth accumulation in late nineteenth-century Europe and the United States was through the inherited amassment of generational wealth, which, as we will see, coincided with a general trend toward increased testamentary freedoms across Europe and the United States.[10] Aphasics were, in aggregate, value-generating to the extent that they contributed to one of the most remunerative forms of economic activity at the end of the nineteenth century.

This chapter, then, examines how through a style of diagnostic reasoning— a style that was economically underwritten—prominent clinicians and researchers were inclined to affirm that aphasia could be entirely compatible with, and thus never an impediment to, financial transfers and the heritable accumulation of wealth; and, therefore, that aphasics ought never to be excluded tout court from the sphere of economic activity. Aphasia was among the gravest neurological and neuropsychiatric disorders—that much was never in dispute; and unlike the various psychopathologies, aphasia normally presented very discernible symptoms even if the nature of those symptoms was quite anomalous and enigmatic. Aphasics were not strictly among the borderland figures that I described in chapters 1 and 2, whose social inscrutability needed to be adjudicated through the lens of economic conduct. Despite all this, aphasia was still perceived to be far more compatible with economic competency, and with far fewer qualifications, than insanity and even entrepreneurial madness ever were. This likely had much to do with the fact that aphasiological appraisals were subtended by a system of wealth accumulation that would have been perceived as far less perilous, and less socially precarious, than, say, business ventures for which the risk of sudden financial gains were countered by an equivalent—and equally pathogenic—risk of sudden losses. Aphasic testation was in many respects an innocuous discourse, a topic largely minimized in histories and theories of aphasia, and one that lacked the controversies and fanfare that accompanied the more audacious forms of enterprising madness that I discussed in the previous chapter. At the same time, however, aphasic testation unified the categories of pathology and economic value, and thereby formalized the idea of pathological value, far more discreetly and comprehensively, with far less contention, than any other morbid psychiatric condition. Guaranteeing the aphasic's ability to write a will was the silent means by which psychiatry safeguarded the generational accumulation of wealth.

From an administrative standpoint, however, the status of an aphasic's testamentary capacity was ultimately a forensic or medico-*legal* concern. That being said, one of most common refrains among aphasiologists, as we will see, was that the law could never quite capture the inconsistent and varied nature of the aphasias, which presented a level of diversity and inexplicability that even the multifarious nature of insanity could not quite emulate. Consequently, the idea of a juridical style of reasoning, which I described in the previous chapter, was far less evident in the discourse on aphasic testation. There was little inclination to preclude aphasics *in advance* from entering the economic sphere. Instead, it was more frequently assumed that aphasics were from the outset capable of carrying out financial bequeathments, acts often designated as the most anodyne form of economic activity, even as inheritance was among the most dominant means, in aggregate, of wealth accumulation at the time.

In other words, while the compatibility between *insanity* and economic competency was a matter of at least some controversy, with respect to the aphasia, the de facto medical position was that aphasics normally demonstrated an acceptable degree of economic aptitude. But this economic style of reasoning about aphasic testation was not necessarily an effort to ensure the legal rights of aphasic patients but, rather, to concretize an economic imperative. It was not a juridical demand but an economic incitement—we might even call it an ideological impulse—that prompted the concern over whether an aphasic could write a will. So, while aphasic testation was undeniably a legal category, the discourse itself was not about recodifying a legal doctrine per se but about formalizing an economic norm. To be clear, norms and laws are not mutually exclusive categories. The historian Jan Goldstein has noted that while it is useful to distinguish laws from what Foucault called the "much more general power" of norms, the two remain, as Goldstein puts it, "symbiotically joined."[11] This line of reasoning is very much Weberian at its core. "Law [*Rechtsordnung*], convention [*Konvention*], and custom [*Sitte*] belong to the same continuum," Max Weber wrote, "with imperceptible transitions leading from one to the other."[12] The real difference between a norm and a law is that a law is simply a norm that can be coercively applied, one whose transgression can be met with sanctioned violence. Consequently, while the discourse on aphasic testation was outwardly forensic, it was in reality an economic encodement.[13] Juridical analyses of aphasic testation were surrogates for a deeper economic rationality, and psychiatric justifications for an aphasic's capacity to write a will were animated precisely by that economic function.

This chapter concludes with an account of why the discourse on aphasic testation disappeared after the end of the First World War, just as precipitously as it emerged in the 1870s. At the end of the nineteenth century, aphasics for the most part comprised a specific demographic of older propertied patients, whose aphasia was most frequently occasioned by strokes and age-based cognitive decline. After World War I, however, aphasia researchers focused their case studies on a whole new population of patients: injured soldiers whose aphasic conditions were primarily the result of war traumas and for whom will-writing was subordinated to the more pressing challenge of postinjury social reintegration.[14] While the specific discourse of *aphasic* testation receded, the problem of *pathological* testation did not. The specific topic of aphasia was instead replaced by a cognate neuropsychiatric disorder: senility and age-based cognitive decline, which functioned as a proxy for aphasia not only because it presented a compatible symptomatology but because it replicated many of the same medico-economic challenges of older propertied patients seeking to transfer their wealth. As a result, senility inherited the same economic style of reasoning that psychiatrists had earlier deployed with respect to the aphasic.

Language Disorders and the Status of the Intellect

By most accounts the neurological study of speech disorders began with the French anthropological anatomist Paul Broca and a hypothesis he presented during a discussion of the relationship between craniometry and intelligence at the 1861 meeting of the Société d'anthropologie in Paris, namely, that a subsection of neuroanatomy enabled language performances.[15] Broca was by no means the first to address the specific neuropathology of speech loss. In France alone, the physicians Jacques Lordat and Jean-Baptiste Bouillaud had discussed cases of the discrete pathological loss of language four decades earlier.[16] But Broca's contribution differed to the extent that he explicitly linked a precise functional impairment of language (motor expression) to a specifically damaged region of the brain (a portion of the left frontal cortex, thereby dubbed "Broca's region"). Broca's essay was published at a moment when European medicine was witnessing the escalation of debates over the anatomical localization of mental functions in the brain—that is, whether and how pathologies could finally link mental activity to cerebral structures—which was itself a culmination of the steady biologization of the most fundamental traits of human nature.[17] Aphasia, then, at least in its earliest discussions, was not only one of the first thoroughly *neurological* disorders, but also the paradigmatic disorder through which nineteenth-century medicine could construct the new anthropological figure *homo cerebralis*.[18]

And yet, from the outset, aphasia possessed a kind of double life in nineteenth-century neuropathological research. On the one hand, it functioned as a clinical expedient for debates concerning the strict localization of mental functions. For many researchers committed to a tradition of pathological anatomy, language loss was defined by the clear-cut deprivation of specific linguistic capacities (lexical memory, articulation, verbal comprehension, the ability to write), capacities that could be situated within discrete cortical and subcortical regions of the brain by virtue of lesions observed and identified in postmortem analyses. Aphasia, in this sense, was one of the most effective strands of pathological research to further corroborate the value of localizational paradigms.

On the other hand, clinicians could not help but admit that aphasia was an especially unusual disorder, a "cerebral symptom of a special nature," as the prominent French psychiatrist Jules Falret would call it.[19] One of the earliest and most surprising characteristics that defined its distinctiveness as a neuropathological disorder was that the loss of language did not simultaneously necessitate the loss of intelligence. Broca himself discussed the case of a man named Leborgne, whom he referred to as "Tan," after the only monosyllabic sound that Leborgne was capable of uttering. Broca admitted, "His state of intelligence could not be exactly determined." While Leborgne appeared to understand everything that was said to him, he had a great deal of difficulty making his own thoughts known, given the constraints on what he was capable of uttering as well as the hemiplegia from which he suffered, which dramatically restricted his movements.[20]

Trousseau was among the first clinical researchers to insist that medical assessments of aphasic patients demanded an investigation into the status of their intellect. "It is important to investigate whether *intelligence is impaired* among aphasics, and to what extent," he explained. "But this not an easy evaluation."[21] The clinician could only hope to identify patients' intelligence by interrogating whatever forms of communicative expressiveness, such as gesture or writing, continued to endure among them. This was, however, made difficult by the fact that, in many cases of aphasia, both writing and gesture were equally impeded. Trousseau adopted the position that thought and speech were distinct faculties, while remaining somewhat ambivalent as to how or to what degree they were linked. He believed that, overall, there was typically impairment to some degree in the intelligence of aphasics, while nevertheless admitting that those with some forms of aphasia exhibited virtually no intellectual deterioration. So while thought and speech were not rigorously opposed, Trousseau seemed to imply that they were not simply conjoined either, functioning instead as two ends of a continuum leading from mind to language.

The broad partitioning of thought and speech was a recurrent motif throughout the early history of aphasia research. In his influential monograph on aphasia, the British physician Frederick Bateman argued that authors like Broca and Trousseau had limited the use of the term "aphasia" to describe cases where intelligence was either unaffected or only "slightly impaired." Bateman did not conflate intelligence with speech but claimed, instead, that while in some cases damage to the brain might debilitate speech alone, leaving intellect intact, there were many cases where the intellect was generally damaged, thereby affecting linguistic capacities. Both cases could be called "aphasia" even though they represented very different kinds of neuropathological conditions. Speech, from a neuroanatomical standpoint, could potentially be impaired without any harm to the mind, whereas a so-called lesion of the mind could not occur without some downstream effect on language capacities.[22]

Bastian too affirmed that "the widest latitude of variation" persisted among the degree of intellectual impairments observable among aphasics. He agreed with Trousseau that while some intellectual weakening was observable in many cases of aphasia, "in others it became much more difficult to ascertain the existence of any general failure in mental power." According to Bastian, "typical" aphasics maintained a degree of intellectual integrity, even though for other aphasics complete dementia was also observed.[23] A British colonial physician, stationed at a medical college in Bengal, corroborated "exactly those [symptoms] observed by Trousseau and others": the loss of language capacities "whilst the intelligence was comparatively good."[24]

Falret tried to sum up these observational inconsistencies, explaining that the relationship between intelligence and aphasia was both "very controversial and difficult to resolve in any absolute manner." It was rare, he proposed, for intelligence to be completely preserved in an aphasic; most aphasics experience some alteration, if not debilitation, of their intellectual capacities. This was not, however, a result of aphasia being an impairment of intellect, with consequential incapacitating effects on expressivity. Aphasia was, in a strict sense, an impairment of language capacities alone; however, without the use of signs, Falret explained, the mind could not typically function in its normal manner. It would be difficult to assert in principle "that intelligence could be completely preserved in a pathological state formed by a brain lesion." In any case of aphasia, the intellect was by necessity "besieged in some more or less profound way." At the same time, however, the degree of intellectual impairment varied dramatically from patient to patient, and the primary clinical challenge was to determine the exact state of intelligence that remained.[25]

The prevailing question of whether the faculty of speech could even be distinguished from the mind was built upon several competing philosophical and medical traditions. "It is debated," wrote the German clinician Adolf Kussmaul, "whether conceptual thinking [begriffliche Denken] is bound up with words (Condillac, Max Müller, Bastian, et al.) or independent of them (Locke, Helmholtz, Maudsley, Finkelnburg, et al.)."[26] Still, Kussmaul too accepted what had become the common consensus among clinicians and aphasia researchers: that words and ideas (Vorstellung) were ultimately different things, and that their "independence . . . is easy to demonstrate."[27] Their distinction was doubtless fueled by demands drawn out of localization research; the validity of the doctrine of discrete brain localization was partly conditional on the presumptive separability and independence of mental functions. It is also possible that the legacy of moral insanity, which I describe in chapter 2, played a role in the conviction that speech and intellect could be partitioned, as it already inclined clinicians to believe that certain genres of mental pathology could effectively incapacitate fundamental attributes of a patient's behavior while leaving intelligence and reason intact.

The French psychiatrist Henri Legrand du Saulle, having just begun his tenure at the Salpêtrière hospital in Paris, defined aphasia as the pathological loss of "speech" (parole) combined with the preservation of "more or less some notable part of intelligence and integrity of the phonetic organs."[28] And yet, what defined intelligence was not, as Trousseau and others had claimed, a patient's ability to comprehend and communicate. Instead, according to Legrand du Saulle, an aphasic signaled intelligence through the preservation of characteristic behavior, by demonstrating "fidelity to his old habits" as well as a continued concern with prior affairs and interests. An aphasic could show

himself to be intelligent, Legrand du Saulle wrote, "when, in a word, he differed little from himself or did not differ at all."[29] Suspicions of diminished
competency would instead arise when patients lost interest in their affairs and
enterprises, when they displayed sudden indifference, apathy, absentmind-
edness, or negligence—when they displayed a marked departure from their
previous personalities. The very desire to put one's affairs in order, to produce
a will and ensure the proper transfer of wealth, could itself function as evidence that an aphasic continued to possess a sufficient degree of intellectual
competence.

For some clinicians, such as the Scottish physician John Wyllie, the best
evidence to support the proposition that the faculty of thought and language
were distinct could be found among deaf populations, whose intellectual capacities remained perfectly intact, Wyllie argued, despite their speechlessness.[30] Deafness was often invoked as an analogue to certain forms of aphasia
as a way of justifying the preservation of an aphasic's intelligence. The discrete separation between language and thought commonly agreed upon by
many clinicians and aphasia researchers, however, did not mean that there
was uniformity in terms of how the two faculties were believed to be related.
For some, language and thought were intimately linked, the one effectively
mirroring the other.[31] For others, however, the difference between the two
faculties was stark and hierarchical, language being simply "reason directing the motor nerves of the tongue, lips, and palate," or "externally expressed
thought."[32] According to the French neurologist Gilbert Ballet, Jean-Martin
Charcot's *chef de clinique* at the Salpêtrière, "Ideas and words are independent of each other. The word is the auxiliary of the idea, it is not its necessary complement." He admitted that while it might not have been possible to
think *well* without signs, it was, nevertheless, still possible to think without
the capacity of language. "We can therefore see from this," Ballet concluded,
"that the relative integrity of intelligence can coincide with the partial or total
abolition of signs, that is to say with aphasia."[33]

By the 1880s, it was rare for aphasiologists not to agree that at least some
degree of intellectual weakening would result as a consequence of aphasia,
even if many aphasics nevertheless could exhibit a remarkable degree of continued intelligence. "Nothing varies," Legrand du Saulle wrote, "like the intellectual disorders observed among aphasics." Pathological disturbances of language might end up having just as much, if not more, of a debilitating effect
on patients' emotional and moral dispositions as on their intelligence; rather
than intellectual impairments, a clinician might very well encounter states
of depression, heightened suggestibility, and limited voluntary capacities.[34]
In any case, Legrand du Saulle concluded, a fundamental duality lay at the heart

of language disorders. While intelligence was undoubtedly weakened by the loss of language, given the indispensable part language played in the brain's normal processes, still aphasia was a disorder that was quite compatible with continued intellectual competence.

"I Am Able to Reason in a Semi-Conscious Fashion"

In one of his unpublished papers, likely from the early 1890s, Jackson presented a short case study of a patient, whom he identified (perhaps with Trousseau in mind) as Mr. X., "a very intelligent and very well-educated man, aged 31." Likely a clergyman based on Jackson's brief biographical descriptions, Mr. X. had suffered a traumatic brain injury that left him suffering from a series of neurological ailments, including seizures, bouts of temporary unconsciousness, and occasional mild paralysis or numbness in the right limbs. It was, however, the attacks of temporary aphasia, of "all clinical varieties," that were of special interest. Mr. X. suffered only temporary states of severe language disturbances, interspersed with extended periods of lucid mental health. The transient nature of his aphasia allowed Mr. X. to recount the experience of his pathological episodes, which he did in a surprisingly cogent, albeit brief, statement that Jackson appended to his case notes. "The account the patient gives," Jackson wrote, "is a very admirable one for clearness and consecutiveness." In his own words, Mr. X. explained,

> What happens to me when an attack [of aphasia] comes is that I may be speaking and suddenly I stop—it may be in the middle of word—and for some seconds I am helpless, and then my faculties seem to rearrange themselves, I can go on, but I cannot always pick up where I left off, and often can't recall what I was speaking about. . . . I have found that the attempt to rehearse the alphabet usually brings me to myself, which shows that while in this helpless condition *I am able to reason in a semi-conscious fashion.* . . . My aphasia is a break in the continuity of my thinking. It is as if my thinking were "switched off" and yet from the above you will see that there is often (and perhaps always) a measure of connected thought.[35]

Mr. X. corroborated, in a way few aphasic patients could, some fundamental tenets of Jackson's theory of aphasia.

Decades earlier, Jackson had observed an especially striking characteristic of the aphasic symptomatology, an observation he attributed to the French psychiatrist Jules Baillarger, that aphasic patients could often involuntarily pronounce words, the very same words, that they could not otherwise voluntarily call to mind.[36] His earliest interpretations of this phenomenon challenged from the outset the emerging dominant trend in the study of speech loss, that language could be partitioned into motor, sensory, auditory, and visual functions and that these functions could be anatomically localized within discrete subdivisions of the brain.[37] Jackson was clear, quite early in his writing, that he "does not attempt to localize language in any limited spot."[38] And, in striking fashion, he further submitted that there was no such thing as a "faculty" of language at all and that aphasia could not consequently be understood to be a discrete disease.[39] Suffice it to say, Jackson's ideas were not initially well received by other researchers of aphasia. It was only after the early 1870s, when his neurological theories began to acquire a degree of international repute, that Jackson's writings on speech loss began to receive a more positive reception.

Jackson proposed that language itself, so its pathology revealed, is best understood according to a hierarchy of expressivity, at the top of which was an "intellectual," voluntary, and "propositional" form of linguistic intelligibility and below which lay an "emotional" or automatic form of affective expression.[40] Jackson sought to make sense of the aphasias much like other neurological and psychological disorders, through the language of nervous evolution and dissolution. As I detailed in chapter 1, all mental and neurological processes were, for Jackson, organized according to a hierarchy of complexity and broke down in the inverse order in which they developed. The so-called lowest levels of the nervous system were the oldest, most organized, and least susceptible to disruption, whereas the highest functions were the evolutionarily newest, least organized, most delicate, and most complex.

In the same way, even the highest language centers, such as the so-called speech center ("Broca's region"), were nothing more than the subdivisions of more complex neurophysiological arrangements. Sophisticated speech functions emerged or "evolved out of" lower and evolutionarily earlier linguistic processes, such as the physiological activity underlying phonemic enunciations. The highly developed speech center designated as Broca's region was simply the "re-representation" of "what we may call primitive and so to say commonplace simple movements of the tongue, palate and lips." Higher language centers developed when older linguistic functions were "compelled into new combinations," a process that Jackson likened to an "evolutionary descent" reminiscent of "a little 'internal Lamarckism.'" When newer nervous arrangements emerged out of older ones, they acquired a substantial amount of complexity and functional independence, thereby representing more than just the physiological processes of tongue and palate movements but also more holistic mental and symbolic functions as well.[41]

Neuropathic dissolutions, on the other hand, as Jackson had long proposed, consisted of a double process: the loss of higher functions and the increase in the activity of the remaining lower functions. Jackson was convinced that the process of dissolution applied in principle to any neuropathic state, including aphasia. The negative symptoms that accounted for the loss of language capacities had to be contrasted to the positive symptoms that remained, such as the "words, syllables or jargon which a partially aphasic person does get out." These positive symptoms corresponded to the healthy nervous processes and continued evolution among lower linguistic functions that nonetheless endured despite the higher-level language loss.[42] The typical symptoms of aphasia, then, "the recurring utterance in speechless men, the erroneous words uttered by those who have defect of speech," were simply the direct expression of lower linguistic functions that had become disinhibited.[43] The neurologist should not, Jackson claimed, simply be concerned with what was lost in aphasia, for the functions that were permitted to increase were just as important in diagnosing the severity of the disorder.

From his earliest writings, Jackson schematically distinguished higher and lower linguistic functions into what he called "intellectual" and "emotional" language.[44] They were, as far as Jackson was concerned, "two inseparable yet distinct forms" of expressivity, the separation of which was displayed only through pathology. In healthy states, intellectual and emotional language were united. In most cases of aphasia, however, it was intellectual language that typically bore the brunt of the injury; "emotional language escapes altogether."[45] For Jackson, intellectual language had a very specific linguistic definition: "To speak," he explained, "is not simply to utter words, it is to propositionise."[46] The propositional nature of intellectual language was crucial for Jackson. A proposition referred to any utterance capable of generating a symbolic meaning that exceeded the simple sequence or summation of words. Not every pronouncement or statement was a proposition in the proper sense. The statement *a jumped over fence the dog* was simply a sequence of words; it only attained the level of propositional meaning, *a dog jumped over the fence*, once the words were properly organized. Propositional language, then, was achieved when words were organized so as to "[refer] to one another in a particular manner,"[47] when the words exhibited a "proper inter-relation" and were "modified by each other."[48] Jackson's propositional theory of language was certainly unique among late nineteenth-century aphasiologists for its refusal to take language as a simple transposition of thought. The most fundamental unit of speech, Jackson maintained, was the proposition itself, a conclusion on account of which the linguist Roman Jakobson would later call Jackson "among the precursors of modern linguistics."[49]

Aphasia, then, amounted to the loss of intellectual language or the capacity to propositionalize fully and completely. Aphasics would resultantly experience a reduction or dissolution to a lower and more *automatic* mode of linguistic behavior, what Jackson characterized as "emotional" language or affective expressivity. Aphasia, then, was yet another instance of pathological automatism. And yet, emotional language or linguistic automaticity was itself a highly common experience among normal, healthy individuals as well. Common instances of ordinary automatic emotional language were observed during moments of heightened agitation. Jackson's primary example was swearing. "Swearing is," Jackson claimed, "strictly speaking, not a part of language." It was a reflexive effort "to add the force of passing emotions to the expression of ideas." Swearing belonged to the category of what could be described as the emotional amplifiers of language, which also included the "loudness of tone and violence of gesticulation."[50] Emotional language and linguistic automatisms amounted to the only linguistic recourse available to aphasics, even though emotional language was normally intertwined with intellectual propositions among healthy language users.

At the same time, however, the reduction to automatic, emotional language did not represent the absolute annihilation of the capacity to propositionalize among aphasics. In the same way a complex act of physical dexterity demanded the mobilization and combined coordination of several automatic motoric sequences, so too did higher linguistic functions rely on a combination of automatic linguistic capacities. "Before a proposition is uttered," Jackson explained, "before voluntary use of words, words must have been automatically revived."[51] While aphasics were stripped of their voluntary use of propositional speech, they continued to retain automatic linguistic capacities that were not entirely devoid of propositional value. "There is nothing strange in supposing that there is an unconscious use of words," Jackson proposed.[52] Aphasics might have been speechless, but they were still *unconsciously* linguistic. They could, for example, initiate linguistic propositions, even though their neuropathology prevented those propositions from ever extending beyond the moment of their initial discharge. As John Forrester writes, in aphasics, "about-to-be-uttered utterances" become "'trapped' within the now automatic functioning" of the nervous system.[53]

This seems to be precisely what Jackson's patient, Mr. X., meant when he claimed that, even in states of pathological speechlessness, "I am able to reason in a semi-conscious fashion." Many aphasics were, in effect, unconsciously linguistic and so, therefore, capable of a degree of unconscious mentation as well. Jackson had long opposed the presumption that thought and language existed separately, refusing to partition as his colleagues did the intellect from linguistic capacities.[54] Speech and thought were fundamentally entwined, and mental activity at any level was always a kind of symbolic cognition. Aphasia, like any neuropathology, was not a subtractive loss but a pathological transformation of symbolic mentation itself. "When a patient loses speech from the disease of his brain," Jackson explained, "he loses not so much a quantity of general mind—if there is such a thing—as a special part of his mind. He cannot reproduce the motor symbols which elaborate thought requires."[55] An aphasic like Mr. X. did not simply lose the ability to speak while retaining the ability to think; instead, his symbolic mentation had been entirely transformed, altogether reduced to a lower and more "semi-conscious" state of both linguistic and intellectual functioning.

Despite his unwillingness to adopt the common clinical practice of separating thought from speech, Jackson's formulations did not entirely oppose conventional clinical suppositions concerning pathological speech disturbances. If anything, Jackson radicalized what the most sophisticated aphasiologists had already sought to articulate. Aphasia, it seemed, was not only compatible with the preservation of some degree of mental competency, but it was even compatible with the continued capacity to symbolize linguistically, albeit unconsciously and involuntarily. What ultimately marked aphasia as different from comparable neuro- and psychopathologies was that its onset did not inevitably amount to the immediate determination of incompetency and thus the forfeiture of civil capacities, even though it was, through and through, among the gravest pathologies of the mind and brain.

Expanding the Freedom of Testation

The seventeenth-century English jurist Sir Edward Coke, despite admitting that the nature of mental unsoundness was both variable and transient to some degree, nevertheless espoused a strongly dichotomized view that the line between competence and incompetence was sharp and discernibly so.[56] In England at least, Coke's sentiments were strongly echoed for the next century and a half by philosophers and jurists, including John Locke, John Brydall, and William Blackstone.[57] Early modern contestations over the economic affordances granted to the insane were much more concerned with the effective administration of property. As such, very little latitude was juridically bestowed on pathological populations who were, for the most part, exempted virtually in principle from the economic field; medical authority was rarely powerful enough to underscore the complex nuances of mental pathologies and potential compatibilities between mental pathology and economic conduct.[58] Blackstone was firm in his conviction that "mad-men, or otherwise *non-compotes*, ideots, or natural fools, persons grown childish by reason of old age or distemper," comprised a clear-cut and divisible category of individuals who were all "incapable, by reason of mental disability, to make any will so long as such disability lasts."[59]

By the nineteenth century, however, the prohibitions on the testamentary privileges of non-normal populations began to ease. In 1823, when discussing the prerequisite competency necessary for testamentary acts, the American physician and medical jurist Theodric Beck cited Blackstone verbatim.[60] Notably omitted, however, from Beck's invocation was the very next line of Blackstone's passage, which extended the prohibition of testation to those "born deaf, blind and dumb."[61] Instead, Beck unequivocally affirmed that verbal and hearing impairments, congenital or otherwise, did not invalidate the intellectual capacity necessary for testation. French physicians made similar concessions. Étienne Georget contended that deafness on its own did not annul the possibility of intellectual development, so long as the individuals in question had "received a complete education."[62] While French jurists recognized the testamentary competency of the educated deaf, they continued to hold that without education, deaf plaintiffs were incapable of producing wills; this legal equivocation particularly distressed Harvey Prindle Peet, the president of the New York Institute for the Instruction of the Deaf and Dumb. As far as Peet was concerned, a formal education was not necessary for a deaf subject to "possess the intelligence necessary to manage his own affairs, to make all civil contracts, to execute a deed or a will, or to give evidence in a court of justice," and the legal failure to recognize this fact was a "defect of the [French] law."[63]

Within the first several decades of the nineteenth century, medico-legal experts recognized that the absence of speech was not in principle a sign of intellectual deficiency. Language incapacity and intellectual incapacity could be partitioned and analytically divided, clearing the way for the interpretative possibility that aphasia, like deafness, did not necessarily imply testamentary incapacity. Although testation had long been foreclosed to pathological subjects, it had become by the mid-nineteenth century something more akin to a site of negotiation, a way of determining the minimal state of operative mental health requisite for civil and economic agency. For Isaac Ray, "wills involving questions of mental condition" presented the most interesting class of cases for medical jurisprudence. Testamentary inquiries complicated by the fact of mental pathology provided a "wider range of inquiry" than did other forensic problems that arose at the intersection of mental illness and the law.[64] Adjudicating a testator's state of mind often involved more rigorous investigation, while simultaneously attributing a degree of latitude to a testator's mental capacities.

In 1864 the American jurist Isaac Redfield discussed the relationship between mental capacity and testation, and while he maintained the general conviction that the unsoundness of mind negated the possibility of testamentary capacities, he also admitted that it was not easy "to lay down any precise rule as to what exact amount of mental capacity is sufficient to enable one to dispose of property by will." Of all civil economic acts, he explained, testation required only a sufficiently minimal degree of mental capacity, less than was necessary in the creation of a commercial contract for the simple reason that testation was made up of a single bequeathing individual, whereas a contract was complicated by the presence of multiple parties with varied interests and demands. According to Redfield, testamentary capacity only required an "active memory" and the ability to make rational judgments. Such requisites did not, therefore, preclude the possibility that a testator could still suffer from substantial pathology. Even the monomaniac, Redfield explained, was "entirely capable of transacting any matters of business out of the range of his peculiar infirmity; and he often manifests considerable sagacity, and forecast, in keeping the particular subject of his delusions from the knowledge of others."[65]

In the few years preceding the emergence of aphasia as a discrete and autonomous disorder, the medico-legal preconditions for testamentary capacity focused not only on the sufficiency of testators' intellectual capacities but also on their degree of intent and expressivity ("in intelligible language," as one particular legal hearing adjudged).[66] Nevertheless, what clinicians and forensic practitioners appeared to verify, even before the emergence of aphasia, was that testation was no longer an act from which pathological subjects should be flatly barred but one in which they could conditionally participate. With respect to financial testation, by the middle of the nineteenth century, there was already a shift away from what I have characterized as a strictly juridical style of reasoning toward a more economic style. Pathological states did not in principle bar individuals from writing a will and bequeathing wealth, and pathology and testation increasingly came to be perceived as theoretically compatible possibilities. The American jurist Edmund Wetmore argued that while "intellectual insanity" invalidated the act of will-making, on account of the delusional states that it occasioned, "general moral mania," that is, pathological changes to the conduct and character of a patient, "has never been held as a sufficient ground for annulling a testament."[67] Some forms of insanity were better than others in fulfilling the conditions of testamentary competence. Thus a style of forensic reasoning was already developing, even before aphasia emerged as a prominent neuropsychiatric disorder.

In France, Alexandre Brière de Boismont went even further, admitting that while delusional mental states often compromised a testator's "mental freedom" (*la liberté d'esprit*), the mere presence of delusions did not in principle invalidate the possibility of testation. Deploying an unmistakable economic style of reasoning, Brière de Boismont claimed that if a patient were "conscious of what he does and directs himself according to the ordinary rules of human wisdom," and all this "despite his hallucinations and illusions," then testamentary rights could not be suspended. Patients could suffer from delusions, but so long as the delusions did not impede the testamentary acts themselves, patients were free to exercise their economic rights.[68] In his discussion of testamentary capacity, Richard von Krafft-Ebing argued that it was impossible to infer the "mental unfreedom" (*die Geistesunfreiheit*) of a testator based simply on the paradoxical (*paradoxen*) or otherwise bizarre propositions written into a will. "The bizarreness and eccentricity of a healthy person should not be confused with the delusions of the mentally ill," he wrote. The contents of a will itself, no matter how strange, could never indicate whether the testator was sufficiently competent or not. Testation could admit of abnormalities in a way that other sectors of civil and economic life could not.[69]

Many clinicians were quick to point out that the criteria for soundness of mind could differ quite drastically in civil versus criminal proceedings. The same amount of mental illness that might legally preclude patients from being allowed to manage their affairs might not be sufficient to exculpate a perpetrator of a violent crime from responsibility. This was, for some, a troubling standard—less because of the extent to which it represented a juridical skepticism toward an insanity defense, but more because it typified an unjust readiness to confiscate patients' wealth in the name of protecting their property. "Our law," wrote Douglas Maclagan, chair of medical jurisprudence at the University of Edinburgh, "is not free from the reproach often directed against it, that it cares more for a man's estate than for his body or his life."[70]

Medical jurists, particularly throughout Great Britain and the United States, argued that while societies might benefit from the juridical skepticism directed at the insanity pleas of criminal defendants, they did not benefit from excessive restrictions placed on testamentary rights. Although insane, wrote a Scottish lunacy commissioner, many pathological testators "are still capable of doing certain things in a sufficiently rational manner, to make it fair to them, and not injurious to society, that these acts should be held as valid. Among these things is the making of a will."[71] The American medical jurist Henry C. Chapman reminded his reader that "strange as it may appear . . . less mental capacity is required by law to make a will than to permit the managing of property or the enjoyment of personal liberty."[72]

While denied to individuals suffering from states of absolute and unequivocal madness, testation was an economic act that could nevertheless tolerate deviations from normal mental health. Under the auspices of testation, madness and economic conduct could effectively coalesce. Throughout the nineteenth century, testation was increasingly made sense of through an economic style of psychiatric reasoning, in an ostensibly systematic effort to ensure that the bearers of wealth were granted the medical authorization to bequeath their fortunes and to facilitate the amassment of inherited wealth. This style of thought was expressed most unmistakably (though by no means exclusively) by American and British clinicians and jurists; it was undoubtedly an expression of broad national differences in the freedoms and restrictions underlying the power of testation.

In England, the freedom of testation enjoyed a comparative lack of restriction from as the early as the eighteenth century, just as it would in the United States, which drew upon British testamentary legal codes.[73] The Wills Act of 1837, amended in 1852, streamlined in both England and Wales the relationship between a testator's wishes and the process of will-making by, among other things, effectively formalizing holographic wills, that is, private wills that require no witnesses and only a signature. By 1858, as a consequence of divorce law reform, ecclesiastical courts had lost virtually all jurisdiction over probate matters, while commentators increasingly remarked upon the "comparatively unrestrained powers of an English will," which emphasized the interests of the testator over the "reasonable claims of close kinship."[74] English testators could essentially disinherit family members and bequeath their wealth to whomever they wished. "The object of first importance," wrote the British jurist Henry Sumner Maine, "is the execution of the testator's intentions."[75] Freedom of testation even became a topic of popular consumption, particularly with the broad dissemination of manuals for will-making intended for both solicitors and the lay public.[76] An American jurist would later conclude that the common attitude shared by both Americans and the British—that one's "belongings should be subject to [one's] volition alone"— had been "fostered by the concession of free testamentary power."[77]

In France, on the other hand, the power of testation was more constrained. Testamentary freedoms had been more or less abolished during the Revolution, only to be reinstituted in limited form in the Civil Code of 1804.[78] By midcentury, however, debates began to emerge criticizing testamentary constraints. While efforts to legislatively expand testamentary rights were defeated, the matter of testamentary liberty nevertheless became a topic of heightened legal and political discourse in France.[79] In both France and England, in other words, there emerged around midcentury a widespread set of discussions and debates concerning the status and function of testation. A similar phenomenon occurred in Germany, whose testamentary freedoms lay somewhere between the French and British models. By 1896, the German Reichstag implemented the right to produce holographic wills, thereby limiting intestate succession and bestowing upon the testator the central decision-making power in the distribution of personal wealth.[80]

Despite its legal standing, the freedom of testation became in the last half of the nineteenth century a topic of social policy, political debate, and popularization in both Europe and the United States, along with a general though certainly qualified drift toward greater testamentary freedom.[81] At the very same moment, will-making itself was increasingly viewed from the standpoint of medical jurisprudence as the civil economic act most tolerant of pathological deviations. And it was precisely in this historical context that aphasia emerged as a recognized medical disorder.

The Aphasic as Will-Maker

While an increasingly economic style of psychiatric reasoning held that some mental pathologies were compatible with testation, when aphasia emerged as a recognized disorder, it eventually came to stand in as the exemplary pathological condition capable of exhibiting the most comprehensive degree of compatibility with testamentary aptitude. With respect to the aphasic's testamentary capacities, clinicians would often claim that aphasic competency had to be considered on an individual basis. The question was precisely, as Jackson epitomized it, not whether *the* aphasic could write a will, but whether *this or that* aphasic could do so. This form of inquiry was not, however, the expression of a commitment to an ethics of individualized care, for this was a very different sort of question than was asked with respect to the insane testator. Insanity constituted from the outset a principled disqualification from civil life, but one that could, nevertheless, be subsequently mitigated through certain extenuations. Clinicians did not ask whether *this* or *that* insane patient could write a will, for to be insane as such meant that will-writing was impossible in principle. The question, instead, was whether the insane person was sufficiently lucid during the time in which the will was written (though even temporary lucidity was not always sufficiently mitigating[82]), or whether they were insane precisely in the ways that would nullify will-writing. An acceptable form of insane testation demanded a qualified and particularized sense of insanity—typically "borderland" and moral pathologies rather than, say, delusional psychosis.

Aphasia, on the other hand, did not have nearly the same level of disqualifying force. The very fact of aphasia did not rule out economic participation; a person could be perfectly aphasic, indeed extremely aphasic, precisely during the time in which the will was written without jeopardizing the validity of the testament. Will-making was simply a dispensation more readily afforded to aphasics than to the insane, and so it was consequently through aphasia, much more so than insanity, that neuropathological subjects could make a valid case for participating in this very specific form of economic conduct. Insanity, as we saw in the previous chapter, was situated at the intersection of two counterposing styles of reasoning, the juridical and the economic, whereas aphasia was predominantly made sense of through an economic style alone. The explanation for this difference likely came down to demographics. Mental disorders were common at any age, whereas aphasia, as Collins noted, "is of so much more frequent occurrence in the aged than in the young."[83] The population of aphasics at the end of the nineteenth century was by and large comprised—or at least believed to be comprised—of elderly patients who were not only more obliged to write a will but also more likely to have accumulated greater wealth over their lifetimes. Forestalling an aphasic's prospects to write a will could potentially interrupt the flow of heritable wealth from a demographic standpoint. In both England and France, as Thomas Piketty has noted, the flow of inheritance by the end of the nineteenth century accounted for nearly a quarter of the national income, before dropping precipitously after 1910 (a pattern consistent with other European nations).[84] Interruptions to flows of inheritance would have been viewed not only as an economic obstruction but, as clinicians reveal in their defenses of aphasic testation, a moral and medical infringement.

The aphasic's economic remit was not always or exclusively confined to testation alone. The prominent American neurologist Charles Hughes proposed that aphasics were capable of a surprisingly rich array of economic activities, including making contracts, taking out mortgages, granting powers of attorney, generating promissory notes and bank checks, and purchasing property, as well as writing wills. So long as the aphasic patient was capable of exhibiting some degree of intelligence, while also demonstrating a proper awareness of "his crippled brain state," then any and all economic freedoms should be afforded. The conditions of competence, even among those suffering from "damaged brains," was simply a recognition of one's pathological state and a willingness to adapt to it accordingly. "A sane aphasic," Hughes explained, "displays his normal character, adapting itself differently to his changed environment."[85]

Nevertheless, the earliest discussions concerning the testamentary freedoms of aphasics did not emerge particularly swiftly or decisively. Although Frederick Bateman first published his monograph on aphasia in 1868, it was not until the second edition over twenty years later that he included a chapter on the forensic ramifications of the disorder—"a subject of paramount importance to the medical jurist," he wrote in the new preface, "and which as far as I know, has not been treated by any British writer." What Bateman meant was that no British medical writer had treated the testamentary question as systematically as their French and German counterparts had already begun to do.[86] Less systematic references do appear as early as the 1870s, though. Henry Maudsley maintained in 1874 that no medical facts had proven that an aphasic was in essence deprived of testamentary capacity, even if some aphasics had simply become far too cognitively enfeebled to engage in any civil or economic activity whatsoever. "Each case," he insisted, "must be decided on its own merits."[87]

The first sustained debates on aphasic testation emerged in the early 1870s. An early polemical discussion was introduced by the Parisian appeals court lawyer Joseph Lefort at the 1872 meeting of the Société de médecine légale. Lefort admitted that while the French Civil Code (specifically articles 489 and 499) "enacted measures to protect the insane," specifically through the legal mandate of interdiction for those deemed incapable of managing themselves or their property, no such provisions had been made with regard to aphasia. This was, Lefort explained, "a lacuna easy to understand since this affection, while not ignored by those who first wrote the Codes, has been scarcely studied and only recently well known."[88] Lefort felt that aphasia was not a mental disorder properly understood, and that it was medically more aligned with deafness than with mental unsoundness.[89] Interdiction was no more applicable or appropriate for the aphasic than it was for the deaf person. Legally aphasia ought to be approached no differently than deafness, that is, according to the dichotomy proposed by French law that civil liberties were accorded only to the literate deaf.[90]

Lefort's paper was followed by a brief reply cowritten by members of the Société, the most prominent of whom were Falret and the Parisian physician Louis-Jules Béhier. They flatly rejected Lefort's blunt equivalence between aphasia and deafness, arguing that the frequency of intellectual weakness among aphasics warranted the conviction that aphasia was, at least in some measure, a kind of mental pathology. Aphasia could be the subject of legal interdiction, though in a way that was not entirely analogous to cases of insanity. The authors instead proposed a tripartite scale of possible juridical interventions. If an aphasic's intelligence were "completely obliterated," they wrote, or if the aphasic were utterly incapable of any form of external expression, "then the patient must be interdicted."[91] If instead the aphasic had not entirely lost intelligence, if intelligence was only incompletely manifest, then the patient must be provided with judicial council. If, however, the aphasic's intelligence were preserved, and sufficiently capable of being manifest through speech, writing, or signs, then "it goes without saying that the patient needs no judicial protection and that he must be permitted free administration of his person and property."[92] The first two conditions were, in effect, no different from how any insane patient might have been treated under the auspices of French law. It was the final condition that represented what was so novel about aphasia, for according to it, aphasics could be at once recognizably ill from the standpoint of neuropathological medicine, yet simultaneously free to manage themselves and their affairs.

Within a few short years, clinical aphasiologists were increasingly inclined to acknowledge that testamentary and other economic competencies were a common feature among aphasics. Relying on numerous case studies of patients who, despite their loss of language, were nevertheless capable of managing property and wealth, Kussmaul concluded, "Mental combinations directed at commercial affairs, that can be put into effect by means of language, can sometimes be excellently performed despite a high degree of ataxic aphasia."[93] In the same year, a précis of a lecture on the jurisprudence of aphasia delivered by the Parisian physician Jean-Théophile Gallard appeared in a prominent French medical journal. Gallard claimed that while most aphasics experience some degree of "intellectual decline [*dépression intellectuelle*]," they nevertheless "have a sufficient amount of intelligence and free will to take an interest in the management of their personal affairs."[94] Gallard then introduced the same tripartite metric described five years earlier by Béhier, Falret, and others—that aphasics should face three possible legal responses to their pathological state, depending on the degree of observable intellectual impairment: interdiction, judicial council, or liberty. Gallard's précis was introduced to English physicians through an article in the *British Medical Journal* that paraphrased and advocated for Gallard's views on the economic liberties of aphasics.[95] An English psychiatrist penned a reply to the piece, calling Gallard's views "instructive," but not without adding one particular point of qualification: that even in cases where an aphasic's intellect remained intact, sometimes language disturbances could so debilitate the mechanisms of expression that testation, while theoretically possible, become practically unfeasible. The sheer technicalities of will-making could sometimes present far too many obstacles for a patient whose means of expression had become fundamentally disturbed.[96]

The aphasic's entry into the sphere of economic life was often perceived as potentially constrained by a series of practical barriers. At the heart of these barriers lay the anomalous nature of aphasia itself, its pathological inimitability, summed up by many clinicians in the persistent refrain that no two cases of aphasia were alike, and that aphasia was wholly dissimilar to insanity and other common mental pathologies. This meant that aphasia could not conform to preexisting juridical standards of competency, even though its consistency with standards of *economic* competency was something upon which clinicians agreed. As Legrand du Saulle explained, "A testamentary examination [for an aphasic] does not even have the same value it would normally have for a lunatic or the ordinarily insane [*de déments vulgaires*]." In aphasia, the clinician had to be skeptical of even the most apparently self-evident manifestations of incompetency. "One must not lose sight of the fact," he wrote, "that the incoherence of writing style and use of words does not necessarily imply a corresponding incoherence in the ideas of the aphasic."[97]

The law, many European clinicians claimed, had virtually nothing to say about aphasia; as difficult as it had been for European legal conventions to adapt to the nuances and complexities of mental disease, aphasia's idiosyncrasies, variability, and symptomatological opacity posed an even graver forensic conundrum. The Austrian physician Emil Frischauer lamented the relative paucity of legal discussions in Austria concerning the testamentary capacities of aphasics. Aphasic testation, he wrote, has "not been discussed in the legal literature; in larger works of Austrian law, it has hardly been touched upon, and even in the literature of common law, one seeks in vain for a scientific discussion." The absence of legal doctrines had much to do with the fact that, with regard to aphasics, "a general rule cannot be established," and unless an aphasic has been rendered "idiotic [*blödsinnig*] or deprived of his reason," then all testamentary freedoms should stand.[98] The German psychiatrist Friedrich Jolly worried that the question about aphasic testation was not whether patients possessed a sufficient degree of intelligence but whether their preserved intellectual capacities could translate into the administrative formalities prescribed by testamentary procedures. Nevertheless, he admitted, even this sort of prerequisite was hampered by the fact that in Germany there was still little national uniformity, even after unification, for testamentary guidelines. Any attempt at uniformity, Jolly argued, must be organized around the fact that aphasia cannot be excluded in principle from the right of testation.[99]

In France, Désiré Bernard, a house officer at the Salpêtrière under Charcot, echoed this increasingly dominant position that the law said nothing about aphasia, due in part to the fact that "this pathological state presents modalities too varied for a common rule to be formulated in all cases of this illness."[100] Admittedly, for some neurological clinicians, like William Gowers, the testamentary question amounted to a relatively cut-and-dried appraisal demanding little more than a simple determination of the subtype of aphasia in question. Pure motor aphasia, according to Gowers, likely corresponded to the ability to make a will with the same degree of likelihood that sensory aphasia corresponded to absence of testamentary capacity.[101] For many others, however, aphasia posed very nearly no grounds whatsoever for legal interdiction. George Savage went so far as to propose, "Aphasia *may exist to almost any degree*, yet there may be sufficient mental power to transact all the business of life sanely and satisfactorily."[102]

Aphasiologists for the most part underscored a conviction that the law was simply unable to capture that language pathologies were almost systematically unsystematic. But it was not simply the inadequacy of legal doctrines that these researchers noted, for what they also observed was that the law could not make sense of what the economic conduct of aphasics appeared much more demonstrably to reveal: that they were largely quite capable of writing wills. As such, explicit medico-legal debates were made up of a combination of declarations concerning the law's shortcomings and efforts to codify an underlying economic norm—namely, that aphasic neuropathology was still consistent with the capacity to transfer wealth and that heritable wealth accumulation should not be medically undermined. Explicit forensic analyses of aphasic testation, I propose, were undergirded by a deeper economic style of reasoning, which consisted not only in justifying the compatibility between aphasia and will-making but also in fashioning technical work-arounds that could function as administrative expedients for the execution of a will.

These work-arounds become especially prominent by the 1890s, in the form of clinical techniques that were intended to perform a double function: first, assessing the actual degree of an aphasic's testamentary capacities and, second, subsequently translating the patient's otherwise incommunicable desires into the terms of testamentary formalities. The American neurologist Charles Karsner Mills argued that a trained clinician might eventually be able to extrapolate from the location, size, and severity of specific brain lesions the nature and genre of aphasia from which any given patient might be suffering; and, in doing so, to conjecture, on the basis of the site and severity of the lesion alone, what testamentary capacities might possibly be retained and whether a further investigation into the patient's testamentary desires would be warranted and possible.[103] Mills advocated for the implementation of interpretative clinical procedures through which speechless patients' pathologies could, in effect, be circumvented. He invoked in particular a probative technique introduced by Moses Allen Starr.[104] The technique was an assessment protocol, comprised of a series of tests intended to uncover the nuances and variations of any given patient's aphasia. Perhaps patients could read but could no longer write; perhaps they could speak but not understand speech; perhaps they could write but only what was dictated to them or only what they copied; perhaps they could understand language but could only speak by repeating spoken words. The variations presented in aphasic pathologies were extensive but not impossible to document if the assessment procedures were extensive enough.

Some clinicians maintained that it was not even particularly necessary for aphasics to be able to display a comprehensive and unadulterated understanding of the details of the will itself. "The average layman," wrote the neurologist Byrom Bramwell, "even when all his faculties are alert and bright, has great difficulty in understanding and clearly appreciating the meaning of a complicated legal document such as a will." If the legal technicalities of testamentary procedures were already difficult enough for an otherwise healthy layperson, then why would an aphasic be expected to exhibit any testamentary proficiency? Technical impediments hardly signaled a problem of competency.[105] A better solution to simply submitting an aphasic to the practical challenge of testation was, instead, to draft a will on the aphasic's behalf but to do so in such a way that the will would contain obvious and palpable errors as to, for example, the nature of the properties and wealth being bequeathed or the names of beneficiaries. Based on the patient's responses, or lack thereof, to those errors, a clinician could swiftly assess the aphasic's testamentary capacities. "The process which has been indicated above," Bramwell warned, "is necessarily slow, and must occupy much time." It was, furthermore, a process that was far too difficult to be independently conducted by "an unskilled observer (a lawyer) who knows nothing of aphasia and the methods of eliciting information from aphasic patients to ascertain the wishes and intentions of the patients."[106]

These sorts of hermeneutic techniques were not intended to accentuate the pathological nature of aphasia but, rather, to extract the remnant and underlying normalcy and intellectual health that most aphasiologists believed lay intact for aphasic patients and, since the patients were not capable of articulating that health themselves, to transpose that normalcy into the realm of everyday civil and economic life.[107] Jackson had proposed such an orientation toward aphasia as much as three decades earlier when he claimed that at the very heart of any aphasic's linguistic "error" lay, still, a kernel of decipherable normality. In 1868, Jackson presented the somewhat incoherent utterances of various aphasic patients and the intended meanings of those utterances side by side, in an effort to demonstrate that a degree of coherent intentionality was still capable of being extracted from the otherwise degraded nature of aphasic expressions (fig. 4.3).

Aphasia, like any other neuropathology, represented the disinhibition of

What the Patient did say.	What the Patient intended to say.
What did you put it in the way for ? Shut the door.	What did you put it in the way for ? Move it.
Whos bottle is this with there money no water in it ?	Whose bottle is this with their beer in ?
If I laid in the next street to you I would let you see.	If I laid in the next bed.
Don't you wish there was no women ?	Don't you wish there was no men ?
What would you charge to write 3 bottons on my shoulders ?	What would you charge to sew 3 buttons on my trousers ?
He eat nearly a whole pig.	He ate nearly a whole fowl.
Can any one pull my knife out for me ?	Can any one pull my tooth out for me ?
I hope I shall be able to eat some breakfast now.	I hope I shall be able to have some tea.
You have not read my breakfast to me yet.	You have not read the prayers to me this morning.
Bring me a quarter of an hour of butter.	Bring me a quarter of a pound of butter.
What am I to say it is a clock ?	What day of the month must I put down ?
He wants to make my hand hot.	He wants to burn my hand.
I have just had the thing taken of my face.	I have just had the blister taken off my neck.
He won't pay me what I owe him. He paid for the stamp.	He won't let me pay him what I think is right. I have only paid for the stamp.
The sister as just drest my boot for me.	The sister has just dressed my blister for me.
What, have you broke yourself again ?	Have you torn your dress again ?
Smoking in the morning did not make my arm bad.	Smoking in the morning did not make my head bad.

FIGURE 4.3. A chart by John Hughlings Jackson illustrating the difference between aphasics' actual utterances and their inferred linguistic intentions. From "Observations on the Physiology of Language" (Queen Square Archives, A20, [QSA/839], 1868), 20.

otherwise normal automatic processes, and there was, as far as Jackson was concerned, no reason to conclude that volition and automatism were diametrically opposed. "I think the experiments which disease makes on the nervous system," Jackson wrote, "show that even the will is not a separate thing [from cerebral processes], and that the bulk of the most elaborate actions are automatic, the 'will' being then their leader."[108] The aphasic was never *not* capable of some degree of volition and intellection; to be fully incapable was tantamount to complete dementia, the absolute loss of mind. Thirty years later, during a short panel devoted to the topic of aphasia and testation at the sixty-sixth annual meeting of the British Medical Association, the Glasgow physician William Gairdner declared that "an aphasic person—I will push it to the full length of saying a man completely aphasic, but not otherwise insane or stupid—has, as regards his inner mind, probably the full capacity of making a will of some kind." Perhaps not an ideal will, he admitted, but a legally sufficient one nonetheless, since, ultimately, the testamentary capacity of the aphasic was *not a general question of law, it is not a thing which can be put into legal or into physiological categories at all.*"[109]

Senility and the Afterlife of Aphasic Testation

By the end of the First World War, as aphasia came to be established in the work of Henry Head, Kurt Goldstein, and other notable neurologists as an exemplary disease state through which the inner workings of the mind and brain could be deciphered in radically novel ways, the testamentary discussions that had littered much of aphasiology since the early 1870s all but disappeared. A new population of aphasic patients comprised of war-injured soldiers provided neurologists the opportunity to radically reimagine the holistically integrated nature of the nervous system and the biological grounds for symbolic experience.[110] The operative administrative question that organized these novel neuropathological inquiries was not whether aphasics could administer their wealth but how brain-wounded soldiers could be healed sufficiently enough either to be returned to the front or, subsequently, to be reintegrated into society. These new directions in aphasia research did not mean that the question of pathological testation had vanished, even as aphasia ceased to function as the conceptual vessel through which those debates were carried out. The question concerning the compatibility of pathology and heritable testation was picked up by a different, though very much cognate disease category: age-based cognitive decline or senility. Senility and aphasia shared a considerable degree of clinical, symptomatological, and demographic overlap; to the extent that testamentary questions were posed in the same ways and with respect to similar populations, the debate over senile testation simply picked up where aphasic testation had left off.

Redfield, for instance, never discussed the topic of aphasia in his *The Law of Wills*; given its publication in 1864, he would not yet likely have been familiar with the medical concept. Nevertheless, Redfield affirmed that no mental pathology was more connected to the question of testation "as that of the imbecility of old age, or senile dementia."[111] For Redfield, the question of age-based cognitive decline and will-making were unavoidably linked, and his discussion of the relationship between senility and testation unfolded in ways that would starkly resemble the discourse of aphasic testation in the decades to follow. He pointed out that while senility was undoubtedly defined most characteristically through manifest cognitive decline, it appeared that according to a great deal of English and American case law, "intellectual feebleness alone will not disqualify one for making a will."[112] From that standpoint, Redfield argued, it ought not be inferred "that mere weakness of understanding, in a healthy sane mind, in a sound body, is to be adduced, as any impediment to the valid execution of a will." Cases of age-based senility, he warned, could never be adjudicated through a generalizable framework. "Each case," he argued, in language that would become common for aphasiologists, "will have to be decided upon its own peculiar facts and circumstances. . . . Hence, the decisions to not wear the appearance of uniformity or consistency."[113]

The peculiarity of aphasia often led clinicians to attempt to make sense of it by way of analogy, through comparisons to proximate or adjacent disease states. As to whether aphasia was, for example, a genre of insanity was one such (often inconclusive) debate. While some psychiatrists claimed that madness "almost never coexists with aphasia," and that "insanity and aphasia are very rarely associated,"[114] others emphasized the "gross apparent analogies between manifestation of madness and aphasia."[115] But the pathological condition to which aphasia was most frequently compared was age-based cognitive decline—either in the form of senility or in forms of dementia loosely associated with age. Often the comparisons were simply passing associations, substantiated on the grounds of the demographic commonality of the two disorders. Legrand du Saulle observed that even among healthy elderly populations, "words are poorly recalled and dates are forgotten, even as memory of facts remains faithful and tenacious."[116] The American psychologist G. Stanley Hall noted that while certain intellectual capacities continue to develop after the age of sixty, it was usually at that point in a person's lifetime when "faint symptoms of aphasia and amnesia begin to show themselves."[117]

Often, however, the comparison between aphasia and senility was explicit and direct. The Parisian physician Alexis Legroux compared the intellectual impairments observed among aphasics to the "senile weakening of intelligence" (*l'affaiblissement sénile de l'intelligence*), or the slow and often imperceptible decline of cognitive capacities in old age. "The aphasic," he cautioned, "cannot be considered a madman [*un aliéné*] or an incapacitated person [*un incapable*]."[118] Others observed that, in addition to intellectual weakening, aphasics often experienced a marked decline in their moral and affective capacities. One physician remarked on the ease and frequency with which aphasics "become disturbed and cry . . . which brings these patients closer to childhood or senility."[119] Hughes was perhaps most explicit about the link between aphasia and senility. "The particular forms of insanity with which real aphasia is most likely to be associated," he declared, "are the insanity of old age—senile insanity as it is called—and the general paralysis of the insane."[120]

In 1910, the French neurologist Maurice Brissot explored the extensive ways in which aphasic symptoms appeared to supervene in cases of senile dementia, and vice versa. According to Brissot, aphasic symptoms often arose progressively over a lifetime, the effect of a regressive dissolution of the brain's functional hierarchy. "In aphasia," he wrote, "words disappear according to Ribot's 1881 general law of regression."[121] His invocation of Ribot (and, implicitly, Jackson) was intended to show that the neuropathology of language disturbances could be mapped onto the cumulative deterioration that characterized age-based cognitive decline. The scholarship linking language disorders and senile dementia, however, was woefully sparse, he lamented. It was possible that the very "process of senility" (*le processus de sénilité*) could give rise to aphasic states. "In summary," Brissot explained, "the patient with senile dementia is not always a dement [*un dément*]; he is sometimes only an aphasic."[122] And yet, Brissot agreed, to the same degree that aphasia could emerge as a complication of senility, certain states of progressive and serious aphasia could often be mistaken for dementia, even in patients who did not otherwise suffer from any cognitive deterioration.[123] The discernible boundaries between aphasia and senility were hardly clear-cut. The neurologist R. Percy Smith likewise noted the "preponderance of senile cases" among many of his aphasic patients.[124] What especially complicated the link between aphasia and senility, he explained, was the added complexity of identifying the pathological boundaries of senility itself.

A central question in the emergent medicine of age-based cognitive decline was how to differentiate normal senility from pathological senile dementia. In trying to make sense of the causes of age-based cognitive decline, George Beard proposed that what nevertheless linked all the causes together was their underlying evolutionary structure. Normal cognitive decline differed from pathological decline not in terms of the structure of decay but rather in the speed and intensity of loss of higher function. "In some of these cases," Beard explained, "the decline was purely physiological, in others pathological; in the majority it was a combination of both."[125] Like Beard, Charles Mercier held that the difference between normal dementia and its abnormal state was simply the degree and rapidity of intellectual loss. The sort of debilitation a person might normally experience in "extreme old age" would be considered pathological if its onset took place earlier in a person's lifetime.[126] Jean-Martin Charcot observed an analogous difficulty in distinguishing normal age-based cognitive loss from its pathological variants. Often, he explained, the physical changes occasioned by old age are such that at a certain point "the physiological and the pathological states seem to mingle by an imperceptible transition, to be no longer sharply distinguished."[127]

By the turn to the twentieth century, it was becoming increasingly clear to clinicians that the porous boundary separating normal, physiological senility—"the ordinary second childishness of old age"—from pathological dementia was not simply a medical conundrum but an institutional problem as well. In 1902, a New York psychiatrist noted that nearly a quarter of American patients committed to psychiatric hospitals from 1890 to 1900 were over the age of sixty; whether the asylum was the proper setting for treating normal age-based cognitive decline could only be addressed once the boundaries between the normal and pathological had been clarified. It had become a medical convention, he explained, to admit that senility was "an inevitable phase of man's existence on earth." What was necessary, however, was simply to identify a standard of senility that could be called normal.[128] "Senile [pathological] dementia . . . presents symptoms which differ from those of normal senility principally in degree, and as already stated it is impossible to define the dividing line."[129]

Senility did not merely denote a disease state but signified the unavoidable normalcy of the brain's eventual pathological turn. The supposed inevitability of cognitive decline meant that precluding a pathological testator from transferring wealth in advance—that is to say, attending to senile testation through a juridical style of reasoning—would be tantamount to declaring that all people eventually become incapable *in principle* of transferring wealth or making bequests. Senility, then, posed an even sharper version of the economic problem introduced by aphasic testation. To bar aphasics from writing wills would have interrupted the flow of inheritance of a subsection of the population of the aged and propertied; but to bar senile patients would in theory have interrupted the flow of inheritance for that entire population, given that cognitive decline was posited to be an unavoidable condition of aging. The economic style of reasoning that subtended the question concerning the economic and testamentary capacity of senile patients was the very same that had been deployed with respect to aphasics.

Neither aphasia nor senility expelled in advance a person from the domain of economic activity, and, in fact, both conditions represented an acceptable allotment of pathology within the economic sphere, a means by which pathological states could be construed as value-generating from the standpoint of inherited wealth. Just as clinicians believed that patients with extreme aphasia could still write wills, so too did they believe that even extreme senility was compatible with a sufficient degree of economic and testamentary competency. The mental conditions necessary for carrying out "the immediate personal needs and natural desires and rights of the individual," Hughes maintained, "may coexist with very considerable pure uncomplicated old age atrophy of brain." That specific form of civil and economic capacity might have been limited, but it was the conviction of Hughes and many other clinicians that the capacity remained undeniably intact.[130]

The aphasic was a paradoxical figure for turn-of-the-century transatlantic psychiatrists and neurological clinicians, at once profoundly ill—the victim of one of the most debilitating neurological disorders—and simultaneously tethered to a medico-economic conviction that neuropathology and testamentary aptitude were reasonably compatible states. Aphasia's purported inconsistency with the law and with simplistic biological or nosological descriptions meant that it evaded efforts to be pigeonholed medically or juridically. Clinicians did not always know precisely what sort of disease aphasia was or how it could be adjudicated under the law; the aphasic's status as a medical or legal subject was complicated and unsettled. What was much clearer, however, was the aphasic's status as an *economic* subject, to the simple extent that any given aphasic was perceived to be very likely able to write a will and transfer wealth. Economic reasoning, in other words, made plain what medical and juridical analyses could not always, for one thing that seemed incontestably decided was that aphasics could partake in some manner of economic activity. We might even say that it was through an economic style of reasoning that aphasia was perhaps most successfully stabilized as a medical concept.

By the beginning of the twentieth century, aphasia had became one of the core conceptual frameworks through which mental and neuropathology could assume a degree of economic value, by concretizing the medical belief that some morbid states could nevertheless perpetuate wealth. We see through the history of aphasic testation the emergence of another form of pathological value, one that deviates from, but nevertheless complements, the pathological value of entrepreneurial madness. In the case of aphasia, the attribution of economic value was prompted at least in part by a tacit economic incitement to ensure that the flow of inheritance be neither interrupted nor curtailed. In that sense, while the history of aphasic testation can certainly be read as a history of the medical defense of patients' civil rights, it must also (perhaps instead) be interpreted as a history of a form of unspoken complicity with one of the chief economic consequences of late nineteenth-century inheritance practices: the immense and starkly unequal distribution of wealth that it precipitated, and, in effect, the facilitation of economic inequality.

5

The Pathology of Work

"La 'Sinistrose'"

At the end of December 1907, several of the most prominent newspapers in Paris reported on the commencement of a civil trial that had begun in the fourth chamber of the Tribunal de la Seine, the Parisian civil court of first instance.[1] The articles uniformly bore the same title: "La 'Sinistrose.'"[2] The legal case concerned a suit brought by a manual worker named Sampère, who, eighteen months earlier, had sustained a workplace fall from a ladder. While he was immediately treated at the Hôtel-Dieu hospital in central Paris, Sampère nevertheless claimed upon his hospital discharge that he had not been sufficiently treated. He subsequently insisted that his injuries had not fully healed, that he was still incapable of resuming work, and that he was due further remunerations from his employer for his incapacity, particularly under the auspices of France's workplace accident insurance law of 1898, which guaranteed compensation to workers who had sustained occupational injuries.[3]

In order to adjudicate the case, the court employed the expert testimony of the neurologist and psychiatrist Édouard Brissaud, a clinician at the Hôtel-Dieu who was already well known in Paris as one of Jean-Martin Charcot's most prized pupils and a member of the inner circle of the *maître*'s presumptive intellectual and clinical heirs.[4] It was actually Brissaud's medical testimony that sparked the interest of the Parisian press. Far from confirming Sampère's own assertions, Brissaud concluded that while the worker indeed suffered from an affliction, it was not due to the injuries that he claimed. Rather, Sampère had unknowingly become victim to an altogether novel disorder—a "new illness," which Brissaud named "sinistrosis" (*sinistrose*). Reproducing Brissaud's court testimony verbatim, the newspaper articles represented the first public announcement of this so-called new disorder.

Sinistrosis, Brissaud maintained in the court proceedings, was a morbid mental condition that pathologically inhibited a worker's voluntary capacity to return to work after having otherwise fully recovered from a workplace injury. The disorder was, he contended, entirely genuine; Sampère was not a malingerer. And it was a unique illness, unrelated to the so-called traumatic disorders such as hysteria, neurasthenia, and psychasthenia that had pervaded French psychiatric medicine for decades. Sampère was not experiencing the residual shocks of his workplace accident. Instead, he had developed, quite independently of his injury, a distinctly obsessive conviction, in the form of a "fixed idea" (*idée fixe*), that "any accident taking place during the course of work constitutes a damage requiring compensation." His pathological fixation on restitution and his incurable conviction that he remained unfit to return to work despite medical judgments that he had entirely recovered from his injuries had nothing to do with his fall from the ladder. The accident did not cause the sinistrosis, Brissaud declared. Instead, "the accident is nothing but a pretext" for a sinistrotic's obsession with reparation. Sampère was not the victim of an accident but, rather, "victim to himself."

Hardly a week after the trial commenced the court ruled that, based on Brissaud's expert testimony, Sampère did not in fact have the right to any sort of indemnity, a judgment that was later reaffirmed in the court of appeals. The January verdict effectively validated a disease category that for all intents and purposes had no diagnostic reality in French medicine at the time of its introduction. The court at once affirmed the socio-legal reality of sinistrosis while simultaneously decreeing that, as a genre of disorder, "sinistrosis could not rely upon the 1898 workplace accident law in order to lay claim to compensation."[5] A "sinistrotic" worker, in other words, could not be protected or indemnified. While sinistrosis appeared to be entirely linked to workers' ideas related to occupational accidents and compensation claims—and thus a disorder that seemed predominantly to afflict industrial laborers alone—it was nevertheless a disease that was judged to be occasioned not by any accident as such but, rather, by a morbidity inherent in the workers themselves.

It was not, to be sure, the popular press alone that caught wind of this new disease. Eugène Quillent, a Socialist labor and health advocate, penned a harsh rebuke of the disease classification and the court decision in the pages of the Communist daily *L'Humanité* shortly after the Tribunal de la Seine ruled against Sampère. Quillent accused Brissaud of trivializing workers' suffering under the cover of medical authority. "The smiles of the insurance lawyers and judges," Quillent wrote, upon hearing of "literary masterpieces" like sinistrosis indicate that they know how to "read between the lines."[6] *L'Humanité* would maintain its position for decades that the category of sinistrosis was nothing more than an "evasion" and "anti-worker [*antiouvrière*] inclination on the part of experts and doctors of insurance companies to prevent injured workers from getting what is due to them."[7]

A little more than a month after the initial verdict against Sampère, Brissaud published his first systematic account of the "authentic illness" of sinistrosis. Of particular concern for the neurologist was the question of etiology, for there was one specific pathogenic source that Brissaud argued was most responsible for triggering the morbid "predisposition" among workers to fixate on feeling injured and demanding remuneration. That source was none other than the 1898 workplace accident insurance law itself, which functioned as a "substratum" fueling workers' false beliefs and fixed ideas. As Brissaud explained, whatever pathological predilections workers might have, they were "both inefficacious and innocuous before the law of 1898," since before the onset of the law, injuries were neither indemnified nor financially backed.[8] Sinistrosis, Brissaud explained, that strange pathological disinclination to return to work as a consequence of a morbid belief that injury still persists, "proceeds exclusively from an erroneous interpretation of the [1898] law and consists in a sort of rational delirium [*délire raisonnant*], based on a false idea of compensation."[9] What Brissaud essentially implied was that virtually every industrial worker in France, whose labor since April 1898 had been legislatively insured against occupational injuries, was now by definition susceptible to this "new illness." Brissaud had, in effect, transformed every worker into a potential sinistrotic.

Given Brissaud's renown at the time (he was elected to the French Academy of Medicine a year after his "discovery"), sinistrosis would likely have developed into a full-fledged medico-industrial doctrine had Brissaud not died unexpectedly in 1909 from an inoperable brain tumor.[10] Brissaud's death, however, did not prevent other clinicians and medical practitioners from continuing to expound upon the nature and symptomatology of this alleged illness. Discussions of the disorder quickly circulated in the pages of prominent journals, immediately becoming a new metric by which physicians assessed compensation claims, and it even found a clinical foothold in colonial Algeria.[11] Hardly two months after Brissaud's first published account of sinistrosis, one physician called upon the *Dictionnaire de l'Académie française* to formally incorporate the term within the French lexicon.[12]

With Brissaud's death, however, clinicians were free to draw upon the initial formulation of the disease, while elaborating on both the symptomatology and the social ramifications of this "essentially modern illness."[13] Sinistrosis was quickly reframed through the language of "psychosis,"[14] a "special psychosis of workplace accidents."[15] And its etiological link to the 1898 workplace accident law was even more strongly reinforced. "The workplace accident law created a new illness, a psychosis," wrote one physician, "that all experts have only confusedly observed, but which was singled out by Professor Brissaud."[16] Sinistrosis was also of particular interest to the French insurance industry, which perceived it as "a disease of the century," deserving "a very special place" among the "many evils" of society. "Relations between employees and bosses are of the gravest concern at the present moment," wrote a commentator in the pages of the insurance organ *L'Argus*, "yet 'sinistrosis' is an illness of the employee against the employer."[17]

The popularization of sinistrosis also became an opportunity for politically conservative outlets to inveigh against liberal economic policies that protected workers against occupational injuries.[18] Readers of the conservative *L'Intransigeant* were told that sinistrotics were little more than malingerers.[19] In the pages of the nationalist right-wing *L'Action Française*, a commentator maintained that sinistrosis was proof that it took longer for workers to recover from workplace injuries when they were insured. Comparing France's workplace accident legislation to comparable laws instituted in England, Germany, and Austria, the commentator concluded, "It is certain that 'pension hysteria' exists in those places as it does for us here, that insured injuries take longer to heal than uninsured injuries." The difference was that, in France, the discussions had become so acute and heightened that it "finally required the creation of a new term," one whose cause was "ascribable to a contemporary law!"[20]

Just four years after Brissaud first introduced sinistrosis, numerous cases of the disorder's utilization in workplace litigation were being recorded throughout France.[21] By 1917, it was claimed in the popular press that sinistrosis, "well known to insurance companies that cover workplace accidents," was on the rise, forcing on insurance companies an alleged 30 percent increase in annual costs.[22] Sinistrosis was a leading "pathology of work" (*pathologie du travail*), a "neuro-psycho-morbid syndrome," which arose when a worker's mind had been pathologically "permeated by the value of his rights," instead of healthily imbued with a sense of the "importance of his duties."[23] By the 1920s and 1930s, however, with its popularity somewhat on the wane, sinistrosis lost much of its medical urgency, eventually stabilizing as a simple synonym for the morbid impulse to commit medical fraud, even as interwar doctors continued to maintain that the "social insurance law gave birth to a category of delirious claims-seekers [*revendicateur*]."[24]

Sinistrosis enjoyed a brief, albeit dramatic shelf life. Much of this undoubtedly was due to Brissaud's untimely death. The disorder must be counted as among the novel psychoneuroses that Charcot's disciples attempted to promulgate as a replacement for hysteria after their mentor's death—diagnostic categories whose institutional repute and dominance were, like hysteria, intrinsically tied to the persona and presumed longevity of their authors.[25] On the other hand, while accident insurance represented one of the most prominent and controversial debates concerning the legislative adoption of social insurance policies throughout Europe at the end of the nineteenth century, as the controversies surrounding workers' insurance cooled during the interwar period, so too did the urgency of identifying cases of sinistrosis.[26]

But what precisely did this strangely transient "pathology of work" represent as a historical and conceptual phenomenon? Of all the topics thus far explored in this book, the subject matter of this chapter may likely be the most familiar to readers, embedded as it is at the well-surveyed intersection of psychological medicine and the political economy of labor. This chapter in part complements scholarship that has explored how firmly entrenched psychiatry was by the end of the nineteenth century within international debates concerning welfare and insurance legislation, occupational pathologies, and worker entitlements, at a moment when, as Rhodri Hayward puts it, mental health was viewed as a "a crucial variable in economic power."[27] However, despite the abundance of research specifically on the French welfare state, the specific place of psychiatry there has not been emphasized quite as strongly as it has in the German and English contexts, and consequently the very concept of sinistrosis has remained virtually unexamined.[28]

And yet at the same time, sinistrosis was more than just a representative case study of the relationship between medicine, insurance, and labor—more than just the French instance of transnational debates concerning so-called pension neuroses, as they were cynically dubbed in Germany and elsewhere. The discourse surrounding sinistrosis not only represented another instance of the deployment of an economic style of psychiatric reasoning, but epitomized far more acutely than analogue German or English debates the coalescence of pathology with economic value. In some respects, sinistrosis revealed that the entire international psychiatric endeavor to conceptualize occupational psychopathologies in relation to worker compensation policies was itself fully part and parcel not only of an internationally diffuse economic style of thought but also of the general economization of madness—or what I have been calling the emergence of pathological value. In contrast to entrepreneurial madness and aphasic testation, economic reasoning with respect to disorders such as sinistrosis was quite overt given that in many cases the ostensible etiologies of these disorders were linked to economic policies themselves. In the case of sinistrosis, injury indemnity was unequivocally believed to function as the source of a morbid condition whose main behavioral symptom was to make false financial claims and thereby burden an entire economic sector.

Sinistrosis, therefore, instantiated one of the most explicit convergences of psychiatric and economic thought. It was a disease that denoted not only the economization of pathology but also the pathologization of economic policy. To that end, we must recognize that the very concept of sinistrosis was as much a medical as it was an economic expression, the psychiatric conversion of a potent anti-labor economic norm. In the previous two chapters, I showed how pathology could be interwoven with economic value in such a way that some pathological states could be subsequently regarded as value-generating. This chapter illustrates the inverse, that economic value could be attached to pathological conditions in such a way that certain pathologies were only ever appraised as value-forfeiting. Sinistrosis, after all, was persistently viewed as costly, dangerous, and detrimental. Its value could be measured according to the same economic metrics used to appraise other disorders, with the one major difference that while some forms of madness were judged to be sources of remuneration, sinistrosis was only ever institutionally deemed to be the source of financial deficit. While for Brissaud and other clinicians, sinistrosis was indisputably real, the disorder was a means by which occupational traumas and worker injuries could be both legitimated and simultaneously devalued. The pathological affordances granted to mad enterprisers and propertied aphasics simply did not translate to industrial laborers. Pathological value is a notion that helps explain how and why abnormality could at times be regarded as a privilege capable of being (economically) defended and vindicated; but it also helps explain how and why some purportedly non-normal conditions, like sinistrosis, were not only discredited but, on the basis of a certain economic rationality, exaggeratedly disavowed.

To explain how sinistrosis emerged as a disorder that was equal parts psychiatric and economic, this chapter provides a genealogical sketch of the origins of the concept, beginning with its derivation from a diagnostic and theoretical puzzle that long confounded clinical psychiatry, namely, the problem of simulation. Michel Foucault noted the extent to which simulation, or the question of whether a patient was authentically ill or simply feigning mental illness, represented a core "historical problem of psychiatry in the nineteenth century," observing, furthermore, how fundamentally interwoven that question was with clinical suspicions concerning economic and insurance fraud.[29] According to the historian Ruth Leys, the concept of simulation was eventually regarded as rooted in the psychodynamic effects of traumatic pathologies and the predisposition to suggestibility that traumas introduced.[30]

But as we will see, medical debates concerning simulation went well beyond the simple dichotomy of whether a patient (and potential claimant for compensation) was lying or telling the truth about their disease. For many medical experts, the question of simulation was palpably complicated after the 1870s by the diagnostic emergence of traumatic disorders—such as hysteria—which proposed that simulative behaviors could very well be legitimate symptoms of postaccident pathological states. The question, then, concerning suspected malingerers was not whether they were simulating but whether the simulations were simply voluntary expressions of greed or, instead, the involuntary expression of a post-traumatic pathology, a so-called simulation disorder. The preoccupation with simulation disorders, of which sinistrosis was a later subvariant, took on a new urgency after the European-wide proliferation of nationalized accident insurance legislation between 1884 and 1906. Ascertaining the truth of simulation eventually became a form of psychiatric reasoning shot through with economic imperatives and overdeterminations, a form of reasoning that was enacted almost exclusively upon the population of industrial laborers.

This chapter, then, examines how sinistrosis conceptually emerged at a moment when clinicians began to postulate that social welfare economic policies and worker compensation laws could involuntarily induce workers to simulate workplace injuries and thus inadvertently seek to defraud the insurance industry and the state. What sinistrosis effectively revealed was that injuries and illnesses related to workplace accidents and, thus, embedded within a political-economic matrix—conditions otherwise viewed as narrowly medical phenomena—were both regarded and experienced as economic forms through and through. Disorders such as sinistrosis allowed medical practitioners to claim that worker compensation laws were virulent agents in themselves, as traumatically pathogenic to the health of the worker as actual workplace accidents were, and that they could engender morbid mental states, idleness, and false symptoms among even the healthiest laborers. The concept of sinistrosis, however, went beyond the mere moral aspersion of worker entitlements. It was a concept through which the perils associated with industrial labor could be reframed and reimagined as a risk to very essence of industrial capitalism itself.

The Menace of the Impostor

A comparative upsurge in the international medical discourse on simulation was framed, after the 1880s, around political and economic anxieties over civil litigation and actuarial compensation. That being said, and as early twentieth-century forensic psychiatrics would attest, the medical preoccupa-

tion with the simulation of illnesses was centuries old.[31] Early modern physicians and medical jurists, such as Johannes Baptista Silvaticus and Paolo
Zacchia, described the fraudulent simulation of disease states, enumerated
which morbid conditions were most often imitated, and detailed techniques
that physicians could adopt to detect impostors. It was commonly held that
the disorder most frequently feigned, and the one most difficult to detect,
was insanity.[32] What was most notable about this early modern discourse on
the relationship between simulation and mental disease, however, was its historical resilience. Much of the tone, if not the very language, of these early
modern analyses continued to be adopted in the writings of medical experts
and medical jurists in the early decades of the nineteenth century.[33]

During the first half of the nineteenth century, the problem of disease
simulation was marked by a different set of apprehensions than would arise
at the end of the century. Here the concern over simulation was prompted
by the increasingly formalized relationship between medicine and the law,
which in the case of psychiatry was integral to the process of shoring up institutional authority for the profession. Feigning an illness was believed in most
instances to be an act of deliberate fraud, but an act not always detectable
by an unskilled observer. Consequently, the underlying rationale for many
earlier nineteenth-century treatises on disease simulation was to make a case
for medical expertise, to argue that it took the keen eye of a trained physician, rather than a jurist alone, to discover an impostor. These earlier writings, therefore depended upon the authority and conceptual footing of long-
standing forensic principles, such as those laid down by the likes of Zacchia,
while simultaneously emphasizing that the rapid proliferation of medical
knowledge particularly in the field of mental pathology meant that simulation was at once a veritable problem for the law and one only the physician
could properly address.

In addition to recapitulating classical forensic thought, early nineteenth-
century writings on simulation introduced novel features into the study and
detection of medical impostors. The first and most prominent was the association of simulation with socioeconomic class. The greatest driving incentive
for the simulator was financial interest, which, as a young Parisian medical
student in the early nineteenth century wrote, "suggests to the poor [pauvres]
a thousand different strategies to arouse our pity."[34] The domain of likely
simulators was quickly expanded to include laborers, soldiers and conscripts,
criminals and prisoners, and women and the youth. Economic motivations
remained, for the most part, an explicit though often diffuse justification for
simulative behavior; the relationship between simulation and financial gain
was not yet being articulated, as it would be by the end of the century, as

integrally and systematically related. The British medical jurist William Guy placed his discussion of feigned diseases in the same chapter as his discussion of life insurance; it was an organizational decision that was uncannily prescient of what by the end of the century would be a thoroughly entangled set of topics, but that in 1844 was perceived as only contingently related at best. "These two subjects are placed in the same chapter," Guy wrote, "for the sake of convenience, and not on account of any great similarity which they bear each other."[35]

Perhaps the most notable characteristic of these earlier writings on simulation was the degree to which they tended to be almost entirely organized around what might be called a logic of suspicion, or what a French medical extern retroactively dubbed the "medicine of suspicion" (*médecine soupçonneuse*).[36] Such a logic, at least in the early half of the nineteenth century, took the form of three noticeable features: the prevalent expression of incredulity on the part of the physician, the clinical supposition that there was a fairly clear-cut difference between an impostor and an authentically ill person, and the belief that detection was, in most cases, achievable thanks to the physician's superior knowledge of diseases. Clinicians advocated for the use of inquests and interrogations to root out simulators and proposed a variety of clinical and observational techniques to do so, a process an early American medical jurist labeled the "administration of medical police."[37] "The pretender," wrote a British physician in 1814, "is therefore always endeavoring to exhibit the extravagancies and act in character the several peculiarities of the malady most commonly of the maniacal form."[38] Simulators were destined, it was readily believed throughout much of the first half of the nineteenth century, to "perform" diseases incorrectly, and through sufficient observation, interrogation, and even manipulation, the fraud would be revealed. In the context of medical inquests, "the insane person will attempt to act sane, whereas the pretender continues to act foolishly."[39]

In the United States, the feigning of mental disorders and particularly of epilepsy was a central problem and common suspicion in antebellum Southern medical discourse, and it was particularly central in debates between abolitionists and pro-slavery advocates. The authenticity of the disease could, for abolitionists, expose the very pathogeny of bondage, whereas advocates of slavery maintained that "shamming" was simply a ruse either to avoid labor or to undermine a sale in a slave trade.[40] Even Northern doctors could not help but reveal their heightened suspicions of feigning among Black patients. Thomas Blatchford, a resident physician at the New York state prison, was quite content with manipulative, coercive, and even violent modes of fraud detection. He reported the case of "a black girl, about twenty-two years of age" who had been incarcerated for "some misdemeanor," and had begun upon her confinement to exhibit morbid symptoms, including rigidity and clenching of the limbs, the rolling back of the eyes, and mild convulsions.[41] Suspecting her of feigning epilepsy, Blatchford ordered his attendees to raise the girl several feet above the floor and to simply drop her, in order to see whether she would allow herself to fall or catch herself, which she indeed did. "A heavy shock of electricity is a test of feigned epilepsy," he later freely and horrifically admitted, "which I have employed with much success."[42]

On the whole, however, faith in the physician's medical knowledge and, conversely, in the impostor's lack of such knowledge remained an integral supposition for early nineteenth-century beliefs in simulation detection. Impostors had a tendency, wrote Isaac Ray, to "overdo the character they assume, and present nothing but a clumsy caricature."[43] Others pointed out that certain physiological indicators of disease, from body temperature and ocular activity to the ideational patterns and fixations in cases of insanity, were simply inimitable.[44] The English psychiatrist John Charles Bucknill, who would characterize the detection of feigned insanity as "one of the most important points in the diagnosis of mental disease," explained that those feigning insanity erroneously believed that insanity corresponded to "utterly outrageous and absurd" conduct, or "the total subversion of intelligence"— extremes that did little but expose the feint.[45]

By roughly 1860, however, subtle transformations in the discourse on simulation began to appear in medical writings across Europe and the United States. Bénédict Augustin Morel described the case of "Derozier," a merchant who attempted to escape punishment for a series of church robberies by feigning madness, but who, instead, overperformed what he believed to be the nonsensical reasoning characteristic of insanity. In contradistinction to the quite explicit medical anxieties over the seeming inevitability of deceptive efforts that marked the earlier decades of simulation discourse, Morel qualified his entire case study of Derozier with an important stipulation: "Cases of the simulation of madness are quite rare."[46]

In an early analysis of malingering in the United States military, Silas Weir Mitchell and two coauthors concluded that exaggeration, rather than outright simulation, comprised their bulk of cases of malingering. The specific concern over the simulation of insanity was, they argued, immaterial. Given the encumbering and unpleasant restraints placed upon a suspected madman, anyone who would venture to feign insanity would by all accounts likely already be a "monomaniac." In any case, the primary problem as it pertained to insanity in the military was not that it was being simulated but, rather, that authentic cases were actually quite numerous. "The number of cases of

insanity in our army is astonishing," the authors wrote. "The assistant surgeon at the insane asylum informed us that the average admissions there from
the army alone were rather over one every day."[47] While acknowledging, like
Morel, the rarity of simulation proper, against the much likelier presence of
exaggeration, Mitchell and his coauthors reframed the true simulator as a
sort of madman while simultaneously conceding the astonishing prevalence
of insanity in military life. As to why the number of insane would be so high,
their discussion of epilepsy provided a useful clue. Many cases of potentially
doubtful epilepsy were, the authors urged, likely real; that "real epilepsy
should be so frequent in the army," they wrote, "need not excite astonishment
when we consider the numberless cases of wounds of the head, sunstroke,
falls, etc., and the immense exposure incident to such a life. When compared
to the number of insane, it is not remarkable."[48]

We begin to see how a different sort of diagnostic picture of simulation
was beginning to take shape after the 1860s. While clinicians never abandoned their logic of suspicion, several noticeable revisions eventually appeared in their diagnostic approach to the problem of feigning. First and
foremost was the fact that the formerly clear-cut distinction between authentic illness and mendacity effectively broke down as simulation came to embody, for a variety of different reasons, a genre of pathology in its own right.
Tinged as it was, therefore, with at least some degree of morbidity, clinicians
were less inclined to observe *pure* and unadulterated cases of medical fraud,
even as involuntary manifestations of simulation rose to purported epidemic
proportions after 1900, particularly after the passage of workplace accident
insurance laws throughout Europe. Greed and vice were no longer the sole
motives attributed to simulative behavior; if a pathological justification could
not be identified, then external causes were to blame. The most recognizable
exogenous source, what Mitchell and his coauthors already alluded to in 1864,
was the notion of the traumatic calamity and the accident.

The Economy of the Accident

Nineteenth-century European industrialization was marked, in the minds of
medical experts and administrators, by the precipitous rise of eruptive and
increasingly recurrent catastrophes linked either to mechanical failures or to
human error (or even chance occurrences) across a broad range of industries,
from railroads and mining to factories and mills.[49] By midcentury, England
was reporting more industrial accidents than any other European country;
it was only by the end of the century that the risk of accident had become a
normalized feature of European and American political economy, a veritable

transatlantic experience.[50] In 1898, Émile Cheysson, statistician, engineer, and inspector general for roads and bridges in France, concluded, "The modern accident does not resemble that of the past; while that one was clear, simple, and isolated, the modern accident is obscure, mysterious, and takes on the magnitude of a collective catastrophe. Now, men killed together in a railway collision or in a mine collapse makes a very different impression on public opinion than a hundred bystanders individually run over in the streets of Paris."[51]

The railway accident was the phenomenon that most contributed to the emergent discourse of industrial calamities, due in part to the rapid expansion of railways throughout the nineteenth century and their subsequent sedimentation within the European imaginary.[52] By 1840, within only the first two decades of England's massive national railway development, civil litigations against railway companies for collisions, derailments, sudden stops, and other mishaps were on the rise.[53] The entire forensic inquiry into accidents concerned how best to characterize injuries in relation to the catastrophes that occasioned them. By the mid-1860s the alleged injurious effects of railway travel remained an object of both theorization and dispute among physicians both antagonistic and sympathetic to railway interests. The surgeon John Eric Erichsen used the term "railway spine" to describe the injurious impact of railway accidents and even railway travel on the neurological constitutions of passengers.[54] A year later, James Ogden Fletcher, the medical officer for the Manchester, Sheffield, and Lincolnshire and Great Northern Railway Companies, responded with the assertion that railway accidents were not nearly as pervasive or dangerous as was commonly perceived. Writing just three years after an 1864 legislative decision that made railway companies financially liable for the health of their passengers, Fletcher proposed that there was actually little evidence to suggest that there was anything especially pathogenic about railway travel. Furthermore, British railways were not nearly as accident-prone as American railways, and per capita rates of accidents were lower in England than they were in France (even though France had fewer total accidents and, like Prussia and Belgium, fewer riders and trains).[55]

Through comparative statistical analyses of mortality rates from a number of common causes in London from 1851 to 1860, Fletcher claimed that railway accidents were far less to blame in the deaths of ordinary Londoners than was cancer, gout, or even drowning.[56] This was the underlying problem that, according to Fletcher, disqualified Erichsen's conclusions about either the prevalence or the likelihood of railway spine. "The whole of Mr. Erichsen's reported experience consists of six cases," Fletcher wrote. "How much better and more satisfactory would it have been if Mr. Erichsen had given us an average of the cases that have been followed by such results?"[57] Fletcher did not deny that railway travel came with an unescapable degree of risk. "There will necessarily be a certain loss of life, connected with railway traveling," he wrote, and "that must be considered unavoidable."[58] The risks were simply low and acceptable.

It was not, however, the likelihood or inevitability of a railway accident in itself that concerned Erichsen. For while railway collisions were comparatively infrequent, it was the sort of injury they occasioned that was crucial to recognize. The railway accident was the pinnacle and exemplar of a novel pathology type: traumatic injuries or shocks, or harm specifically tied to the violent contingencies of civil life and often experienced in the form of falls, blows, and even horse-and-carriage accidents. While traumatic shocks were regarded as increasingly common hazards of modern metropolitan living, their severity, indeed their frequent severity, was in no context better typified than among "persons who have been subjected to the violent shock of a railway collision."[59]

Admittedly, the category of traumatic shock was never exclusively tied to railway or industrial catastrophes alone. Natural disasters were also believed to elicit similar sorts of pathological effects. Earthquakes, for example, were often invoked for their particularly traumatic intensity.[60] The French neurologist Georges Guinon listed earthquakes and lightning strikes as among the various *agents provocateurs* of hysteria.[61] Emotional upheavals could also induce similar states of morbid "mental shock."[62] What linked trauma to industry had less to do with the specific sort of shock that the industrial accident could provoke; in that sense, a railway collision was not fundamentally unlike the gravest natural or emotional calamity.

The decisive difference, however, was that industrial accidents were met with a very different kind of interpretative response, a different sort of judgment in the eyes of the public. In 1901, the French commerce minister (and eventual French prime minister and president) Alexandre Millerand claimed in a ministerial report that industrial accidents "are appraised not only according to their number, but also, and above all, according to their gravity." It was not the likelihood of the industrial accident that stirred up the social imaginary but its lethality.[63] Even Fletcher had noticed the extent to which railways accidents were typically set apart from other sorts of disasters as examples of "gross carelessness, and culpable indifference to the destruction of life."[64] What was so singular, then, about the industrial accident was the extent to which it was tied to notions of detrimental negligence and, thus, to indirect liability, to an element otherwise "not found in ordinary accidents, namely, *the question of compensation*."[65]

When the English railway surgeon Herbert Page declared, "Every case of

railway injury is more or less the subject of medico-legal inquiry," he was re-
marking on the extent to which railway accidents after midcentury were perpetually coupled with a subsequent, and often successful, civil suit—litigation
that more and more relied legally on the category of organic shock.[66] Railway
workers were typically disproportionately more frequently injured than travelers. Among British workers, the number of fatalities between 1860 and 1873
increased from 121 to 773.[67] Injury rates for American railway workers were
as high as 1 in 12, while mortality rates hovered around 1 in 117.[68] A British
railway engineer affirmed that though some of the fault of railway accidents
could be attributed to human error, "there may have been mitigating circumstances, such as the omission of the railway company to provide a safeguard."
In such cases, "blame was cast on signalmen that should have lain at the door
of other officers and directors of the company."[69] To the extent that whole
companies could be held culpable, postaccident litigation was often quite lucrative, and by the mid-1860s, railway companies resigned themselves to the
belief that they would likely be on the losing end of such litigation.[70]

The industrial accident became, therefore, a sort of elementary unit in a
comprehensive economic circuit, so to speak—the antecedent stimulus to a
compensation-seeking response. The accident, however, was not simply *incidentally* monetary. Rather, the very concept of the modern, nineteenth-century
accident was economic in principle, a feature of a new liberal market ideology
whereby the contract form became the overwhelming paradigmatic lens, the
grid of intelligibility, through which accidents were increasingly perceived and
interpreted.[71] Not every disaster, therefore, was an accident, recognizable as a
breach of either implicit or explicit contractual expectations, for which one
might assign accountability without ascribing fault. Nevertheless, an "accident" could extend as far as the contractual paradigm would tacitly allow, and,
therefore, across an array of commercial, economic, and social phenomena.

While injuries occasioned in the workplace were undoubtedly regarded as
breaches of implied occupational expectations, disease in the broadest sense
could be viewed as an interference in the general ability to accumulate wealth.
Any given illness could be, to some minimal extent, observed according to
the economic logic of the accident. It was not, therefore, that industrial accidents were assigned an economic value. Rather, to perceive an industrial
accident was increasingly a form of economic perception; and, likewise, a
traumatic injury itself soon became a form of economic experience. Before
becoming inextricably bundled to the actuarial logic of nationalized accident
assurance, as they would be by the end of the nineteenth century, accidents
and injuries were, among European clinicians, already being suffused with
economic meaning.[72]

In contrast, then, to the earlier logic of suspicion—an undoubtedly juridical style of medical thought—psychiatric approaches to industrial accidents and to the authenticity of the injured claimant were precipitously transformed into tacit economic appraisals. As psychiatrists further broached the injurious effects of accidents, they not only adopted an economic rationality for making sense of traumatic pathologies, but did so while simultaneously affirming the possibility that simulation and imitative behaviors could be genuine symptoms of post-traumatic states. As we will see, against the backdrop of the economization of accidents and traumatic injuries, the very meaning of simulation underwent a transformation of its own, shifting from a form of conduct assessable in primarily juridical terms to a form of conduct saturated with economic possibilities and therefore monetarily appraisable as something to be either clinically impugned or medically defended.

Neuro-Mimicries

As economic meaning and value were increasingly associated with occupational accidents, greed was often cited as a primary motive for litigating plaintiffs. Fletcher relied on the German medical jurist Johann Ludwig Casper to emphasize the persistent likelihood of fraudulent simulation in accident compensation litigation.[73] And yet, at precisely the moment when litigating plaintiffs were being increasingly suspected of simulating illness for pecuniary gains, the category of simulation itself underwent a transformation as clinicians and medical theorists began folding simulative behavior into the diagnostic features of an array of mental and organic diseases.

By the early 1860s, prominent psychiatrists were describing cases of "morbid imitation," forms of mental disorder characterized by a pathological "disposition to exaggerate" and "tell absurd and motiveless lies."[74] Pathological forms of imitation offered an epidemiological explanation for why certain mental disorders, such as hysteria, chorea, and even epilepsy, often spread like contagious diseases.[75] At the same time, the idea of a natural faculty for mimicry, a concept promulgated for centuries by philosophers, was being revisited from the standpoint of neurophysiology and physiopathology. Jules Luys defined imitation as among the core unconscious and automatic functions of human reflex physiology.[76] The neurophysiology of imitation explained the underlying mechanisms of socialization and education, while simultaneously explaining how and why certain pathological phenomena, such as crowd psychologies and mob mentalities, could so easily approximate the "*contagium* of infectious diseases."[77] The difference between normal physiological imitation and morbid simulative behavior was simply a matter of degree. Soon prominent philosophers and social theorists—Herbert Spencer and Gabriel Tarde among others—were writing about the zoological, physiological, and social foundations of imitative behavior.[78]

The concept of "nervous mimicry," introduced by the prominent English surgeon James Paget, represented an influential contribution to the upsurge of medico-philosophical interest in both the natural and morbid manifestations of mimetic behavior.[79] Paget had been preoccupied with what he observed to be the symmetrical coincidence of disease states in the human body, that organic afflictions appeared to affect both sides of the body simultaneously.[80] Organic pathologies, Paget reckoned, could manifest in the body in the form of affinities at a distance. His concept of nervous mimicry, or "neuromimesis," derived from his research on these symmetrical affinities. In cases of nervous mimicry, Paget wrote, a nervous disorder could imitate or mimic local organic diseases, often with such a degree of resemblance that it was virtually impossible to discriminate between what was real and what was simply imitated.[81] Nervous mimicries might include simulated pains, paralysis, or other sorts of apparent organic disorders.

Such mimetic phenomena, Paget admitted, approximated what many clinicians had long believed to be among the central diagnostic criteria of hysteria. Among the litany of symptoms attributed to the hysteria diagnosis over the course of the nineteenth century, one particularly persistent refrain concerned the extent to which the illness could "[appear] under so many forms, and [exhibit] so many different characteristics";[82] that "simulation itself is a symptom of hysteria, though not an inseparable one";[83] that "essentially characteristic of the hysteric is the spirit of duplicity and lying";[84] that "there is scarcely a disease that [hysteria] does not at some time simulate";[85] and that hysteria "may produce symptoms that closely simulate those of organic disease."[86] In hysteria's ability to counterfeit a variety of other illnesses, forensic experts warned that the difference between the hysteric and the impostor was a difficult line to draw.

For Paget, however, nervous mimicry was an altogether distinct disease state, not to be confused with hysteria, though he admitted that instances of alleged feigning long attributed to hysterics were likely mistaken cases of neuromimesis, which, Paget argued, was nothing more than an organic disorder of the nervous system; for that reason, there was no organic disease that could not be pathologically mimicked.[87] At the same time, however, psychological disturbances brought on by "sudden mental distresses, emotion, disappointment, long anxieties, or exhaustion by overwork" could easily induce nervous disorders with marked neuromimetic symptoms.[88] Paget admitted that "in nearly all mimicries a mental influence may be discerned."[89] Nevertheless, given the authenticity of neuromimesis as a veritable organic disorder, Paget was adamant that neuromimetic phenomena were fully authentic disease states.

The very possibility that imitation could be reimagined as itself a kind of disorder meant that certain core assumptions about simulators had to be revised. Chief among them was the assumption that patients espousing, for example, local pain in a part of the body that did not otherwise present organic damage must be lying. Paget was even willing to propose that patients who intentionally feigned illnesses were not necessarily engaging in acts of malicious deceit but were, instead, acting "like children, who almost involuntarily imitate diseases." It was an impulsive weakness of will that led to most inventions and exaggerations of symptoms, not an avaricious intent to defraud. In the end, it was very difficult "to distinguish between the frauds of the willful and those of the will-less." Still, there was never any reason to believe that willful fraud was in any way frequent "among those with nervous mimicry or hysteria," that is, among patients most liable to be suspected of deceit.[90]

Traumatic Simulations

The increasing interest in mimetic behaviors was accompanied by the gradual neuropsychiatric legitimacy of the diagnosis of hysteria. It was mainly through the medical affirmation of hysteria after 1870, particularly in terms of its wide-ranging symptomatology and its link to a precipitating trauma, that simulation was able to crystallize into a viable pathological symptom.[91] While clinicians like Paget sought to maintain a difference between hysteria and mimicry, it was instead as a symptom of hysteria that simulation obtained its status as a legitimate pathological expression. Midcentury railway physicians had already proposed that hysteria was one likely consequence of the traumatic effects of railway collisions, which could often yield symptoms that mimicked organic nervous disorders.[92] Among traumatized patients there was often an observable "unconscious exaggeration of symptoms," a "simulation or nervous mimicry" that was "common to all hysterical people."[93] Many railway physicians, however, never fully abandoned their commitment to an organic approach to traumatic pathologies and remained polemically committed to upholding a strict divide between somatic and psychical etiologies.[94]

Scholars have instead looked to the English railway surgeon Herbert Page as the first trauma clinician to explicitly emphasize the purely psychogenic nature of traumatic shock, that is, the first to fully "psychologize" the effects of railway collisions.[95] It might, however, be more accurate to claim that Page did not so much psychologize traumatic injury as envision traumatic injuries as part of a complex psychosomatic economy.[96] It was a topic Page was not the first medical expert to broach; clinicians had observed for some time the mutual influence that the mind and body had on one another, or how excessive strains on the mind could translate into somatic disease states.[97] In 1883, Page published his influential forensic manual on traumatic railway accidents gathered from his observations as surgeon to the London and North Western Railway Company, where he had been employed for a decade.[98] While written for railway physicians and, ostensibly, in the interest of the railway industry, the book went a long way to shedding light on and medically legitimating simulative behaviors as pathological effects of traumatic injuries.

Simulations were not necessarily unintentional contrivances of the mind's influence on the body; nor were they simply cases of intentional medical fraud. One of the major effects of the "purely psychical causes" of shocks associated with railway collisions were cases of the "nervous mimicry" of "functional disorders" throughout different parts of the nervous system. Although he invoked Paget's descriptions of neuromimesis, Page offered a very different explanation for the mechanisms underlying these psychically induced simulated pathologies.[99] They were nothing more than the automatic emergences of disinhibited lower nervous centers, which arose because higher and otherwise inhibiting neurological processes had been "put *hors de combat*" as a consequence of traumatic shock.[100] The expression *hors de combat* was part of a passage that Page excerpted from the work of the English psychologist James Sully, who was himself drawing on the work of John Hughlings Jackson.[101] To whatever degree Page advocated for a psychogenic view of traumatic injury, he appeared to do so through a Jacksonian theory of neuropathology. Page would later describe how the entire diagnostic gamut of traumatic neuroses was an effect of the "dynamic disturbance" of the highest governing nervous centers, the disorder of which could take place "without the existence of *gross* structural lesion."[102] With the "dissolution" of the highest centers, Page claimed, the lower "more automatic [centers] are stable and abide."[103]

From an unmistakable Jacksonian perspective, Page claimed that traumatic neuroses, and consequent simulative behaviors, were simply cases of pathological automatism. The sheer "terror" of a railway collision was sufficient to diminish the regulatory power of higher cerebral processes, resulting in a "hypnotic" condition of the brain, which Page described as a state of complex automatic behavior, wherein the traumatized individual "may act in a seemingly purposive manner," yet nevertheless unconsciously much like a "post-epileptic automaton."[104] In such an automatic "hypnotic stupor," a patient would become highly suggestible and likely to unconsciously adopt physical symptoms that had no actual organic source.[105] Psychical shock alone, then, was capable of disinhibiting automatic processes, which could present as mimicked symptoms, simulative behaviors, and mild exaggerations.[106] "Exaggeration," Page insisted, "may not be, nay very often is not, altogether willful or assumed. Exaggeration is the very essence of many of those emotional or hysterical disorders which are so common in both sexes after the shock of collisions."[107]

By the 1880s, psychiatrists across the North Atlantic were increasingly willing to recognize that psychical trauma could function alone as the pathogenic source of a hysterical state, one of the core symptoms of which was the onset of simulative or exaggerative behaviors. Despite hysteria's proximity to other functional nervous disorders, such as neurasthenia, it was normally only the hysteric who was thought to be most likely to simulate.[108] Only under the auspices of the hysteria diagnosis, in other words, could simulation appear as an authentic pathological symptom. While many psychiatrists affirmed that traumatic phenomena could trigger hypnotic dissociations and states of heightened suggestibility, Ruth Leys describes how the history of trauma research nevertheless oscillated between a "mimetic" and "anti-mimetic" assessment of traumatized patients. The difference revolved around how closely the clinician felt patients remained immersed within the trauma and its perpetration and whether patients could separate themselves from the traumatic precipitant sufficiently in order to properly remember and narrate it.[109]

The "mimetic" debates that followed industrial traumas, however, were arranged according to a slightly different dichotomy, though one nevertheless organized around the presumed veracity of the post-traumatic condition. These debates were about whether post-traumatic simulations were authentically pathological manifestations and whether they could be causally attached to a precipitating accident or catastrophe; both inquiries were ultimately reducible to the same underlying question: Does the simulation warrant financial compensation? For if, as Page maintained, the simulator was nothing more than an unconscious automatist, then it would make sense (as we saw in chapters 1 and 2) that the propriety of simulation would be adjudicated in economic terms—that is, for the medical value of simulation disorders to become thoroughly affixed to and even confounded with their presumed economic value.

The economic stakes of simulation disorders were particularly conspicuous for Jean-Martin Charcot who not only, and perhaps most famously, shored up the connection between hysteria and simulation but did so by situating their pathological affinity within the political economy of industrial labor. Charcot proposed that one of the pathogenic effects of trauma was that it induced hypnotic and suggestible states among the injured, as a result of which patients would likely become unconsciously susceptible to ideas that could "settle in the mind like an isolated parasite"—including the idea that they were physical injured even if they were not actually.[110] Charcot was quite explicit that his views on traumatic suggestion affirmed and augmented the conclusions reached by his Anglo-American colleagues, even affirming at one point that his views were in keeping with Page's quasi-Jacksonian understanding of psychogenic trauma as the dissolution of higher cerebral functions and the disinhibition of lower processes.[111] Traumatized patients were, therefore, eminently suggestible, likely capable of finding themselves, possibly from a doctor's prompting or from some other inadvertent source, mimicking a mental illness or physical ailment.[112] As Sigmund Freud would later succinctly proclaim, "All neurotics are simulators [*Simulanten*]; they simulate without knowing it, and this is their sickness."[113]

But it was precisely in the context of his turn to male hysteria, after 1882, where Charcot most strongly linked simulation to post-traumatic hysterical states.[114] Charcot himself conceded that he was not the first clinician to observe hysterical symptoms among men who had suffered traumatic shocks.[115] Fletcher was among the first to associate traumatic pathogeny with the masculinization of the hysteria diagnosis, noting the "well-known fact" that "men 'active in mind, accustomed to self-control, addicted to business, and healthy in body,' do occasionally, after severe emotional excitement, become affected with a train of symptoms known as 'hysterical.'"[116] What Charcot recognized was that the masculinization of traumatic hysteria was, among other things, an economization of the very question of the hysteria diagnosis and the ancillary problem of simulation.

Male hysteria was not simply a medical anxiety but a financial one as well, for it posed an economic challenge in a way that female hysteria ostensibly did not, at least not so evidently. It was the paradigmatic presumption that industrial laborers, including railway workers, were mostly men.[117] And as even Charcot's "American colleagues" in all their "practical spirit" had already observed, it was predominantly men who in aggregate were most susceptible to violent occupational traumas such as railway accidents and who normally claimed damages for the psychogenic traumas they sustained—the evidence of which was presented in the form of hysterical and simulative symptoms. Very often, as Charcot conceded, "thousands of dollars are in the balance," not only in the form of compensation but in the interruption to labor, as postaccident nervous disorders frequently prevented injured men from returning to work for several months, if not years.

As Freud would later point out, male hysteria not only signaled the likelihood of a traumatically instigated catastrophe, but it was almost always inter-

preted as an "occupational disorder" (*Berufsstörung*), that is, a medical contestation where money was ultimately on the line.[118] Traumatic simulation was
invariably an economic affair, its pathological validity being inextricably tied to
the question of financial reparation and the political economy of labor power.

The Competing Truths of the Simulator

After the early 1880s, it seemed that simulators of all sorts abounded, in the
clinics as much as in the workplace, and psychiatric researchers and clinicians
in both Europe and the United States found themselves having to contend
with a gamut of simulative possibilities, from presumably deceptive actors
to involuntary mimetics and even psychotically impulsive liars. The growing concern that simulation was on the rise was not necessarily consistent
with the assumption that imitation was merely an act of deception and fraud.
Alleged simulators were analyzed according to two counterposing medical
positions: on the one hand, the continued deployment of a logic of suspicion
that presumed that the simulation was nothing but fraudulent deception and,
on the other, the medical validation that simulative behavior was an authentic
symptom of hysterical and post-traumatic pathology.

An increasingly common trend among many notable psychiatrists and
neurologists was to downplay the presence of fraud entirely. "Simulation is
rare and still rarely successful," wrote Richard von Krafft-Ebing. Nervous disorders were simply too complex, and too well understood, to be duplicitously
feigned with any degree of success.[119] "I now look back on a rather long series
of cases in which I had to examine accident neuroses [*Unfallnervenkranke*],"
wrote the German neuropsychiatrist Paul Möbius. "I have never in any way
found a case of pure simulation," that is, medical deception.[120] The more a
clinician claimed to observe impostors, Möbius explained, the more likely
it was that the clinician simply lacked conversancy in psychogenic injuries.

In making sense of accident trauma, Möbius had partially adopted the
category of "traumatic neurosis" from neurologist Hermann Oppenheim.[121]
Oppenheim was at pains to distinguish the category of traumatic neurosis
from, in particular, hysteria, to which it was nevertheless often compared,
and eventually subordinated.[122] Still, Oppenheim also admitted that "the occurrence of simulation in traumatic neurosis has been greatly overestimated
in the past."[123] "It is dangerous," he warned, "to focus the examination solely
on the detection of fraud [*Betrug-Entlarvung*]. In the first instance, it is important to prove the existence of the disease."[124] It was not uncommon for
clinicians to emphasize the infrequency of full-fledged duplicitous feigning,

over and against the higher occurrence of milder and normally involuntary exaggerations.[125]

This did not, however, imply that forensic concerns over the possibility of simulation had simply subsided. In keeping with a feature of the earlier treatment of simulation, clinicians in the latter decades of the nineteenth century continued to identify certain populations as more likely to simulate illnesses, particularly insanity. Criminals, for instance, and those who stood accused of crimes bore the brunt of such medical suspicions.[126] A vigilant distrust of the authenticity of purported symptoms was a similarly common practice among prison medical officers.[127]

Not all experts, furthermore, shared the belief in the general scarcity of medical simulation among the broader population. The prominent forensic physician Ambroise Tardieu proposed that medical knowledge of wounds and injuries should be supplemented with knowledge not only of how injuries might be sustained but also of how they could be imitated.[128] Victims of accidents, especially given the compensatory possibilities that inhered in industrial disasters, were marshaled into a new class of suspected simulators. Reports were published recounting the utility of electrical stimulation in detecting feigned symptoms particularly among people claiming injuries after railway collisions. Advocacy for coercion, intimidation, and fear to both detect and deter simulators continued as part of the clinical arsenal for decades.[129] Gabriel Tourdes, a professor of legal medicine at the medical school at Strasbourg, claimed that the study of simulation was nothing more than the "history of fraud in disease," and, moreover, "the existence of evil."[130] Dedicated textbooks, of the sort observed earlier in the century, continued to be published on the topic of disease simulation, with applications across a range of institutions, from legal medicine to military administration.[131] Against the older medical confidence that disorders were too complicated to be blatantly feigned, some clinicians warned that impostors could study the symptomatologies of, particularly, traumatic neuroses well enough to fool even seasoned physicians.[132] Such a scenario might especially be likely among "poor workers" (les travailleurs indigents), for whom, according to one Parisian medical student, "existence is a daily problem."[133]

The discourse of simulation by the end of the nineteenth century was marked by a morally bifurcated vision of the simulator as either villain or victim, organized around a contestation of positions over whether simulators were rare or common (and in what form), fraudulent or pathological (and to what degree). It is particularly telling that the renewed medical skepticism toward likely medical impostors was conversely met with novel theoretical

elaborations regarding precisely how simulation could be a legitimate manifestation of genuine pathology. The strongest justifications emerged, on the
one hand, out of expansions of the suggestibility thesis and, just after the turn
of the century, the theory of psychopathic personalities.

The notion of the latent suggestibility of the hysteric, popularized through
the work of Charcot, legitimated simulation after the 1880s in a way that comparable medical diagnoses, such as Oppenheim's traumatic neurosis, could
not.[134] Charcot's student and house physician Georges Gilles de la Tourette
continued to hold that suggestibility was the primary mental state of the
hysteric.[135] Like his mentor, Gilles de la Tourette upheld the belief both that
trauma was a central *agent provocateur* of hysteria, and that certain populations were simply more hereditarily predisposed to traumatic injury.[136] But
Gilles de la Tourette especially underscored the "essentially perverse" and
deceptive nature of hysterical mentalities, that hysterics were "extremely im-
pressionable beings." Some hysterics were so suggestible that, even in waking
life, they could be compared to "pure automatons" and "hypnotic somnambulists";[137] that their passivity was like a "photographic plate" that merely recorded and repeated the impressions it received, sometimes in an amplified
form. Gilles de la Tourette was careful to distinguish the hysteric, who was
merely "a slave to suggestions," from nonhysterical, but still morbid simulators, a different though not unrelated population of "perverse beings, liars,
with malformed brains—degenerates as they are called today, or better *unbal-
anced* as Charcot called them."[138]

Another Charcot disciple, the neurologist Joseph Babinski, rejecting
Charcot's general picture of hysteria, transformed the hysterical symptom of
suggestibility into an autonomous diagnostic category of its own, which he
dubbed "pithiatism."[139] The pithiatic was simply a former hysteric, reduced to
the one set of symptoms that, according to Babinski, got to the very essence
of the disorder: that ostensibly hysterical symptoms arose through suggestion and could disappear through the influence of persuasion.[140] Suggestibility
and auto-suggestibility were the most crucial symptoms for Babinski, a diagnostic criterion too specific to warrant the nosographic encumbrance of the
general category of hysteria.[141] Pithiatics were often prone unconsciously or
"subconsciously" to imitating conduct that they passively experienced. Babinski even argued that the force of suggestion was so strong among pithiatics
that aspects of their voluntary behavior could be affected by it as well.[142] But
ultimately the difference between the malingerer and the pithiatic was unambiguous. The ordinary fraudster acted in a completely deliberate manner,
whereas the pithiatic was, at best, a *demi-simulateur*, who only "plays a role, in

an unconscious or subconscious way," an indeliberate *acteur*.[143] The pithiatic rendering of hysteria became a useful hermeneutic tool in forensic analyses of traumatic psychopathologies following industrial accidents.[144]

The utility of pithiatism was, furthermore, notoriously exhibited in the writings of French colonial psychiatrist Antoine Porot for whom pithiatism became the preferred nosographic rendering of the North African Muslim's so-called primitive mentality.[145] As Porot professed, in his entry on "simulation" for his 1952 *Manuel alphabétique de psychiatrie clinique et thérapeutique*, the fundamentally pithiatic nature of "the native mind" meant that Muslims and indigenous North Africans were, in his estimation, simulators through and through.[146] Pathological pretense, then, became the framework through which the apparently transgressive behavior of colonized populations was ultimately interpreted—behavior that soon turned quite singularly toward political unrest. As Frantz Fanon would point out, the primary function of the colonial diagnosis of political insurgency as pithiatic pretense and simulative fantasy was to neutralize and delegitimate the ineluctable and, as Fanon saw it, necessary violence of Algerian decolonization.[147]

Somewhat implicit within the broad thesis of pathological suggestibility was the assumption that even voluntary simulation was pathogenic in its own right—that, to quote one French physician, "certain cases of madness that are simulated for a long enough period during the course of a forensic examination could degenerate into real insanities."[148] The American psychologist Morton Prince claimed that simulation possessed an almost infectious virulence, writing, "It is well known that many neuroses, although they may have originated in volitional attempts to deceive, nevertheless pass in time beyond the control of the will and persist as true pathologies."[149] These views were in line with an attitude that increasingly imagined simulation less as the symptom or unconscious externalization of an underlying hysterical or pithiatic disorder, and more as a disease in its own right, pathological even in its *conscious and volitional* form.

This medical orientation toward the consciously pathological simulator was most saliently formalized in the context of early twentieth-century German writings on psychopathy, beginning with the 1904 edition of Emil Kraepelin's psychiatric textbook, where he included in his newly organized chapter on "psychopathic personalities" the category of the "morbid liar and swindler" (*die krankhaften Lügner und Schwindler*).[150] For some psychiatrists, not all conscious simulations were effects of a morbid compulsion to lie but were instead the effects of "pathological fantasies" (*pathologischer Phantasievorstellungen*) grounded in organic modifications and hysterical affections.[151] For others, however, like Eugen Bleuler, the nature of voluntary simulation (a rare occurrence, according to Bleuler) was quite clear: "Those who simulate mental illness with some skill are almost all psychopaths [*Psychopathen*], some of them are really insane." For a clinician to ascertain that a patient had been consciously and even willfully feigning did not, therefore, prove that the patient was mentally sound; indeed, the very opposite could likely be the case.[152] German clinicians in particular deployed broad categories of psychopathy, degeneracy, and psychosis to reevaluate behavior like "conscious feigning" (*bewusster Vortäuschung*) no longer simply as wrongdoing. In many cases of deception, it was "impossible to judge where simulation—that is, conduct formed through purposive consciousness [*Zweckbewusstsein*]—stops and disease begins."[153] Conscious simulation could be simply a prodromal state of dementia praecox;[154] or it could be the behavioral manifestation of what psychiatrist Ernst Kretschmer called "hysterically inferior predispositions" (*hysterisch minderwertiger Veranlagung*).[155]

By the first decade of the twentieth century, simulation had become a genre of conduct thoroughly shot through with pathological overdeterminations. In his work on shell shock, the English physician Charles S. Myers noted that even the "simulators" who came to his attention had nevertheless been "feeble, nervous subjects themselves."[156] Some medical experts tried to make sense of the range of simulative disorders, lately classified through novel designations such as "pathomimia" or "mythomania," by describing them as disordered expressions of an otherwise normal and healthy phylogenetic adaptation; that the evolutionarily selected inclination to lie or deceive for survival purposes had become, in the course of disease, disinhibited and exaggerated.[157] *Pseudologia phantastica* was the term that one forensic psychiatrist settled on to characterize cases of pure pathological mendacity, a psychopathic forerunner of so-called factitious disorders.[158]

"The simulator," wrote the Montpellier physician Albert Mairet, "in the great majority of cases is a predisposed person [*un prédisposé*]," susceptible, that is, to a likely "cretinism" (*un tare*) within his nervous system.[159] Even the deliberative, fraudulent impostor likely suffered from some sort of "psychical instability" or "mental insufficiency." Although the conscious deceiver was technically sane, nevertheless, insisted Mairet (quoting the Parisian forensic physician Paul Garnier), "he is an irregular [*un irrégulier*], an abnormal being, deprived of all balance and levelheadedness [*pondération*]."[160] It had become quite possible in the early decades of the twentieth century to unite the two discursive trajectories that had throughout the previous century defined medical analyses of simulation: the logic of suspicion and the pathological elaboration of simulative behavior. Rather than partition the authentic patient from the impostor, as was common in the early half of the nineteenth century, clinicians after 1900 increasingly proposed that a patient could at once be a fraudster as well as an "abnormal being."

What had most strongly differentiated these two approaches to simulation was not any particular attribute of the behavior itself, for even conscious and deliberative feigning—formerly the trademark of obvious deception—could turn out to be the behavior of a psychopath. It was, instead, the rationale, the motive prompting the conduct, that tended to best distinguish the real malingerer from the neuromimetic patient. Was the behavior prompted by illness? Or was it driven by greed?[161] What the unification of the logic of sus-picion and the pathological analysis of simulation meant, however, was that economic desire and madness could cease to remain mutually exclusive con-ditions through which clinicians could make sense of simulation. Financial incentive and pathology could converge as a twofold foundation for simula-tive conduct, and it is here that the concept of sinistrosis came to life.

The Perils of Remuneration

In his prominent monograph on simulation with specific relation to work-place injuries, the Belgian physician and social medicine advocate René Sand argued that any attempt to diagnostically interpret postaccident simulations had by necessity to consider the institutionalization of compulsory accident insurance throughout Europe. It had been over two decades since Germany instituted its own social insurance law in July 1884, and although Sand ad-mitted that the German context "offers us, from the point of view of simu-lation, the largest field of study," he nevertheless cataloged the profusion of similar legal reforms throughout Europe: Austria (1887), Norway (1894), Fin-land (1895), Great Britain (1897), Denmark, Italy, France (1898), Spain (1900), Netherlands, Greece, Sweden (1901), Luxemburg (1902), Russia (1903), and Belgium (1903, 1905).[162]

Sand corroborated that indemnification had explicitly become an un-avoidable new grid of intelligibility for making sense of simulative behavior. "At the outset," he wrote, "it was feared that universalized insurance would cause a larger number of workers to feign morbid traumatic disorders; some even foresaw a 'general demoralization' of the working class."[163] But Sand re-mained unconvinced by these worrisome prognostications. He admitted that the mounting likelihood of potential fraud was met equally with the growing prospect of actual accidents and, accordingly, to legitimate simulative pathol-ogies arising from cases of traumatic hysteria and neurasthenia.[164] Compensa-tion legislation had even become the catalyst for entirely novel forms of imita-tive conduct, such as the "honest" simulation of when a worker, faced with the worrying possibility of not having his injuries taken seriously, deployed "another psychological process," one in which he "imitates the salesgirl who, knowing the customer is going to haggle, will propose from the outset a much higher price than the one they have in mind."[165]

Still, Sand was adamant that simulation was not specific to industrial workers alone but was found in every country and among every social class, including the wealthy, who were in his eyes equally guilty of attempts to defraud railway and life insurance companies.[166] Other physicians and actuarial assessors echoed Sand's observation that simulation appeared to be uniformly present across social and economic classes, equally distributed among workers and the elite.[167] Sand was also not the only physician to take at least a moderately sympathetic and medicalized approach to the apparent simulation crisis; one German psychiatrist observed how readily "simulation-like" (*Simulationsähnliche*) many genuine mental illnesses actually were—that "conscious exaggerations, even falsifications" were genuine symptoms of mental pathologies among patients who could not simply be ignored or cast aside as a so-called pension hysteric (*"Renten" Hysteriker*).[168]

The idea of pension hysteria was part of a systematic wave of moral condemnation in response to the increasing acceptance of simulative pathologies.[169] In Germany in particular, as soon as the category of traumatic neurosis gained actuarial coverage under accident insurance legislation in 1889, it quickly acquired among skeptics the derisive moniker "pension neurosis" (*Renten Neurose*). Even though the actual number of post-traumatic pension applications comprised only a minuscule percentage of accident insurance claims, the backlash was clearly an expression of animosity against Socialist and workers' rights political movements.[170] Pension neuroses, or pension hysterias as they were sometimes called, were caricatured insinuations that patients' "afflictions" were nothing more than expressions of pecuniary interest and the desire to avoid work;[171] or, at best, a kind of pathological weakness of the will, a compulsive self-deception ultimately (and perhaps unconsciously) motivated by greed alone.[172] The social stigma of claiming a pension as a result of a debilitating workplace injury was itself so disreputable that it too could cause a degree of psychological injury, what was eventually designated as "pension struggle neurosis" (*Rentenkampfneurose*).[173] Scholars like the historian Greg Eghigian have noted how deeply "neurosis and welfare state bureaucracy implicated one another," yet even such a conclusion may not fully appreciate how fundamentally intertwined pathology and welfare-state economic policy actually were.[174]

There were certainly clinicians who were critical of bad faith efforts on the part of medical examiners to explain away a patient's seemingly illusory symptoms as nothing but the effects of the struggle to attain a pension (*der Kampf um die Rente*).[175] And yet, there were many who could not help but view simulative behaviors through the bureaucratic, and thus skeptical lens, of actuarial administration. French physicians became increasingly convinced that "the number of accident cases [had] increased alarmingly" since the passage of the 1898 accident insurance law.[176] Medical insurance assessors offered gloomy warnings based on observations from Germany and Austria that through the institution of compensation legislation many workers sought "a source of illicit profit from the effects of a faked or exaggerated accident";[177] or, at the very least, that cases of exaggeration and simulation had simply become more frequent.[178] It was not an uncommon medical conclusion that among the "strange consequences" of the adoption of insurance legislation was "a curious development of 'simulation.'"[179]

Such sentiments were as prevalent in Britain as they were in France and Germany, where doctors often found themselves taking up the task of "policing malingering in the interest of insurance companies or the state."[180] One medical practitioner claimed that since the institution of the British Workmen's Compensation Act, not only had the number of accidents and claims to compensation increased, but so had the amount of fraud, "from the out-and-out fraud down to the unconscious fraud." While "unconscious malingering" was a "contradictory term," it nevertheless accurately described a "common type of case" where workers malingered involuntarily—not with the intent to defraud, but as an unconscious effect of being overly stimulated by the incentives of compensation. "With all due respect to the British working man," the physician wrote, "he is, so far as exaggeration of symptoms is concerned, no better than the hypochondriac of the upper classes. The one exaggerates for compensation, and the other for compassion."[181]

According to John Collie, a city medical examiner in London, the incitement to feign illness amounted to the "moral degradation" of the laboring classes, a fault he attributed not only to their "perverted mental outlook" but to socialism itself, which, Collie argued, has "run riot" throughout the world. Given how likely workers claiming compensation were to exaggerate their symptoms, even unconsciously, the safest approach according to Collie was to "assume that the plaintiff is often untruthful."[182] Collie also claimed that the costs of insuring against industrial workplace accidents had exceeded the estimated actuarial costs of the Workmen's Compensation Act. The skyrocketing number of accident claims, he suggested, could not possibly be justified simply as effects of the growth of industry and the number of laboring bodies. "Malingering and dishonesty," he concluded, "must have had an influence in raising the figures to their present abnormal height."[183] These dishonesties, however, were not simply effects of "deliberate wickedness," for many of these feigning laborers were themselves actually "victims" of the effects of the Workmen's Compensation Act, and other related laws.[184] Workers, Collie believed, would likely exaggerate, either consciously or unconsciously, for fear that their purported illnesses would simply be minimized.[185] In fact, Collie went so far as to suggest that while functional nervous disorders were effectuated by accidents, the prolongation of simulative behaviors and other neuromimetic symptoms were actually the effects not of the accident itself but of the workmen's compensation laws, which functioned as a kind of psychical incentive for the perpetuation of injuries. The possibility of compensation, he proposed, could psychically enable the will to imagine injury, thereby making the boundary between intentional fraud and involuntary simulation difficult to demarcate.[186]

Collie's remarks highlighted an inconspicuous characteristic of postaccident traumas and their resultant simulative states, particularly given how clinicians implicitly wrestled with the veracity of these conditions. Simulative pathologies were increasingly consigned to a kind of transactional reality by the turn of the century, that is, a reality conditional on the extent to which economic instigations—like accident insurance laws—could be sutured into the explanatory chain that linked pathogeny to symptomatology and, ultimately, to prognosis and cure. The possibility of compensation, and the associated strains of litigation or actuarial verification, became integral to the very legibility of simulative disorders. Most of the Anglo-American medical architects of the theory of traumatic shock detailed the effects that compensation litigation had on either the increasing severity or therapeutic reduction of subsequent hysterical symptoms. Erichsen was certain that post-traumatic symptoms would persist for the duration of litigation but would dissipate as soon as legal proceedings had ended, in the form of a recovery so rapid as to lead to suspicion that "the whole of the sufferings were purposely simulated, and the patient was a malingerer."[187] Page concurred that recovery from traumatic psychogenic injuries could only truly begin once litigation proceedings had ended, adding that just as litigation could exacerbate symptoms, the settlement of a compensation claim could function as "potent influence" on recovery and convalescence, even if the influence were "wholly unconscious." Because of how interconnected workplace injuries were with compensation claims, once an injury had been sustained, the very knowledge of the likelihood of remuneration became "part and parcel of the injury in the patient's mind."[188]

The prospect of compensation, and the bureaucratic efforts to procure it, became interwoven into the very experience of the occupational injuries themselves, compounding and intensifying them in ways that were perhaps only "simulative" in an effort to ensure a financial return. Page warned his reader not to misinterpret his meaning; he did not propose that a settlement could in itself be a "curative agent." Compensation was only a kind of inverse catalyst for recovery, a negative therapeutic precondition, insofar as it did not necessarily incite recovery by its presence but, rather, delayed it by its absence.[189] Page would elsewhere offer the reassurance that he was never "inveighing against the right or justice of compensation for injuries sustained." His point was merely that compensation functioned as a kind of unconscious inducement to an injured party to "keep up the invalid state," and to avoid "[escaping] from it as soon as even he can," a phenomenon observable even among honest and authentically injured victims of an accident.[190]

Prominent international clinicians described how litigation proceedings, or other administrative hurdles that perpetuated the sheer "waiting for compensation," or which kept the "specter of misfortune perpetually before [a patient's] eyes," could aggravate or induce symptoms and depressed mental states that were, if not unreal, than at the very least embellished.[191] Litigation of any sort, according to one American psychiatrist, was "the worst possible thing to which the patient with traumatic neurasthenia can be exposed."[192] It was liable to induce what another forensic specialist called "litigation psychosis."[193] In the medical examinations of workplace injuries, Oppenheim observed, clinicians were often challenged not only by the fact that they had to calculate a patient's "earning capacity" (*Erwerbsfähigkeit*), but also because prognoses concerning the resumption of work were often drawn out as an effect of "the struggle for compensation." Prognoses were always less favorable the more the worker's psyche (*Psyche*) was detrimentally undermined by remunerative hurdles.[194]

Sinistrosis and the Pathology of Indemnification

By the end of the nineteenth century, a widespread psychiatric conviction held that compensation could function as an involuntary psychical influence, an external intercession into the mental associations through which a patient otherwise experienced a traumatic affliction. It is here, then, that we return to sinistrosis in order to appreciate how this "new illness" secured a much more profoundly causative relationship between compensation and pathology. Sinistrosis was a disorder that incorporated the role of worker compensation legislation into its nosological makeup so completely that it was as much an economic concept as it was a psychiatric one: it denoted the economization of psychopathology just as strongly as it did the pathologization of economic policy.

And yet sinistrosis was not necessarily singular in this regard. The medical attitude affirming the pathogeny of insurance remuneration was embodied as much in expressions of medical skepticism (e.g., "pension neurosis") as in affirmations of the viability of disease categories such as sinistrosis itself. Writing in 1910, the German psychiatrist Alfred Hoche wondered, given the "fluid borderland [*flüssige Grenzgebiet*] between mental health and mental illness," which phenomena of the modern world should be counted among the pathogenic triggers of insanity. He underscored the prevalent belief in traumatic neurosis, a "national plague" (*Volksseuche*), he called it, which, unheard of decades earlier, had become a "cancerous sore to the entire organism of our whole workforce." The fault lay, however, specifically in state-based accident legislation, beginning with Germany's own workplace accident insurance law of 1884. "The law has," he wrote, "on this there is no doubt, engendered the disease."[195] For Hoche, economic law itself was a pathogen that infected the body politic. "People are actually sick but, strangely enough, they would be healthy if the law did not exist."[196]

The French clinical architects and disseminators of sinistrosis were not the only European medical experts and administrators to explicitly formulate the pathogenic nature of economic policy. They were not the only ones, as historian Andreas Killen has argued, to view the simulator as a general reflection of the economic bad faith in welfare politics and anxieties concerning the stability of the self in fluctuating money markets.[197] French clinicians often noted how "closely linked" sinistrosis was to German "pension hysteria" and that "whichever term is used, the phenomenon is the same."[198] While sinistrosis was certainly perceived to be "a first cousin of pension hysteria," its difference lay in the fact that it was simply a much more discursively explicit and elaborate rendering of what Hoche and other German psychiatrists sought to express. The French physicians Émile Forgue and Émile Jeanbrau quickly latched onto the diagnostic value of sinistrosis as a pathological "state of mind" that was "caused and sustained by the possibility of obtaining the strongest possible indemnity." The disease was not a consequence of an accident itself; rather it was "the law of 1898, which so notably modified the pathology of trauma, [that] created a new illness."[199]

Sinistrosis was not, therefore, unique so much as it was exemplary—paradigmatically so—of the supposition that compensation laws acquired what the industrial accident had on its own already possessed: a pathogenic virulence. The cognate category of pension neurosis among German physicians was a looser and more diffuse term, a medically vernacular form of moral rebuke rather than an internally coherent psychiatric concept. Sinistrosis was, furthermore, a concept that, unlike some of its British analogues,

placed the pathogenic vector almost squarely on the side of economic legislation rather than, say, in a biological predisposition among the workers
themselves.[200] Sinistrosis, then, was exemplary because it was a nosologically
formalized and conceptually coherent disease state that authorized medical
experts to assert unilaterally and unequivocally that psychosomatic exaggerations were instigated by social insurance, that is, by the very safeguards put
in place to protect against traumatic injuries. At the International Congress
of Social Insurance in Rome in 1908, a Genevan physician typified, through
an analysis of the Swiss worker's accident law of 1906, a largely Pan-European
concern—that accident laws incited workers to become "inclined to exploit
insurance." And while the physician was insistent that the "influence of legislation itself is undeniable" in motivating cases of unadulterated fraud, "a place
must be reserved for Brissaud's 'sinistrotic,'" he wrote, as the archetypal case
study of the "unforeseen and immoral" consequence of workplace accident
legislation.[201]

The symptomatology of sinistrosis was perhaps the most paradigmatic expression of what counterpart disorders, such as pension hysteria, attempted
otherwise to depict. Again, sinistrosis was defined primarily as injured
workers' purportedly erroneous belief that they were unequivocally owed
recompense. The belief not only resulted in a kind of "paralysis of the will,"
an inability to resume labor, but also in an involuntary impulse to simulate
and exaggerate symptoms.[202] The psychiatrist Maxime Laignel-Lavastine
proposed that some degree of simulation "is almost inevitable among the
wounded (no matter their moral or social rank)."[203] Sinistrosis meant that
simulations were an inevitable reaction to workplace injuries, and pathogenic
incentives for compensation were perceived to be potent enough to induce
morbid simulative symptoms among injured workers as well as to bring about
the avaricious impulse among the healthy simply to lie. Sinistrosis effectively
flattened the very difference between the unconscious exaggerator and the
willful impostor.

The disorder also highlighted a crucial epidemiological division among
presumably afflicted populations. As Laignel-Lavastine explained, "Given the
same injury, there is a colossal contrast been the normally rapid improve-
ment among the uninsured and the abnormal slowness of the same improve-
ment among the insured." Not all workers, in other words, were susceptible
to sinistrotic symptoms; only indemnified workers were. But since the 1898
law meant that virtually all workers were de facto susceptible, the only way to
curtail the rapid spread of sinistrosis, Laignel-Lavastine declared, was to stop
indemnifying workers altogether.[204]

For Laignel-Lavastine and many other clinicians, sinistrosis was dangerous primarily because of the economic threats that it posed. A neuropsychia-trist in Lyon, for whom this very "modern disease" was "born on the day the right to accident compensation was proclaimed and then enshrined into law," was frank about the disorder's supposed economic damage. "It cost railway companies millions," he explained, "and after the implementation of work-place accident laws, first in Germany, then in France, and most recently in Belgium, it has seriously threatened the balanced budgets of insurance com-panies."[205] The only way to rid the national workforce of this "special pathol-ogy," he argued, while also ensuring the economic well-being of nations and corporations, was to revise the law of 1898.[206]

With the onset of the First World War, sinistrosis was eventually deployed in such a way as to exceed its original remit as a "claims psychosis" or "claims delirium."[207] The category of "war sinistrosis," a French variant of British shell shock, gained enough traction that by October 1915, sixty-seven cases had already been reported.[208] Yet even with its application to warfare, sinsistrosis could never fully relinquish its economic connotations. In their discussion of war sinistrosis, Laignel-Lavastine and psychiatrist Paul Courbon noticed "a large analogy between the reaction of many soldiers treated since their mobi-lization and the reaction of many workers since the 1898 law of workplace accidents."[209] The pathogenic resemblances between soldiers during times of war and workers during times of peace were due in part to the general simili-tude between war and enterprise as well as to the fact that soldiering and in-dustrial labor were equally perilous undertakings. War, the authors explained,

> presents itself more and more as an industrial enterprise. Every battling coun-try is a factory, of which war is the industry, the State is the boss, and the mobilized are the workers. The right to an invalidity pension is similar to that conferred by the 1898 law on disabled workers. The injury or illness contracted by the mobilized in commanded service, either on the battlefield or at the rear, is therefore a real workplace accident.[210]

The crucial difference, however, between the soldier and the worker was that the soldier was regarded as fulfilling a moral duty to the nation. Therefore, a soldier's right to compensation for a battlefield injury was morally justified, they explained, in a way that a workplace injury was not for a worker. The authors refused to recognize industrial labor as possessing any kind of moral value that would warrant remuneration in excess of a simple wage. It was, the authors declared, the very conviction that the worker is owed anything more from his employer or the state that makes the worker into a "delusional claim-ant" and "a constitutional psychopath."[211]

Although sinistrosis eventually transitioned into a simple nomenclatural proxy for medical fraud, the sinistrotic nevertheless remained a perpetual surrogate for the industrial worker.[212] Or, perhaps more precisely, the sinistrotic was both a stand-in for the worker and a re-embodiment of the medical simulator. The industrial laborer had become not only a likely traumatized victim of workplace accidents but also the potential dupe of accident insurance legislation. This was a bifurcated moral position, however, because this alleged victim, as a consequence of disease, could not help but simulate, exaggerate, and therefore exploit insurance companies and the state. The entire idea of a remunerative pathology, exemplified by the sinistrotic's simulative propensities, successfully transformed the industrial worker into both an economic victim and an economic perpetrator.

Ironically, the very effort to mitigate the psychogenic trauma of accidents created, as far as many clinicians saw it, not only new psychopathologies but also new pathogenic sources in the economic domain. Economic policies inaugurated to protect against the financial losses incurred from workplace injuries were accorded a pathogenic power comparable to a literal traumatic accident, a power capable of inducing pathologies primarily among industrial laborers. Through sinistrosis, the worker was situated at the nexus of a variety of precarious psycho-economic vectors—workplace injury, on the one hand, and compensation legislation, on the other (to say nothing of psychogenic traumas of industrial labor itself). Pathologies were as economically disastrous as economic legislation was pathogenic, and nothing represented this entanglement of industrial pathology and political economy better than the figure of the sinistrotic.

As a concept, sinistrosis was more than just a medical diagnosis and represented more than simply the expression of an economic style of psychiatric reasoning. For a brief period, "the sinistrotic" cohered into what Georg Simmel called a "social type," an amalgamated social reality or "form" constituted through the reciprocal dynamics of interpersonal relations.[213] In contrast to "the stranger," Simmel's most familiar social type, "the sinistrotic" coalesced as both a social and an institutional reality, fashioned at the intersection of psychiatric conjectures and economic norms before pervading the medical imaginary as a hermeneutic lens through which virtually all workers in France were to some degree viewed. Perhaps the reason the sinistrotic enjoyed such a robust, albeit temporary, ontological fullness was because its social truth hinged on the purported dangers that it posed. The pressing social reality of the sinistrotic was an expression of its alleged perilousness to the economic order. For in contrast to the enterprising madman and the aphasic testator, the sinistrotic was a figure whose pathological value was regarded as a ruinous and costly hazard rather than a lucrative privilege, calamitous to the health and wealth of the social body, and unlikely ever to be excused, endured, or redeemed.

Social types, as Simmel has argued, are revelatory to some degree, insofar as they unveil something singular about the individuals who inhabit those forms. Sinistrosis, therefore, revealed something about the meaning and value of the laborer within the economic order of late industrial capitalism. François Ewald has argued that, in France, the idea of "professional risk" (risque professionnel)—the so-called principle upon which French lawmakers relied as "the cornerstone of their major accident insurance laws"[214]—was an administrative technology that attended to the inherent perils of industrial labor in two distinct ways: first, as an actuarial calculus to determine whether certain occupations could in fact be insured; and second, as a remunerative calculus to determine whether certain injuries could be compensated and, if so, by how much.[215]

But what the concept of sinistrosis exposed is that industrial labor introduced a very different, somewhat more elusive, and certainly more disruptive kind of risk than what even Ewald observed. The sinistrotic was simultaneously a surrogate for the industrial laborer as well as a way of reframing the historical menace of the simulator, the impostor who continued to pose an ever-present hazard for medical administration. As the simulator was recast into the newly sinistrotic "soldier of industry," the dangers of simulative behaviors were compounded, engendering hitherto unknown kinds of "risks" into the medicalized political economy of industrial labor. Sinistrosis meant that the "professional risk" of industrial occupations was not limited to the threat of accidents alone, since in the eyes of medical experts, insurance laws and workers themselves became the bearers of novel pathogenic hazards. Precisely because compensation laws were deemed potentially virulent in themselves, the worker was suddenly posited as always at risk, not only of sustaining an injury but of being unduly influenced by the infectious force of economic legislation; one way or another, an injured worker was bound to exaggerate or simulate symptoms, inevitably as an effect of being manipulated by the promise of remuneration.

Through sinistrosis the industrial laborer acquired a new legibility, viewed as perpetually in danger of becoming traumatically wounded from all sides. The worker's "professional risk" embedded a hidden hazard for the administration of medico-economic bureaucracy itself, since the covertly sinistrotic worker was in principle always at risk of becoming a risk to the de facto economic order. The question was not whether workers were latent sinistrotics and therefore potentially disruptive to the institutional intersections of economic policy, state bureaucracy, and medical administration; that much was assumed. The question was simply whether the latency of their sinistrosis would actualize and, if so, to what degree. What sinistrosis in the end staged was the idea that workers could be regarded as essential constituents of industrial capitalism while simultaneously, and precisely through their function as workers, a lurking threat to the very operation of the economic order.

6

Appraising Eccentricity

The Value of an Eccentric

Over the course of this book, I have sought to illustrate how madness came to be regarded as compatible with economic forms of conduct and interpretable from the standpoint of pecuniary benchmarks. As madness was increasingly subjected to an economic style of reasoning, psychiatrists were able to attribute financial and even moral value to an array of pathological conditions such that some mental disorders could be lauded as financial assets while others could be disavowed as economic liabilities. Insanity signaled neither the absolute delimitation of financial or commercial possibilities nor a simple expulsion from the market economy; mental pathology was not economically disqualifying so much as something to be appraised, and therefore rejected or redeemed on economic grounds. The sinistrotic, for example, was not simply excluded from the economic sphere; if anything, what the institutional renunciations of sinistrosis exemplified was that the disease was an economic phenomenon through and through, the alleged embodiment of financial disaster and monetary loss.

As I described in the opening chapters, it was the diagnostic impasse introduced by the borderland typology that first set the conditions by which psychiatry would opt to deploy an economic style of reasoning in the first place. By affirming the medical and social reality of borderland figures, psychiatry effectively introduced a hermeneutic problem it could only resolve by modifying its diagnostic aims. Clarifying the social status of performances situated at the boundaries of socio-medical legibility did not require ascertaining the truth of a patient's mental state so much as assessing the propriety of their conduct according to an otherwise profoundly legible arithmetic of economic value: Does the behavior in question generate or forfeit wealth?

This turn to the appraisability of conduct inaugurated an economic style of psychiatric reasoning that continued to pay dividends, as it were, with respect to other psychiatric and neuropsychiatric puzzles, such as language pathologies, traumatic occupational injuries, and the psychoneuroses related to business enterprise. More than just the effect of injurious economic forces, madness itself was fashioned into an economic form, an emergence that I have been calling pathological value.

While mad enterprisers, will-writing aphasics, and compensation-seeking sinistrotics certainly displayed key affinities with the borderland typology, still none of these psychiatric personas quite replicated the nuanced ambiguities of the paradigmatic borderland figure, namely, the eccentric, the psychiatric problem that first instigated this entire analysis. The eccentric, after all, was not a nosological category, for despite the psychiatric anxieties that it introduced, eccentricity could never be transposed into a systematic compilation of determinable behavioral traits. It merely denoted conduct that evaded generalization, and had slipped into the recesses of behavioral legibility. The eccentric was the object of interpretative consternation and contestation, a problem of how to assign meaning and value to behaviors that eluded categorical decipherment. For this reason, any eccentric could be both repudiated and venerated, which is precisely why eccentricity was the exemplary psychiatric problem to which an economic style of reasoning could be most acutely applied. For it was the tacit financial merit of eccentric behaviors that ultimately determined how idiosyncratic conduct would be adjudged.

As this book opened with the problem of eccentricity, it seems only appropriate to end with it as well, to see how the conclusions developed thus far might be redirected back to the hermeneutic crisis with which this book began. But unlike the subject matter of some of the preceding chapters, eccentricity was not a consolidated diagnostic classification, nor was it always the object of a uniformly organized medical discourse. There is, in other words, no such thing as a general eccentric, of eccentricity *as such* outside of a specific socio-behavioral context. Consequently, the very notion of eccentricity must be approached according to the logic of the case study, as an instance of "thinking in cases," as John Forrester put it.[1] Concluding this book with an exploration of eccentricity, then, means identifying a case study relevant to, and capable of typifying, what this book has sought to illustrate—namely, an economic style of psychiatric reasoning and the historical emergence of pathological value. And we find just such a case in the life of an all but forgotten onetime minor celebrity of the American Gilded Age, the turn-of-the-century "eccentric" businessman and self-proclaimed clairvoyant John Armstrong Chanler.

Chanler was born in 1862 to a family in the upper echelons of New York's financial elite. He was the great-great-grandson of the German émigré and New York real-estate mogul John Jacob Astor, and his family ancestry was linked to other prestigious American families as well, including the Livingstons and Stuyvesants. By the age of fifteen, Chanler had become an orphaned millionaire, the oldest of ten (seven surviving) siblings, most of whom would, by the end of the century, count themselves part of the so-called New York Four Hundred, a semiofficial and highly exclusive coterie of the city's de facto aristocracy.[2] The United States and New York in particular, as the historian Sven Beckert has claimed, was arguably "the most bourgeois of all nineteenth-century societies."[3] Chanler's early life was therefore set in the unofficial capital of one of the North Atlantic's most coherent and emblematic expressions of elite class identity. That setting afforded Chanler a particular privilege: to cast himself as a heroic man of affairs, an intrepid protagonist of a harrowing life story, and above all as an oracular virtuoso in possession of a prodigious mental endowment. For in his early thirties, Chanler "discovered" that he was in possession of a parapsychological power, which he dubbed his "X Faculty," insisting that it was akin to an entirely autonomous personality. This X Faculty, Chanler claimed, granted him a set of remarkable new talents, including the gift of prophecy, a prescient business acumen, a sudden talent for creative literary composition, and the ability to commune with the spirit of Napoleon and the devil himself. Chanler's eventual notoriety owed less to his purported talents than to the fact that they belonged to a man connected to families who were best known for furnishing society with captains of industry, politicians, and socialites, not self-proclaimed prophets.

What makes Chanler such an illustrative case study, as well as a fitting topic for this culminating chapter, is the extent to which his style of life conduct and general self-performance embodied the dynamic relationship between madness and enterprise I have sought to describe throughout this book. His behavior and general style of self were almost consistently made up of performances of pathological privilege that played out, and were repeatedly justified, in economic terms. What we can observe most especially about Chanler is that he adopted an internally directed economic style of reasoning; he was a man who recognized himself as psychologically abnormal, but interpreted his abnormality, indeed fully extolled and redeemed it, as a lucrative financial prospect. This is therefore a case study of an economic stylization of the abnormal and idiosyncratic self—a stylization grounded on an entitlement to view oneself, and to be viewed by others, as the source of immense pathological value. Chanler essentially fashioned himself at the intersections of creative madness and financial virtuosity decades before actual economists touted the perspicacious foresight and the daemonic irrationalism of the prosperous enterpriser and business luminary.[4]

Chanler's life was punctuated by two significant events. The first was his involuntary institutionalization, engineered by his family, at the Bloomingdale asylum in upstate New York for a period of four years, which ended when he escaped the asylum grounds. The second was a legal battle in a Virginia courthouse that took place subsequent to his escape, where he was not only declared competent but celebrated as a scientific and business prodigy. In both instances, his purported X Faculty took center stage. In New York, in the eyes of preeminent physicians and his doctors at Bloomingdale, his belief in his prophetic aptitude was merely the symptom of a paranoid delusional disorder. In Virginia, however, based upon the expert testimony of prominent psychological researchers, including Joseph Jastrow and even William James, Chanler's X Faculty was exalted as the expression of a mediumistic capacity that likely arose from a profound mental aptitude.

This bifurcation of interpretations actualizes a point that Charles Mercier only hypothetically proposed, and that I discussed in chapter 3: that business venturers are retroactively viewed as insane when their inventions and undertakings fail but are considered virtuosos and mavericks when they succeed.[5] Chanler's institutionalization in 1897 was precipitated by a series of potentially risky investments in southern textiles. Apprehensions by his family that he was likely squandering his inherited wealth were only exacerbated when they caught wind that his speculations were prompted by what he claimed were his newfound prophetic abilities. After the four intervening years, however, Chanler's financial prognostications were eventually corroborated. Southern textiles became a lucrative industry, and while his incarceration at the Bloomingdale asylum prevented him from directly reaping any of the rewards for himself, the sheer success of his prospective speculations reframed the value of his otherwise pathological behavior. It was not Chanler's style of conduct alone—his belief in and performances of the X Faculty—that divided subsequent medical evaluations of his competency and social worth. It was, instead, the constellation of economic interpretability within which his conduct was situated that made him seem like a financial risk in one moment and a shrewd and prescient entrepreneur in another. The *very same* expressions and behaviors that in one instance were perceived as genuine insanity could, against the backdrop of economic success, be reframed as the very engines of enterprise.

Although Chanler was labeled an "eccentric" and "eccentric millionaire" in the popular press for many years, we might be inclined to regard his unwavering belief in his fantastical abilities as a sign of a serious mental disorder.[6] But taking such a stark view potentially impedes an important analytic perspective on Chanler, which is that his abnormality was not simply the passive object of decipherment, diagnosis, and arbitration by friends, doctors, lawyers, and judges. It was also something akin to a "technology of the self," that is, a moldable style of conduct that Chanler practiced and cultivated with a strong degree of intentionality and awareness, betting on and in many cases profiting from the possibilities that it could occasion.[7] He remade himself, we could say, in the very image of his abnormality when at the age of forty-six, he commemorated his unconventionality by christening himself anew, altering the spelling of his last name from Chanler to "Chaloner"; it was the more authentic spelling, he averred to the press, "the ancient form of the name," and the one I will hereafter employ.[8]

Chaloner was furthermore quite capable of demonstrating a tactical shrewdness with respect to the performance of his abnormal self. He mitigated public pronouncements of his X Faculty with an exaggerated scientific skepticism, deploying a lay and self-taught knowledge of transatlantic psychology in order to eschew occult explanations of what he experienced in favor of experimental confirmations of his mental prowess. While his actions and pronouncements could often seem outrageous, he was adept at balancing his apparent absurdity with a requisite normality. His eccentricity was not simply at the hermeneutic mercy of others. It was also a practiced performance, a technique of self-expression. And yet in all instances, however and by whomever it was regarded and assessed, his eccentricity was inevitably routed through an economic grid of intelligibility, the final measure against which Chaloner's medical health and social value was appraised.

This chapter is not the first recounting of the case of John Chaloner.[9] Prior renderings, however, have tended to rehearse a uniform and somewhat hackneyed motif, that Chaloner was a victim—a man who though peculiar was not necessarily insane, and whose rights and liberties were trampled upon, who was taken advantage of by unscrupulous friends and family members, and who after a long struggle finally regained his economic freedoms. This is exactly the story that Chaloner gave about himself.[10] And while we need not discount how unscrupulously his family did behave in their efforts to institutionalize him, nevertheless framing Chaloner's life as a tale of victimization almost entirely ignores the vectors of financial, racial, and gender privilege that he, an immensely wealthy white man, was generally afforded and that, as we will see, he was quite aware of possessing. While privileged persons can certainly be victims, a story of victimization can become a way of evading the role that social entitlements play, for these entitlements were precisely what enabled Chaloner and others to believe that his abnormality could be value-generating. That he could come to enjoy the privilege of yielding pathological value is the feature of Chaloner's life that is almost completely disregarded when all that is at stake in the story of John Chaloner is his victimization.

Therefore, the purpose of this chapter, this case study in eccentricity, is not to recuperate nor to redeem this admittedly unique historical actor. The goal is not to bring Chaloner to life, to let his voice be heard, to make him familiar, translatable, or comprehensible. In many respects, Chaloner made such a task quite inaccessible, as the data of his life and personality were consistently awash with cryptic ambiguity. His personal writings, for instance, were not only the expression of his own voice but also that of his purported X Faculty, which he claimed functioned as the primary "author" of many of his essays, letters, and compositions. His personal accounts of events were uneven, inconsistent, and often at odds with what others reported. Throughout most of his life, he was just as readily pathologized as he was esteemed. Chaloner was, to put it plainly, interpretatively opaque—the quintessential eccentric. For this reason, efforts to fully decode Chaloner or to restore his supposed authenticity on ethical grounds will likely fail or, at best, yield a troublingly incomplete picture, just as the narrative of his victimization overlooks the many ways that he was very much *not* a victim, certainly not in any structural sense.

The function of this chapter, then, is not to arrive at the truth of Chaloner, not to wrest *him* from the obscurity of the past, but to elucidate something *behind him*, as it were, namely, the matrix of meaning within which Chaloner's self-performance and style of conduct became legible both to himself and to others.[11] For while Chaloner's experiences and behaviors were undoubtedly distinctive, they were embedded in a condition of possibilities—a constellation of values and reasoning—that not only substantiated those experiences and behaviors, but validated analogous performances and stylizations of the self that likely continue to this day. This chapter, therefore, considers what it means for an individual to internalize an economic style of psychiatric reasoning, and how such a style could function as a hermeneutic strategy for self-intelligibility. The economic style was not simply an institutional lens for psychiatric medicine, but it was also a diffuse interpretative grid that could be personally adopted and deployed as a framework for subjectivity, through which some advantaged individuals could practice, stylize, and experience themselves and their apparent eccentricities, but only if they possessed the structural privileges to do so. While Chaloner was in the business of making money, his strongest aspiration was not actually to accrue wealth but to convince the world that he should be regarded in precisely the same way he saw himself, as an anomalous prodigy who was able to harness and direct his creative abnormality, his pathological originality, toward financial and business success.

Prophetic Speculations

Chaloner became the effective heir to the family estate after both his mother and father (the latter being US Congressman John Winthrop Chanler) died during his adolescence. Chaloner's initial wealth was derived in much the same way as it was for most New York elites, through inheritance.[12] Chaloner's early life followed the customary track of any wealthy New Yorker of his social standing. He received his primary education at the prestigious Rugby School in England and otherwise grew up between the family's Hudson River estate and New York City, the Astors being, in effect, the proprietors of lower to mid-Manhattan. After graduation from Columbia University and a brief stint in Paris, Chaloner met and hastily eloped with the American author Amélie Rives, shortly after the publication of Rives's semiautobiographical novel, which depicted her lurid and only barely fictionalized relationship with Chaloner.[13] The novel earned Rives a degree of literary notoriety but, together with the elopement, provoked the ire of the Chanler siblings, who blamed Chaloner for having "hurt the feelings of his entire connection on this side of the water."[14]

Chaloner and Rives divorced barely six years into the marriage, but not before Chaloner purchased a farm near Charlottesville, Virginia—close to the Rives family estate—where he would come to permanently reside, even after his divorce, much to the chagrin of his New York family and friends. In 1894, shortly after withdrawing from New York, Chaloner quickly became preoccupied with an apparent investment opportunity in cotton textiles developing just south of the Virginia border, at the northern most point of the Roanoke River in North Carolina.[15] There, a former Confederate soldier and state senator, Thomas Emry, along with some additional investors and engineers, were seeking to harness and profit from the waterpower of the Roanoke to try to reclaim, as one newspaper reporter put it, "the great money value that was going headlong down the stream never to return."[16] Emry was not the first or only investor hoping to take advantage of the many canals that had previously been dug and built into the Roanoke River.[17] And Chaloner's turn to southern textiles was also not an especially novel enterprise either, as wealthy New Yorkers had been steadily investing in and profiting from southern cotton since the 1850s.[18]

Nevertheless, just as Emry's financial prospects were grinding to a halt as a result of the 1893 financial panic, Chaloner arrived in time to invest in and absorb the struggling venture.[19] He established a holding company that acquired control of Emry's newly constructed power station, through which he could oversee the production of knitting mills along the river. Chaloner's ambitions, however, extended beyond the production of a few mills; he wanted to establish a textile factory town, what would eventually become the city of Roanoke Rapids. And by the end of 1895, he was well on his way to doing so. "A new manufacturing town," a local newspaper reported, "has sprung into existence like magic," underwritten with a "big pile of capital," and with Chaloner positioned as the town's "chief promoter."[20]

Chaloner likely took a great deal of pleasure in how he was described in the local press—"one of the cleverest and most intelligent gentlemen I ever met," as one reporter put it.[21] The speedy and ostensibly successful development of Roanoke Rapids became something of an aspirational beacon of economic optimism. Chaloner's enterprising ambitions were swept up into a regional economic zealotry, and soon he and his associates were extolled as "southern industrial heroes," successful in ways that few industrial venturers hitherto had been. "Within a few more short months or years at most," a reporter opined, and by virtue of people like Chaloner, "the Southern made cotton fabric will contest the claims of those made anywhere else in the world"; it was only a matter of time before the "South will dominate" the textile market.[22] In Chaloner's mind, these public accolades stood in stark contrast to the quiet admonishments and disapproval that he endured from his New York family. It was a pattern that he would see repeated just a few years later, when a New York court would deem him legally insane even though a Virginia court would subsequently and overwhelmingly rule in favor of his competence. Eventually, Chaloner would fully disidentify as a New Yorker and come instead to regard himself as a bona fide gentleman of the American South, not only legally and economically but, as we will see, in the strongest racialized terms.

Nevertheless, a year into the venture, without any indication of tangible returns, Chaloner felt considerable exuberance over his belief in the imminent success of his southern investments, even though the project was demanding a significant amount of underwriting, more than Chaloner had at his disposal. It was not just a matter of constructing factories but also of laying the groundwork of an entire labor town, from workers' housing to general infrastructure. So, in addition to his own heavy investments, facilitated in part by borrowing against his Astor-inherited New York real estate, Chaloner convinced several of his siblings to buy into his project.[23] Still, his enthusiasm remained unabated, and was even amplified by his confidence in several novel schemes he had prepared, which he was certain would radically augment his textile enterprise.

The first, and the one about which he felt most optimistic, was a self-threading device, an invented prototype, for sewing machines.[24] Having bought the invention just in time to have the prototype exhibited at the World's Columbian Exposition in Chicago in 1893, Chaloner would go on to claim that he received four thousand orders of a device that, at the time, could not yet be mass-produced.[25] "As regards the Self Threader *I am a prophet*," he wrote to his ex-wife in August 1896. Chaloner was convinced his device was compatible with an assortment of sewing machine brands, including the Singer brand, which was almost single-handedly responsible for making sewing machines one of the first mass-produced consumer goods to dominate both domestic and international markets.[26] The self-threader would not only yield incredible profit on its own, but it would revolutionize his factory town by substantially augmenting production. "In [self-threader] affairs," he professed, *"I'm a prophet who is willing to back his prophecy by cash."*[27]

A second "scheme" concerned a Chanler family charity, the St. Margaret's Home for orphaned young women in Red Hook, New York, which Chaloner had become responsible for overseeing.[28] While the home normally prepared young women for domestic service, Chaloner hoped instead to train the women for and subsequently redirect their labor into his textile factories. He claimed that transforming St. Margaret's from a "servant's nursery" into a "skilled operative training establishment" would guarantee for St. Margaret's graduates a social and economic independence that they would never otherwise attain in the confines of a "bourgeoise ménage."[29] Chaloner's paternalistic view of the factory as a site of social and economic refuge for young women, however, masked an ulterior economic motive: to extract a return on investment from the charitable obligations of educative philanthropy—that is, to recoup in the form of rent and labor value what was spent on the care and education of the young women.[30] Ultimately, Chaloner remained convinced of the anticipated lucrativeness of his combined southern speculations. "The price of land in [Roanoke Rapids]," he declared, "has advanced 50% in the past twelve months!"[31] It seemed that nothing could disrupt the pending success of his business ventures.

And yet, as precipitously as Chaloner had initiated his enterprise in cotton textiles, it seemed to many that after less than two years of financial development, he abruptly abandoned all of his efforts. A severe falling out with his siblings, particularly his brother Winthrop, who retained partial control of the factory town, had only exacerbated his already strained relations with other members of the governing board of the holding company.[32] In January 1897, Chaloner suddenly relinquished control of his Roanoke holdings and secluded himself within the confines of his Virginia estate. From an outsider's perspective, it appeared as if Chaloner had recklessly abandoned his investments, despite the money that was still so precariously on the line. Behind closed doors, however, what he was at the time not yet prepared publicly to disclose was that in late December 1896 Chaloner had stumbled upon a discovery—a world-historical breakthrough, he believed, that demanded not only his full attention but that accordingly necessitated he leave in the hands of perfectly reliable proxies the management of what he simply assumed was the otherwise inevitable success of his businesses. Something more profoundly rewarding was afoot.

Chaloner had come to believe that he had inadvertently accessed a hitherto unknown and uninvestigated psychophysiological function of the brain—one that could substantially augment his normal mental capacities with a newfound level of creativity, inspiration, and even prophetic aptitude. It made him, as he would declare some years later, "the sole, only, and unique individual in the history of civilization, who has, so to speak, *harnessed said Unknown-Faculty of the brain*."[33] From Chaloner's perspective, however, what he had come across was as much a scientific discovery as it was a singular invention, a novel mental technology that presented significant financial prospects. When he first encountered it, Chaloner would later explain, he examined this unknown mental faculty as one would investigate a patent, by exploring the "state of the art" of the invention. "I found that it was necessary for me to make investigation beyond the interpretations of facts of psychology," he explained.

> In law the legal phrase expresses exactly my meaning—in patent law the phrase is "the state of the art." That means when a man wants to make an invention, say a steam engine, the first thing the patent examiner does is to find out "the state of the art," that is, how far has invention gone into the art of the steam engine.[34]

The X Faculty, as Chaloner would come to call it, was more than just a psychological aptitude; it was a technical innovation with clear economic implications. He approached his X Faculty just as he did the self-threader device, as a potential source of value, and in that sense Chaloner had not recklessly abandoned his commercial endeavors as his friends and family believed. His withdrawal from his Roanoke ventures at the end of 1896 did not mean that he had renounced the spirit of enterprise. Rather, he had simply redirected that spirit toward a new undertaking, the investigation and appraisal of his X Faculty. His economic ambitions had shifted away from the simple profitability of cotton textiles toward a potentially far more lucrative possibility in the form of an exceptional mental prodigiousness, a singular abnormality capable of generating boundless pathological value.

The X Faculty

Chaloner's most thorough account of his discovery of the X Faculty was never made public. It is found in a nearly eighty-page typewritten affidavit, completed in April 1901, for use in his defense in the Virginia trial to reassert his legal sanity after his escape from the Bloomingdale asylum. The document was titled "The X Faculty," with an addendum, which Chaloner wrote by hand: "Being a Setting Forth of the Experiments in Advanced Experimental Psychology Which Led Us to the Discovery of 'The X Faculty.'" Not all the medical witnesses who testified on Chaloner's behalf at the Virginia trial were personally acquainted with him, and so the affidavit functioned as the key piece of evidence on which those experts relied to generate their testimonies. The affidavit was therefore written for a specific strategic purpose, and must be read accordingly—as a means by which Chaloner could mitigate suspicions of insanity by making a case for the legitimate possession of a parapsychological capacity, and to portray himself in the most reasonable and acceptable terms possible.

The document was written in an autobiographical style; it provided an extended account of how Chaloner learned that he possessed the X Faculty, how he experimentally confirmed its authenticity, and what its scientific implications might be. But what most punctuated the document was how the discussion of the X Faculty was almost entirely underwritten by a tacit form of economic reasoning. The economic logic is evident straightaway, for Chaloner had made sure to indicate, just below the affidavit title, that the document was "copyrighted" (fig. 6.1). It was not just the affidavit alone that Chaloner was attempting legally to protect, but the faculty itself, over which he sought to claim exclusive legal rights, believing himself to be the sole proprietor and beneficiary of this new psychological innovation.[35]

Chaloner opened the affidavit by quickly situating the X Faculty within a recognizable lineage of psychophysiological and psychiatric research. "The above is the title," Chaloner wrote at the outset of the document, "which I have given to a force at present unknown to science, but which has been noticed by the following eminent psychologists, to wit, Dr. Carpenter, the originator of the phrase 'unconscious cerebration,' which is Dr. Carpenter's way of explaining the said unknown force, which I call 'The X Faculty.'"[36] Carpenter was one of the earliest architects of the concept of automatism, which, as I detailed in chapter 1, refashioned the category of human conduct, and the divisibility between mental health and pathology, according to an essential ambiguity. Chaloner hoped to avail himself of precisely this ambiguity, to convince readers that while a lay observer might view the X Faculty as a pathological

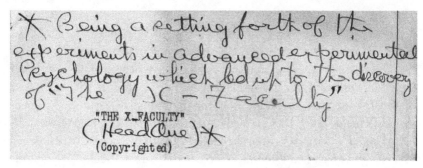

FIGURE 6.1. A detailed close-up of the title of John Chaloner's (copyrighted) "The X Faculty" affidavit. From John Armstrong Chanler, "The X Faculty," John Armstrong Chaloner Papers, 1876–1933, Box 6, "assorted manuscript drafts and notes, 1905," David M. Rubenstein Rare Book & Manuscript Library, Duke University.

condition, it was in reality a genuine parapsychological aptitude. And to do so, Chaloner relied upon not only the language of scientific psychology, to claim that his X Faculty could be experimentally validated, but also an economic style of reasoning, to propose that his X Faculty was tantamount to an innovation capable of accumulating wealth.

To pad his bona fides, Chaloner positioned himself and his exploration of the X Faculty within a constellation of research that had in the final two decades of the nineteenth century sought to establish a "scientific" program for the study of psychical phenomena; in addition to Carpenter, Chaloner invoked the writings of German philosopher and psychologist Max Dessoir and American and British psychical researchers such as Frank Podmore, Thomas Jay Hudson, and Frederic W. H. Myers. Advances in neuropathology and psychophysiology, particularly explorations of subconscious behavior and secondary personality traits, suggested that abnormal psychical phenomena were not occultist fixations but, rather, legitimate explorations of as yet unknown processes of the brain.[37] More than just a dabbler, Chaloner fancied himself a citizen scientist at the very least, eventually declaring, categorically and unreservedly, "I am an expert alienist." His investigations into his X Faculty, combined with his institutionalization and a self-taught "year or two course in Krafft-Ebing and other world-renowned Psychiatrists" had sufficiently prepared him, he believed, to generate valid knowledge claims about psychological medicine.[38]

Chaloner first encountered the X Faculty in 1893, he wrote in the affidavit, when he began to experience an array of acute emotional states—depression, exhilaration, and serene calm—which seemed to anticipate and even portend fluctuations in his business affairs. After nearly two years of observing this

strange "system of daily premonitions," and its utility in directing his business to "a most prosperous condition," Chaloner was convinced that "he had
got hold of something entirely new."[39] These premonitions lasted for another
two years, and by the middle of 1895, it had gotten so that Chaloner "could
tell on waking in the morning whether he was going to receive an important
business letter that day—a letter containing either good or bad news which
required instant attention—or whether any occurrence of any nature good or
bad was to take place within his knowledge that day."[40] The X Faculty made
its earliest appearance as a quasi-prophetic gauge for assessing, predicting,
and directing the state of his business affairs, a general "barometer to navigate
[one's] daily course through life."[41] It was then that Chaloner realized he was
in possession of something completely novel, "beyond the farthest reach of
any branch of science."[42] Chaloner dubbed this newly discovered mental capacity, this psychological innovation, the "X Faculty" in homage to Wilhelm
Röntgen, who in 1895 similarly stumbled upon an invisible natural force, also
using the symbol X to mark the as yet unidentified nature of the X-ray.[43]

Chaloner began to experience his premonitory episodes during the same
period that he developed his Roanoke investments. The strength and apparent
predictive accuracy of the premonitions peaked at precisely the moment that
his business venture seemed to herald a favorable financial outcome. Between
1895 and 1896, Chaloner had become convinced that the success of his factory
town, along with his self-threader patent and other financial undertakings,
was all but guaranteed—a success he felt was due in large part to his "system
of daily premonitions." When Chaloner proclaimed, "I'm a prophet who is
willing to back his prophecy by cash," he was not being metaphorical; the
statement was an expression of an emergent belief that he was genuinely in
possession of an oracular ability that enabled, among other things, the capacity for financial prognostications. The Roanoke factory town and the self-
threader patent were not commercial *speculations*; they were, as far as he was
concerned, *prophecies* in the most literal and authentic sense.

By Christmas 1896, Chaloner continued to explain in the affidavit, his X
faculty displayed another novel trait. While engaged in a bit of note-taking,
Chaloner observed that "his hand, gently, but decidedly, increased its grasp
of the pencil and without the slightest conscious cerebration or volition on
his part began to write." The X Faculty, it seemed, was communicating with
him directly, and, as Chaloner put it, "to say that the present investigator was
surprised is putting it rather mildly."[44] No longer simply an affective premo-
nition, this strange mental process began to take on the characteristics of an
autonomous mental agency, "mental force, both volitional and cerebrational,"
which communicated with Chaloner, at least initially, through the medium

of automatic writing.[45] A "written dialogue" with the X Faculty ensued, and during that first conversation, the X Faculty "directed [Chaloner] in writing to put its prophetic powers, its powers of forecasting the future to the following test": to "go at once to New York City and buy a certain well-known stock on a margin."[46] It was a particularly fitting experiment, Chaloner believed, for a number of reasons, not least because "no more intricate, uncertain, or unknown quantity exists than the fluctuations of the stock market," particularly short-term fluctuations in fractional stock prices.[47] As he would later explain, a "successful prognostication" of a short-term fractional fluctuation would therefore confirm "bona fide prophesy [sic], forecast, or reading of the future."[48]

Chaloner was compelled to travel immediately and surreptitiously to New York in order to purchase $700 worth of stock on a three-point margin. And within forty-eight hours, he had cleared a profit of well over $500, affirming the "successful stock-gambling operation in Wall Street under the direction of the X faculty," which he would recount for years to come.[49] This "strictly practical and not unprofitable test of prophesy [sic]" only strengthened his conviction that the X Faculty was fundamentally linked to commercial and financial success.[50] If Chaloner could somehow learn to wield this newfound power, he would "have got hold of something as Scientifically startling as it surely was financially gainful," a "celestial gift of prophesy [sic]" that he could bestow on mankind as "a benefactor greater even than Prometheus himself."[51] It was precisely at this point, in the final week of December 1896, and after what seemed like the most categorical corroboration of his newfound prophetic aptitude, that he decided to extract himself entirely from his business affairs in order to focus on private investigations into the X Faculty. His departure from the Roanoke ventures, he wrote, was neither hasty nor negligent, as he had "arranged his financial affairs so as to be in a position to be effectively managed by intelligent and honest agents," before embarking upon an adventure into the "unknown sea of the mind."[52]

Chaloner went to great lengths in the affidavit to emphasize the care he took in the subsequent weeks to subject the X Faculty to rigorous experimentation, which he approached not only as a self-avowed student of psychology and philosophy but also as a practicing Episcopalian—and therefore with the explicit and ambitious hopes that the X Faculty could achieve what Chaloner believed was a long-standing aspiration in modern Western thought—namely, to unite scientific psychology with the philosophy of mind and religious doctrines of the immortal soul. As soon as his private investigations had begun, Chaloner reported that his X Faculty began to communicate with him directly, no longer through an intervening graphic medium but through direct access to Chaloner's unconscious, what he would experience as mental "inspirations."[53] These inspirations were akin to hypnotic suggestions, he reported, with the one main difference that they never compromised Chaloner's consciousness or mental autonomy (he would later use the phrase "anti-hypnotic-subconscious-suggestion" to describe the process).[54] He would always maintain that at no point was he ever subdued or manipulated by the X Faculty. "There is no contest between you and me," his X Faculty reportedly communicated to him. "It is understood that we each recognize that the other is supreme in his own sphere. You cannot make me act. I cannot make you act."[55]

What followed in the early months of 1897, Chaloner reported, was a set of revelations bestowed upon him by his X Faculty, increasingly baffling, though one way or another anchored to a pronounced economic rationale. The first was the surprising disclosure that numbers were individually gendered, much like grammatical gender, but in a manner that was obscured and difficult to detect. If Chaloner could eventually come to apprehend "this secret system of numbers," then he would be able to discern the "foundation of all the bewildering complexities which surround games of chance," including the most seemingly random sorts of games, like roulette or faro.[56] Unfortunately, Chaloner lamented, the theory of numeric gender was simply too opaque for him to fully understand.

So, instead, the X Faculty presented a second revelation, something a bit more straightforward but which began with a startling request, that Chaloner gather and transport hot coals from his fireplace to his window using only his bare hands. Chaloner cautiously complied, and only after four agonizing rounds of carrying out the request, his hands noticeably burned, did his X Faculty inform him that he could stop. What could possibly have been the point of such a request? After some deliberation, Chaloner concluded that what was being revealed to him was a moral lesson—in this case, on the importance of the role of willpower as the single most vital criterion and prerequisite for financial achievement. "Since the present," Chaloner wrote in the affidavit, "is essentially the business age of the world's history, the age whose key note, aim, and inspiration is business success, the study of what underlies business success is surely a legitimate one." Such an investigation was especially warranted given than "over ninety per cent of all men who go into business on their own hook, of whatever nature of business whatsoever, fail ignominiously."[57]

Through this strange and painful experiment, Chaloner explained, his X Faculty was able to reveal to him that the "secret key to success in business life is will-power," and that cultivating the art of business meant nurturing the power of the will, a faculty that bestowed upon the man of affairs the ability to resist reckless temptations and to wait for the right and truly profitable ventures.[58] Chaloner wrote that although his hands were badly burned, they healed quite quickly, likely owing, he surmised, to the curative aid of the X Faculty. Furthermore, as a reward for his troubles, the X Faculty conferred on him a gift, he reported, in the form of a physical transformation, to symbolize his psychical conversion. It informed him that in the following days, his eyes would change color from brown to dark gray, a transformation that Chaloner reported he witnessed over the course of three hours. "No explanation of the above change was given by the X Faculty," he wrote, "nor has any ever been given."[59]

By February 1897, Chaloner continued, the X Faculty appeared to be displaying a mind of its own, engaging in schemes to which Chaloner was not always privy. It asked him, for example, to invite an unnamed neighbor—a mathematician—to his home, whom Chaloner informed, by way of a "vocal automatism" purportedly guided by the X Faculty, that he had resolved the ancient geometric problem of squaring the circle, even though it was known at least among professional mathematicians that Ferdinand von Lindemann had demonstrated fifteen years earlier that *pi* could not be transposed into algebraic terms, ultimately proving that squaring the circle through algebraic means was impossible.[60] The mathematician unsurprisingly left mystified and somewhat irritated by the incident; Chaloner too was quite unsure as to the purpose of that particular ruse, and he recounted the event in the affidavit with a strong degree of narrative circumspection, to signal to the reader his own bewilderment at what would likely have appeared as a potentially troubling display of odd, perhaps mentally afflicted behavior.

It was then that the X Faculty made its most startling revelation, Chaloner explained, by notifying him that "he must prepare for trouble."[61] At the time, Chaloner did not understand what the revelation meant; only later did he realize that he was being forewarned of his impending incarceration at the Bloomingdale asylum, which would begin in barely a month's time. Chaloner did not linger in the affidavit on the details of his confinement, but he did describe one final development in the evolution of the X Faculty, which took place three months into his institutionalization, when he received another visitation from his X Faculty. It was then, Chaloner wrote, that the X Faculty for the first time revealed its true identity; it was none other than Satan himself, an admission that Chaloner presented once again with the circumspect neutrality of unbiased reportage, so as to mitigate the potentially pathological overtones of his concession that he was effectively in communication with the devil.

The X Faculty "in the Role of the Devil" was not, it turned out, the maleficent creature Christianity had long portrayed, since, as Chaloner reported in the affidavit, it sought only to assist him in his dire predicament, first by assisting him with his legal efforts to engineer his freedom from the asylum, and second by helping him pass the time by bestowing upon him a newfound literary proficiency.[62] Through the automatic mediumship of the X Faculty, Chaloner discovered that he could suddenly, and for the first time in his life, generate an array of literary compositions, from novels and plays to rhyming couplets and sonnets. His earliest compositions, he explained in the concluding lines of the affidavit, were so strikingly well-crafted that he believed he had made yet another lucrative financial discovery—that he had "so to speak 'struck oil'—struck poetic oil."[63] (Chaloner would eventually publish his literary compositions in the years that followed, though it is doubtful that they were in any way remunerative.[64])

What we can observe in the final analysis, then, is that from Chaloner's perspective his X Faculty was unfailingly coupled with the possibility of lucrative enterprise, unremittingly value-generating, it seemed, one way or another. Even in its most unsettling displays, whether as a request to carry hot coals by hand or in the ostensible guise of the devil, what the X Faculty consistently revealed about itself was that when it was not functioning as the direct source of business innovation, it was at the very least facilitating the conditions for financial success. Chaloner's perception and interpretation of his X Faculty were shot through with an economic style of reasoning, a will to recognize his self-acknowledged abnormality, which already appeared to border on the morbid, as a perennial source of value. Chaloner ultimately came to believe that his imminent institutionalization interrupted a venture that was as important scientifically as it was financially, that he had discovered the preternatural key to profitability in the form of a prophetic aptitude that could counteract what for others were simply the inevitable risks of an unknowable market, and would allow him to act with prudence and shrewd foresight rather than with the mere luck of a gambler.

Mad Ravings

Chaloner noted in the affidavit that during his early investigations his X Faculty cautioned him not yet to reveal this newly discovered truth about himself to anyone. Eventually, however, the X Faculty would become part of the public record, through news reports of trial proceedings as well as through a short pamphlet that Chaloner would publish on his X Faculty in 1911. The advice to keep the X Faculty concealed was undeniably astute, Chaloner pointed out, as lay observers would likely misinterpret what its presence signified, and instead of expressing admiration for Chaloner's new psychological talent, as he hoped, they would likely express concern over his state of mind. The latter was the outcome most anticipated by psychiatric experts. "Some persons exhibit eccentricities of thought, feeling, and conduct," wrote Henry Maudsley, "which, not reaching the degree of positive insanity, nevertheless make them objects of remark in the world, and cause difficulty sometimes when the question of legal or moral responsibility is concerned."[65] In some cases, wrote George Savage, "the conduct of the eccentric person is such that for his own sake he needs especial care, and to this extent he may be regarded as insane."[66]

Chaloner's abrupt departure from his Roanoke venture had already been a source of confusion and exasperation among his siblings, who were both financially and managerially invested in the project. When they were made aware of his seemingly erratic behavior and pronouncements, they decided to intervene. In late February 1897, Chaloner received a visit from his family friend, the architect Stanford White, whom Chaloner had recruited to design the town of Roanoke Rapids and who in Chaloner's absence acted as his legal proxy in the venture. White was accompanied by the New York physician Eugene Fuller, whose real identity was concealed. They convinced Chaloner to travel back with them to New York, under the false pretense of a social calling. Posing as something of an enthusiast in psychical research, Fuller prompted Chaloner, once they had arrived in New York, to tell him more about the "discovery" that Chaloner had made. It was a request with which Chaloner was more than willing to comply, and he even had another "experiment" ready to present, which he felt would be particularly appropriate for the occasion.

"I was informed by my 'X Faculty' that it would like me to go into a Napoleonic trance," Chaloner later explained. "It gave me to understand," he wrote, "that I would represent the death of Napoleon Bonaparte by doing so, and that my features, when my eyes were closed, and face, would resemble strongly those of the dead Napoleon Bonaparte."[67] It was doubtless an odd "experiment"—to determine whether, in a trance state, Chaloner's facial appearance would come to resemble Napoleon's death mask. And the oddity was, at least to some degree, not lost on Chaloner, who worried that the entire experiment, the "most daring" he had conducted thus far, would simply "prove abortive and ridiculous." He therefore treated the provocation to enter this so-called Napoleonic death trance as a test, he reported, simply to have Fuller confirm whether "the prognostications of the 'X Faculty' regarding my face and features strongly resembling those of Napoleon Bonaparte in death were accurate or false."[68] At the same time, Chaloner's self-association with Napoleon was not quite as innocuous as he initially and so judiciously and guardedly made it out to be. Some years later he quite publicly elaborated on what precisely he believed his X Faculty was attempting to communicate to

FIGURE 6.2. A composite rendering of John Chaloner as Napoleon, "as he would appear in the garb
worn by the famous conqueror." From "John Armstrong Chaloner Thinks His Resemblance to Napoleon
Will Aid in Proving His Sanity," *News Leader*, June 8, 1912, 1. Image courtesy of the David M. Rubenstein
Rare Book & Manuscript Library, Duke University.

him through that Napoleonic experiment. "[The X Faculty] told me there was
an extraordinary destiny before me—an almost impossible destiny, and that I
was the reincarnation of Napoleon Bonaparte."[69] Chaloner became so increasingly convinced of his growing resemblance to and affinity with Napoleon
that he sent photographic evidence of his Napoleonic likeness to newspapers.
One paper took advantage of the opportunity to create, apparently in good
faith, a composite rendering, which Chaloner treated as a prized possession
(fig. 6.2).[70]

Chaloner's prolonged self-identification with Napoleon evinced the same
ambiguity that framed public appraisals of his X Faculty. Napoleonic allusions

were noteworthy for the flexibility Napoleon embodied as an idealized arche-
type. On the one hand, he was the quintessential "Romantic genius."[71] As I
discussed in chapter 3, Mercier identified Napoleon as the paradigmatic "man
of business," and so it is not difficult to imagine how a self-styled "man of affairs" might seek to invoke the image of Napoleonic greatness.[72] On the other
hand, throughout the nineteenth century Napoleon had been a recurrent
object of pathological self-identification, the prototype for delusions of grandeur. As the historian Laure Murat has argued, Napoleon epitomized not only
an omnipotent and domineering "superman," but someone whose victories
were earned, achieved, through the power of a prodigious and even abnormal
will—that is, a conquering genius who was, by some accounts, more than a
little mad. Napoleon was, Murat writes, "the perfect example of the American
ideal of a self-made man," a virtuoso who straddled the boundary of singularity and abnormality.[73] To identify with Napoleon was already to identify with
an ambiguous figure, with someone who was indisputably grand but whose
grandeur was simultaneously delusional. In that sense, it was unlikely that
a Napoleonic self-identification could ever be *healthy*, that is, untinged by
some degree of morbidity. The question was not whether such an identification was healthy but simply whether that identification, and the pathological shade it carried, were warranted; and as Mercier's idea of the Napoleonic
"man of business" indicated, the entire metric of acceptability had come to be
determined on economic grounds. Chaloner's Napoleonic self-identification,
much like his X Faculty, would only be deemed an acceptable pathological
privilege if it could be viewed as tied to business success and commercial
prosperity; only then could such an abnormal trait be redeemed as a vessel of
pathological value.

But based on Fuller's befuddled and troubled reaction to the trance experiment, we can surmise that Chaloner's proposed affinity to Napoleon was very
much deemed an expression of delusion rather than of grandeur. When Fuller
returned, he was accompanied by the prominent New York neurologist Moses
Allen Starr, who examined Chaloner on the pretext that he was also an interested occultist—a "masquerading 'occultist,'" as Chaloner later called him—
and after witnessing yet another demonstration of the X Faculty, declared
Chaloner mentally unsound.[74] The self-proclaimed prophet was taken into
custody by order of a legal certificate of lunacy on March 13, 1897, and quickly
dispatched to the Bloomingdale asylum fifty miles north of Manhattan.

Bloomingdale itself had a long record of housing enterprising madmen,
as it were. By 1844, when ranked by profession, merchants and traders comprised the most numerous occupational group among the patients, followed
by clerks, and then "men of leisure and young men without employment."

This fact was observed by the New York physician Pliny Earle, author of several asylum histories in the 1840s and 1850s. "Without explanation," he admitted, "the necessary inference must be far from flattering to these classes." The groups in question "constitute no unimportant proportion of the population of the commercial and wealthy city of New York." However, what the data ultimately implied, Earle observed, was not that wealthier classes were more liable to illness but that they were simply more likely to "resort to this institution" when suffering from mental illness. "The great majority of persons whose pecuniary resources are limited," he wrote, "are taken to places where the expenses are less."[75] The inflated presence of merchants, traders, and other wealthy individuals on the patient rolls signaled merely that these groups could sustain the cost of their stay. That being said, although New York's wealthier commercial classes were, per Earle's estimation, no more prone to illness than any other inhabitants of the city, still their heightened presence in private asylums certainly bore out the psychiatric imaginary, fueling the conviction in the pathogenic effects of business and commerce that would become dominant, especially in the United States several decades later, while also, perhaps more implicitly, normalizing the relationship between pathology and enterprise in all its permutations.

Chaloner was diagnosed with "primary delusional insanity," the second most common diagnostic category at Bloomingdale, after general paresis.[76] From the perspective of his doctors, Chaloner's behavior at the asylum only confirmed their diagnostic suspicions. At the trial to determine his legal competency, which was instigated by the Chanler family and took place in New York City two years after his initial confinement, over the course of one afternoon on June 12, 1899, the Bloomingdale medical superintendent Samuel B. Lyon claimed that Chaloner's strong tendency to exaggerate his "prominent talents" had grown so strong that "he has now the delusion that his mental powers are almost supernatural." Two other Bloomingdale physicians also testified to Chaloner's peculiar conduct, particularly his bouts of "morbid mental excitement," when he would become "excited, effusive, and vociferous . . . , sometimes shouting at the top of his voice."[77] Their testimony was damning to be sure and did not even include the fact, as they were not aware of it, that Chaloner believed that through his X Faculty he was in direct communication with the devil.

Chaloner, on the other hand, was as confident in the status of his mental health as he was convinced that his confinement was the product of a grand conspiratorial plot, of which he was the victim, orchestrated by his family to acquire full control of his Roanoke venture. The asylum's governing board, Chaloner would later attest, consisted of Astors, Livingstons, Stuyvesants,

and Vanderbilts, family friends and even actual family members; several of
the hospital governors who ran in the so-called New York Four Hundred
doubtless rubbed elbows with his brothers and sisters over the years. In addition to the asylum's governing board and his siblings, Chaloner believed the
conspiracy against him went all the way up to the New York governor's office,
and that ultimately Bloomingdale was "run purely for money, purely on business principles," in order to extract revenue from patients with wealth.[78] He
reported that the institutional fees alone to cover the costs of his incarceration
amounted to $20,000, and that was to say nothing of the fact that his entire
estate had been put into committeeship.

As Chaloner saw it, Bloomingdale was a modern Bastille, and his personal
plight a veritable US Dreyfus affair (never mind the mixing of historical analogies).[79] The asylum, he claimed, was not only where New York elites deposited their unwanted relatives, sources of scandal and familial embarrassment;
it was also a weapon through which the wealthy could ruin their competitors.
Chaloner was not the first former patient to underscore, through invocations
of the Bastille, the threats US lunacy laws posed to the spirit of American re-
publicanism.[80] "There exists in the Metropolis of the United States," he wrote,
"in the center of wealth, and alleged culture and alleged knowledge, there
exists in New York City today an organized band of, we shall not say rob-
bers, but we shall say robber barons." They posed, as Chaloner saw it, such a
fundamental "menace to life, liberty, and property . . . over the heads of each
and every sojourner in the City of New York" that New York would soon
likely lose its financial edge in attracting investors, who would opt for less
treacherous cities, such as Philadelphia or Baltimore.[81] His incarceration was
ultimately effectuated, he alleged, for no other purpose than to dispossess
him of his wealth. "It seems that I am doomed to pass the balance of my life
among lunatics," he wrote in one of the few letters he was able to mail outside
the walls of Bloomingdale. "It appears as I understand it merely a matter of
friends or no friends rather than the fact of sanity or insanity."[82]

Lyon, unsurprisingly, interpreted Chaloner's conspiracy theories as part
of his progressive and paranoid delusional insanity, which included "suspicions of persecution by enemies."[83] Lyon told the New York court during
Chaloner's 1899 competency hearing that from the moment Chaloner arrived
at the asylum, two years earlier, he "kept copious notes; manuscripts two or
three inches in thickness," which Chaloner would vehemently guard, claiming
"it contained his whole case."[84] Chaloner, on the other hand, was convinced
that the testimony of the expert medical witnesses had simply been bought by
his family, and that the entire hearing, which did little more than reaffirm the
initial lunacy certificate, thereby returning him to the asylum where he had

already spent two years, was a sham and part of the very conspiracy he was accused of irrationally concocting.[85] Chaloner admitted that while the bouts of morbid excitability that the medical staff witnessed and testified to were real, they were not the result of madness but simply an effect of the "wear and tear" that the two years of "excruciating mental torment" in that "Hell-upon-earth" had aroused.[86]

There were aspects to Chaloner's institutionalization, however, that did not entirely cohere with his account of Bloomingdale as a "Hell-upon-earth." One crucial feature was "the utmost liberty" he was reported to have been granted. A *New York Times* exposé claimed that by 1900, he enjoyed a private asylum ward all to himself, complete with horses and a carriage attended to by a private servant. Chaloner may not have been fabricating the exorbitant costs of his incarceration; he may simply have been concealing the reasons why he was paying such a hefty fee, that is, in order to maintain a lifestyle within the asylum similar to what he had known before. It was a luxury that the asylum was able to provide, confirming Earle's earlier assertions that Bloomingdale catered to predominantly wealthier New York patients. "He had the freedom of the great estate which surrounds the asylum buildings," the article claimed, "and was allowed to go and come pretty much as he pleased."[87] Bloomingdale was not an enclosed penitentiary, and Chaloner was hardly considered a violent patient. He quickly earned the trust of the medical staff sufficiently enough to be able to take unaccompanied walks, even beyond the asylum grounds. It was on one such walk, in late November 1900, that Chaloner withdrew from Bloomingdale, never again to return.[88]

The Bloomingdale annual reports made no mention of Chaloner's escape; his patient record simply listed him as "eloped and . . . not found on expiration of 30 days."[89] It was initially presumed that Chaloner had fled to Roanoke Rapids.[90] In reality he took refuge in the private sanatorium of John Madison Taylor, physician and chief assistant to Silas Weir Mitchell.[91] Chaloner was familiar with Mitchell's clinic and knew Taylor personally, as Taylor had attended to Rives, when Chaloner and she were married, treating her bouts of nervousness with Mitchell's famed rest cure.[92] Spending an additional six weeks at a private clinic about twenty miles away under the care of another physician, Chaloner traveled south to Virginia at the end of the summer. He cannily foresaw that, insofar as he was an asylum fugitive, his sanity would never be believed, let alone legally reinstated, without substantial medical corroboration. With the aid of a newly acquired lawyer, Chaloner petitioned the Albemarle County Court in Charlottesville to adjudicate his mental competence in order to free himself, at least in some measure, from the judgment that he had received four years earlier in New York.

Restoration of a Prodigy

Conspicuously absent during the 1899 court proceedings in New York to determine Chaloner's mental competence were any discussions concerning the
status of his financial investments in Roanoke Rapids. His supposed insanity
was a foregone conclusion, overdetermined by the fact that he had already
spent two years in an asylum. But as it turned out, Chaloner's decision to invest in southern textiles proved, only in hindsight, to have been an immensely
worthwhile undertaking. The economic development of a powerful textile industry in the postwar and postagrarian so-called New South began in earnest
in the early 1880s, growing quickly over the course of the subsequent two
decades, particularly in North and South Carolina, Alabama, and Georgia.[93] It
was the flow of northern venture capital—like Chaloner's—that first bolstered
the southern textile economy, particularly in the 1890s when returns on investment could often peak above 15 percent.[94]

Textiles became an integral means by which southern states could shift to
a dramatically new industrial economy and establish a degree of economic
independence at a level comparable to their northern industrial counterparts.
By 1900, southern states had managed to eclipse in terms of growth their New
England factory rivals, sounding the death knell for New England mill towns
and garment manufacturing, and becoming the de facto center of domestic
cotton textile manufacturing, with a growing foothold in international markets.[95] In another two decades, southern mills would account for half of all
national textile production, increasing to nearly two-thirds by 1930.[96] Systematic anti-unionization efforts kept the costs of factory labor low, which meant
that northern companies would economically benefit by claiming a stake in,
if not altogether relocating to, southern markets.[97]

The relative economic success of the Roanoke venture functioned as an
implicit backdrop at Chaloner's Virginia trial, which began on November 6,
1901. The trial itself was strategically organized as an exhibition of character witnesses and new, sympathetic expert testimony effectively showcasing
Chaloner as the gifted man of affairs he always imagined himself to be.[98] The
trial was also an opportunity for Chaloner to personally mitigate rumors of
his supposed insanity with an overly dramatic public display of his innocuous
normalcy. He was "the picture of physical health" the press reported on the
day of his own court testimony, which he delivered "in a clear, concise decisive
manner and his bearing during the entire case was one of quiet dignity and
close attention in the details in hand."[99] An array of witnesses, which included
Roanoke investors, business associates, and factory workers all testified that
Chaloner was "in a perfectly normal condition as far as business matters are

concerned," that he was "a very careful man, a very prudent man, and managed everything well," and was ultimately a "shrewd, sharp businessman."[100]

The trial was additionally punctuated by a small selection of notable medical experts, some of whom were among the most renowned figures in American psychology at the time. Each attested in some form not only to Chaloner's sanity but to his extraordinary aptitude as both a businessman and, even, an investigator of psychical phenomena. Taylor, in whose sanatorium Chaloner took refuge after fleeing Bloomingdale, testified that while Chaloner possessed a nervous temperament, nothing better exemplified his mental competency than "the fact that he has always been a man of business," in the context of which, Taylor continued, Chaloner displayed his capacities in a way that was both "successful and judicious."[101] Taylor was in large part rehearsing the lingering traces of a professional dispute brought on by clinical neurologists against asylum psychiatrists on the matter of professional overreach and the tendency among asylum clinicians to excessively institutionalize patients who, until they proved to be real and viable dangers, must be allowed to enjoy "certain rights of liberty of speech and action."[102]

Four additional depositions, which attended specifically to the matter of Chaloner's purported X Faculty, were introduced into evidence. The first was from the Philadelphia physician Horatio Curtis Wood, who had in his own work criticized the medico-legal tendency to confuse prodigious intellectual aptitude with insanity.[103] Wood concluded, after an in-person interview, that Chaloner was a "very active businessman" and "always kept up to some extent his interest in psychology." Though Chaloner was "certainly mentally peculiar," in Wood's estimation, he "[did] not see on what ground [Chaloner] could be at present called insane, and would therefore testify as to his present sanity."[104]

A second deposition came from the psychologist Joseph Jastrow, whose 1905 book, *The Subconscious*, would garner the praise of Théodule Ribot, and who would later admit to having been strongly influenced by "the Freudian point of view."[105] For Jastrow, the crucial feature of Chaloner's X Faculty was its automatism—that it was an autonomous mental agency that expressed itself through forms of conduct such as writing. It was precisely this feature, Jastrow claimed, that most substantiated Chaloner's mental *health*, or at least most offset suspicions of insanity. Jastrow adhered to the belief that everyday life was permeated with a degree of normal automatic behavior. These automatisms, however, amounted to more than just expressions of habit or routine but a "mild dissociation in the normal consciousness." Even under "ordinary circumstances," that is, in states of perfect health, everyday forms of "dissociated consciousness" nevertheless managed to find avenues for expression in normal conduct. While automatic writing and speech were the least common forms of the ordinary "dissociations" of the mind, nevertheless such acts were "not to be regarded as a presumptive concomitant or indication of a mentally impaired or diseased condition."[106] "Mr. C.'s automatic writing," Jastrow concluded, "is the expression of a mental condition differing only in a very slight degree, and not in an easily recognizable manner, from his ordinary normal condition or the waking conscious condition of any normal individual." Hence the X Faculty was, as far as Jastrow was concerned, the expression of a "normal mentality," nothing more than a "mental peculiarity," which ought not be pathologized.[107]

In the context of the professional boundaries that defined turn-of-the-century American experimental psychology, Jastrow's deposition in particular came as some surprise. Jastrow had long been hostile to psychical research, and particularly averse to its investigation by amateur researchers, championing instead the psychophysiological underpinnings of professional experimental psychology.[108] Why Jastrow would have elected to defend the psychical investigations of someone like Chaloner is not entirely obvious. A plausible explanation is that Jastrow's opposition to psychical research was largely professional; policing the profession was not the same thing as condemning a private individual to an asylum. Another, somewhat more speculative, explanation concerns Chaloner's socioeconomic status. Jastrow may have been more inclined to endorse and even valorize Chaloner's idiosyncrasies—and if not endorse, then at least tolerate them—because Chaloner was a man of affluence. We can observe such a dispensation play out in a short correspondence that Chaloner initiated with Jastrow a decade after his trial. Chaloner reached out to Jastrow in the hope that the psychologist would once again validate another, and in this case recently exhibited, feature of the X Faculty. Chaloner had come to believe that he had developed the ability to move objects, in a limited fashion, using only his mind and sheer force of will. In newly resumed "experiments" with the X Faculty, Chaloner felt he now possessed "the power to draw heavy curtains and make them remain in a given position for hours by merely passing his hands over them."[109] In his reply, Jastrow was neither reproachful nor dismissive; his gentle skepticism was mitigated with an almost deferential politesse. He suggested to Chaloner that a material explanation was likely the cause—an undeniably mild and generous response to an otherwise outrageous claim.[110]

A third deposition in Chaloner's Virginia trial—perhaps less *expert* in this case—came from Thomson Jay Hudson, an American journalist and civil servant by profession but generally known as an amateur psychical researcher whose first and most prominent book, *The Law of Psychic Phenomena*, was published in 1893 (and somewhat scathingly reviewed in its academic coverage).[111] Hudson, who like Wood also conducted a personal interview, framed his testimony around what he took to be Chaloner's judicious and careful experimental inquiries into his mental powers. "In a word," Hudson explained, "the very antithesis of insanity is the cautious conservatism of the inductive scientists."[112] There could be no way, Hudson insisted, for the X Faculty to be the expression of madness and simultaneously the object of lucid and reasonable examination, epitomized especially by Chaloner's rejection of spiritualism in favor of psychophysiological explanations.

Whereas Taylor and others regarded business acumen as the sensible antidote to allegations of madness, Hudson believed the remedy to such allegations lay in the conduct of the experimental scientists, which, when performed by the nonprofessional, amounted to the actions of a cautious citizen scientist. According to Hudson, "Scientific conservatism constitutes the best possible evidence of perfect sanity."[113] Citizen science and business expertise were equally capable, it seemed, of mitigating suspicions of madness. Moreover, what Hudson's testimony seems further to imply is that scientific innovations could very well possess the same redemptive power that financial innovations did, the power to retroactively find value in an allegedly pathological source. The mad enterpriser may have had as an analogue the figure of the mad scientist; for both, the boundary between reverence and revulsion, virtuosity and pathology, was defined ultimately by the remunerability of the discovery or invention at hand. As Chaloner repeatedly insisted in his affidavit, the X Faculty was as much a valuable scientific discovery as it was a profitable financial innovation.

The final and perhaps most prominent deposition came in the form of a short statement written by the American philosopher William James, unquestionably an impressive testimony to obtain, likely secured through the distant acquaintanceship between the James family and the Astors, and linked somehow to James's brother, Henry, who was familiar with the work of Chaloner's ex-wife, Rives.[114] James did not conduct a formal interview with Chaloner, relying instead on the testimonies of the other expert witnesses as well as on "The X Faculty" affidavit in order to generate his deposition. Deeply influenced by French experimental psychology and psychical research, James like Jastrow affirmed the normalcy of Chaloner's automatic states, admitting "some of my best friends, of superior sanity and mental power possess this faculty. It is impossible to deduce from its presence any conclusions whatever detrimental to the sanity of the person who manifests it."[115] James admitted that automatisms were behavioral expressions that straddled the boundaries between the normal and the pathological, a fact that generated a great deal of

clinical confusion with respect to the psychiatric status of so-called automatists, that is, self-proclaimed mediums, psychics, and other paranormals. "It
is clear already," James wrote just a few years before Chaloner's trial, "that the
margins and outskirts of what we take to be our personality extend into unknown regions."[116] Whether those unknown regions were healthy or not, that
is, whether "we can rightly rank the automatist, either as insane, or as merely
a person in whom subliminal uprushes are unusually facile" was a matter that
could only be adjudged on an individual basis, and based solely on "the actual
messages given."[117]

What James observed in Chaloner's case, and of his X Faculty in particular, was simply a rarer instance of "mental automatism," that is, a normal
feature of the "unconscious" part of the personality "making irruptions into
the conscious part" in ways that were still not entirely understood by psychological science. James went so far as to declare that Chaloner appeared to possess "a strongly 'mediumistic' or 'psychic' temperament," one that he believed
Chaloner commendably investigated in a critical and systematic fashion, and
without the troubling invocations of spiritualism or the occult.[118] Chaloner
would invoke James's acknowledgment of his "mediumistic temperament,"
and use it for years to come as license to self-identify as a confirmed and
genuine automatist and medium, "upon the authority of the late Professor
William James."[119] "Psychology would be more advanced," James concluded
in his deposition, "were there more subjects of automatism ready to explore
carefully their eccentric faculty."[120]

More noteworthy than the sheer fact of James's involvement in the court
proceedings was how his appraisal of Chaloner's "eccentric faculty" reinforced
the tacit economic logics that suffused many of James's most well-known psychological theories. As scholars have noted, in *Principles of Psychology* James
depicted selfhood as a form of self-ownership, a proprietary overlap in self-
consciousness between what is "me" and what is "mine." Elements of Jamesian psychology recapitulated a preoccupation with "propertied individualism" and a market-contractual model of the self.[121] At the same time, James
also articulated a moral epistemology that some have interpreted as a form of
advocacy for the value of unpredictability and chance against the constraints
of a hard determinism, which James put forth not only by endorsing scientific probabilism but also by implicitly affirming the potential profitability of
"fortuitous contingency," or the economic value of risk-taking and financial
speculation.[122]

These diverse economic readings appear to converge in James's commitment to psychical research and to the exploration of abnormal mental
states. Uncertainty and contingency were fundamental characteristics of the

irruptive spontaneity of unconscious psychical processes. James argued that psychological investigations into "automatic performances" were valuable because of "the questions they awaken as to the boundaries of our individuality." Automatic states and behaviors emerged, James explained, "from a personality other than the natural one," thereby exposing the "foreignness" at the center of oneself.[123] To further elucidate the nature of these automatic and unconscious irruptions, James drew upon the writings of Frederic W. H. Myers, particularly Myers's extensive descriptions of what he called the unexpected "subliminal uprush" of nonconscious mental processes, which normally took the form of "an emergence of hidden faculty," a novel and hitherto unprecedented behavior or mental function that demonstrated an unpredictably "new and complex adaptation."[124] For James, the "subliminal uprush" was precisely what linked the phenomenon of genius to psychopathology; as I described in chapter 3, it was believed that such uprushes were responsible for the "new combinations" observable by both geniuses and madmen alike.[125]

The prospect of unknowable and unpredictable subliminal uprushes did not, however, countermand a proprietary conception of selfhood as self-ownership. In his reappraisal of the Lockean theory of person, James explained that the early modern epistemological fiction of the same man having different incommunicable consciousnesses had become a psychiatric reality. A person "may thus live two alternating personal lives with a distinct system of memory in each," James explained. What psychological medicine had appeared to confirm was that "one human body may be the home of many *consciousnesses*, and thus, in Locke's sense, of many *persons*."[126] Instead of emphasizing a single self-possessive individual, James would stress the merits of thinking in terms of "my total individuality," composed of a spectrum of consciousnesses and personalities that any given "self" could nevertheless be understood to possess.[127]

The Jamesian self could still be viewed according to the contractual model of self-ownership insofar as a total individual would also come to possess the products of their psychical alterity, the yield of their subliminal uprushes. The Jamesian self could be both a contractually oriented market actor and a reservoir of irruptive potential, a subject essentially composed, and therefore in possession, of unpredictable unknowns. What James perhaps unwittingly perceived in Chaloner, what he perchance gleaned from the X Faculty affidavit in particular, was that Chaloner was in many ways the emblematic Jamesian self. From James's standpoint, Chaloner was a mediumistic eccentric, positioned just at the borderland of madness and virtuosity, who not only experienced the "subliminal uprush" of a prophetic and enterprising X Faculty, but who understood those unpredictable irruptions as precisely his own, as belonging to him in the form of a proprietary right. After all, what James would likely have observed, marked as it was at the top of the very first page of Chaloner's affidavit, was that the X Faculty—whatever else it might have been—was in the very first instance "copyrighted."

In Search of Lost Investments

The Virginia court affirmed Chaloner's legal competency two weeks after the commencement of his trial, though it would be another eighteen years before his legal competency was established in the state of New York. For nearly two decades Chaloner found himself in the strange predicament of being legally sane in one state while still considered incompetent in another. Chaloner's path to legal competency in New York could not have been more circuitous, and only further reinforced his "eccentric" public persona. In 1909, a nearby neighbor sought temporary shelter in Chaloner's Virginia home, to flee the domestic violence of her abusive husband. When the husband pursued her, a clash ensued during which Chaloner shot the man dead.[128] Although a coroner's jury quickly exonerated Chaloner, the press was far less forgiving.

"Eccentric Astor Millionaire Slays a Virginia Wife Beater," read a headline in the *New York World*.[129] The article claimed that the Chanler family had institutionalized Chaloner because "they had been told by alienists that he was liable to commit murder while in a violent mood." The *New York Evening Post* was even more scurrilous in its coverage. Fresh in the mind of most New Yorkers was the recent murder of Chaloner's duplicitous business associate, the architect Stanford White, at the hands of Harry Kendall Thaw, the wealthy heir of a railroad and coal industrialist and likely sufferer of a serious mental illness, during whose murder trial the newly classified category of dementia praecox was first introduced into US legal and public discourse.[130] While the nation was rife with wealthy men who seemed to get away with murder, the *Evening Post* declared, Chaloner at least "had the rare foresight to have himself declared insane before he shot his man."[131]

Chaloner sued the *Evening Post* for libel, and in their defense, the paper's lawyers claimed that they genuinely believed that Chaloner *was* in fact insane, which had the inadvertent effect of making Chaloner's psychiatric status a matter of legal concern for the second time in the state of New York. Although the trial did not take place for another ten years, in 1919, when the day of Chaloner's testimony finally arrived, he took the stand just as he had done in Virginia nearly twenty years earlier and provided another deft public display of his innocuous and sympathetic normalcy. It took the jury less than an hour to find in his favor, which meant that the New York federal district court had tacitly recognized Chaloner's sanity, which the New York Supreme Court would go on to uphold in July of the same year.

In the years between the Virginia and New York rulings, Chaloner directed much of his energies into protracted legal efforts to recoup what wealth he could, which was difficult given how much of his family inheritance was tied up in legal committeeship in New York. It had become clear as soon as the Virginia trial had concluded that while Roanoke Rapids proved on its own to be a rather successful enterprise, Chaloner's particular holdings sustained considerable losses during the period of his institutionalization.[132] While he often maintained that his textile assets remained profitable, there were considerable rumors that his wealth was all but depleted.[133] And yet his confidence in what he believed would have been the certain success of his prophetic venture into southern textiles never diminished, and Chaloner seemed to offset the actuality of his shortfalls with fantastical accounts of his financial triumphs. In 1915, he published a fictionalized autobiography written as a three-act dramatic play titled *Robbery under Law, or the Battle of the Millionaires*—a self-glorifying portrayal of Chaloner's imagined heroism in the face of adversity and a romanticized and counterfactual telling of the financial triumph that he believed he *would* have achieved had he not been unjustly incarcerated. The main character, Chaloner's alter ego, was Hugh Stutfield, a "millionaire art patron and law writer," whose love interest, Viola, expresses surprise in discovering that Hugh was "such a variegated man of affairs . . . of such wide scope."[134] In one brief scene, Chaloner reimagined his Roanoke venture according to how he felt it should have played out. The character Hugh, like Chaloner, patented a self-threader device for a cotton textile factory that he owned. But instead of losing the patent, as Chaloner did when his committee sold it as well as some of his Roanoke holdings to pay for his medical and legal bills, in the play, the self-threader proved a financial success. "They went like hotcakes," Hugh reports to Viola, not only in New York and in North America but also throughout "the whole of Europe," "South America, Central America, and Mexico," as well as "India, China and Japan, New Zealand, Australia and South Africa."[135]

Chaloner remained embroiled in some form of litigation for virtually the remainder of his life. When he was not trying to reclaim his inheritance or sue newspapers for libel, he was immersed in a vociferous campaign to publicize the conspiracy of which he believed he was the victim, frequently outlining the details of his siblings' plot to institutionalize him and to take over his inevitably successful business by fixing and devaluing his company's stock value. His faith in the reality of his victimization was all but fortified after one of his first public declarations of the supposed conspiracy, which he delivered in the very town of Roanoke Rapids only a few years after the Virginia trial, where he was met with an alleged "storm of applause," the apparent "idol of the working people of that industrial city," as one newspaper put it.[136] His personal plight, he believed, was tied to the systematic state of legal deprivation to which nearly all psychiatric patients were subjected; and to substantiate his case, he assembled a 350-page compendium of international case law and newspaper reports detailing the violences perpetrated by asylums such as Bloomingdale. While he posed as a champion for the rights of psychiatric patients, Chaloner nevertheless felt that the greatest threat that asylums posed to patients was financial, that because of psychiatric abuses "the greatest millionaires may in a day's time be reduced to the level of paupers."[137] When one considers the many cruelties patients experienced at the hands of early twentieth-century psychiatric practitioners, from the removal of teeth and other organs to the inducement of fever states, injections of barbiturates and insulin, electroshock therapy, and lobotomies, the mere loss of immense wealth hardly seems a generally applicable or especially grim medico-legal transgression.[138]

Still, the legal vindications in Virginia and later in New York continued to sanction Chaloner's increasing turn to public life, even as those legal victories had the added effect of exacerbating conduct that became only increasingly anomalous and erratic as time went on. Some of the most unorthodox behaviors took the form of novel forays and projects. Perhaps the most alarming was a newspaper Chaloner launched in 1910, in the town of Roanoke Rapids, inauspiciously titled *The Confederacy and Solid South*, a periodical seemingly devoted to the vicious tenets of southern white supremacy and thankfully contained to only a single inaugural issue.

In his editorial introduction to the newspaper, Chaloner proclaimed that the purpose of the periodical was "to preserve and keep" an enduring memory "of the Confederacy, and a magnificent, unbreakable, victorious political machine as represented by the Solid South."[139] It was a peculiar venture for a native New Yorker to advocate, given that one of the primary political objectives of the solid congressional voting bloc of post-Reconstruction southern states was to secure Jim Crow policies and systematically disenfranchise and politically exclude Black citizens.[140] What could have possibly motivated Chaloner to want, through his newspaper venture, to defend and exalt the "glorious deeds" of the "defenders of the South"? On the one hand, the newspaper was clearly an overwrought display of affinity with a region of the nation that Chaloner felt had been a sanctuary after his institutionalization, among whose residents he found protection and empathy, and through whose legal institutions he first attained absolution. The newspaper symbolized through the rhetoric of regional animosity the rift he felt between himself and his New York family.

On the other hand, and more importantly, the paper was an unambiguous display of Chaloner's perceived racial self-identity. Chaloner had always been aware of the status of his affluence and knew himself to be "a man of ample means," one who "does not live by his hands," and "should he by any chance lose his fortune he has a profession—that of the law—to fall back upon."[141] What the newspaper venture reveals is that, much like his affluence, Chaloner was quite aware of his whiteness as well, that it too was a property he understood himself to be in possession of, an attribute he could perform and stylize as acutely, prohibitively, and with as much entitlement as possible, that is, according to the logic of American white supremacy. "We are a different race from those living north of Mason & Dixon's line," he wrote, and by which he meant to signal more than just the political disparities that divided northern from southern states. "The war between the States was a righteous war," he explained, and "we content ourselves with saying that the Confederacy was historically right in the struggle." With pronouncements such as these and others—as when he declared "we are opposed to indiscriminate immigration"—it becomes unmistakably clear that Chaloner experienced himself and his enterprising prodigiousness not only through the lens of his inherited wealth but through the lens of his whiteness, which, in the Jim Crow South especially, was a lucrative form of property precisely in its anti-Black and anti-immigrant accentuations.[142] The newspaper venture was one of the few undertakings in his life that revealed just how much white self-identity underwrote Chaloner's sense of pathological privilege, his determination and facility to view his abnormalities as value-generating attributes rather than economic liabilities.

And yet Chaloner's seemingly abrupt public embrace of overt white supremacy was met with an almost equally precipitous about-face when, in the following year, he published his first public account of his X Faculty, a short, twelve-page pamphlet titled *The X-Faculty or the Pythagorean Triangle of Psychology*—a text so different from and so harmless in comparison to his newspaper that it seems almost to have been written by a different person. *The X-Faculty* was a pithy philosophical manifesto, a mere fraction of the length of the 1901 affidavit and devoid of any autobiographical nuance. It was an effort on Chaloner's part to take back control of the rumors that had circulated in the press about his purported psychic abilities by recasting them as a short and easily digestible theory of the mind. "Mediums" and "psychics" all possess a peculiar quality, wrote Chaloner, namely, "a fourth automatism, which is variously regarded as the subconsciousness in action, the subliminal self in action, or the X-faculty in action." This "X or Unknown-Faculty," he explained, was present in all people, though most people were incapable of properly harnessing it. If people could successfully channel the power of their X Faculty, their subconsciousness, much in the way that Chaloner believed he had, then they could bring about a harmonious balance between the three dominant faculties of the mind—reason, morality, and desire.[143]

Yet much of *The X-Faculty* amounted to little more than a dense and abstruse attempt to engage in a level of philosophical speculation that Chaloner was neither prepared for nor particularly adept at. The discursive precision and circumspection that had characterized his earliest accounts of his X Faculty had begun to unravel into what increasingly felt like a rambling free association of ideas and reflections. Privately, while completing *The X-Faculty*, Chaloner illustrated the mental harmony that he described in the short book through a series of unpublished cardboard diagrams that, in various incomplete states, depicted the different faculties of the mind converging by way of an unmistakable X pattern formed by the intersection of drawn lines at the center of the diagram (fig. 6.3).[144]

While Chaloner continued to conduct private experiments on his X Faculty, filling up countless pages of writing pads with notes on his experimental observations, the nature of those experiments had become far less coherent than what he had described in the winter of 1896. One typical entry, dated May 26, 1909, reads:

> Began 5:28pm. Standing . . . swinging arms from side to side in front of body, arms rising higher as if raising said by a sailor pulling on a rope, suddenly an X force stiffens and pulled back body and raised the hands extending over the top of the head. . . . Stopped 5:48. Began 5:58. Short tips up and down—electric tingle in the right forefinger—"juggling" of the knees. . . . Stopped 6:15.[145]

Nevertheless, even as his expositions of the X Faculty were becoming more enigmatic and obscure, Chaloner was increasingly emboldened to publicize his psychical prowess more candidly than he had ever done, not only through published writings but through press conferences and self-organized public talks. Much of these public pronouncements, however, did little more than reinforce the fact that Chaloner's eccentricities were beginning to lose their prodigious luster. In one ad hoc press conference in Alexandria, Virginia, in 1912, to which he invited members of the Washington press, Chaloner announced that he had been receiving posthumous messages from a former acquaintance, an ex-Confederate officer named Thomas "Uncle Tom" Jeffer-

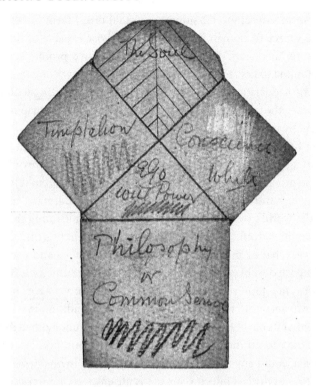

FIGURE 6.3. One of several similar cardboard diagrams that John Chaloner created circa 1911 depicting the different faculties of the mind. From John Armstrong Chaloner Papers, 1876–1933, Box 8, "photographs and postcards, 1902–1932 and undated," David M. Rubenstein Rare Book & Manuscript Library, Duke University.

son Miller. Miller, it turned out, resided in Hell, but through the aid of his X Faculty, Chaloner claimed that he had generated nearly sixteen pages of automatically transmitted and recorded communications from his departed friend—revelations about the nature of Hell and Satan.[146] Prominent newspapers, including the *Washington Post*, the *San Francisco Chronicle*, and the *Chicago Tribune* reported on Chaloner's "revelations" with equal parts cynicism and mockery.

Over and against these extravagant and religiously inflected admissions, Chaloner continued to extol the enterprising advantages and economic value of his X Faculty and its crucial role in prompting and sustaining business success. A year after his New York state competency was finally confirmed, and a week shy of twenty years to the day that he escaped from Bloomingdale, Chaloner delivered a public lecture on the Upper West Side of Manhattan on the topic of the subconscious and how it could open a path to worldly success.[147] The subconscious, Chaloner's eventual proxy term for the X Faculty, was the source of human inspiration, and those capable of harnessing its powers were "mediums," he insisted, endowed with a prophetic aptitude. Chaloner admitted to his audience that shortly after war was declared in early August 1914 he had written to the *New York Herald*, "forecasting with mathematical accuracy the political and geographical changes in the map of Europe to be wrought by the war."[148]

Chaloner hoped that his subconscious accomplishments might function as a source of inspiration. "My subconsciousness is nothing more or less than a Prophet," he proclaimed. "My subconscious can and has foretold the future to me day by day and for years."[149] (He was, he would maintain only a few years later, the world's only true modern prophet.[150]) Cultivating the powers of the subconscious would unveil the "Secret of Success in this life," Chaloner stated, a success that he once again formulated in financial and commercial terms. He imparted to his audience the very lesson he claimed his X Faculty taught him twenty-four years earlier. "Every man that ventures into business on his own account," he professed, "starts into a condition in which over ninety percent of its daring devotees meet absolute and unqualified defeat!"[151] Only by learning to attune to subconscious inspirations, and by fortifying one's willpower, could anyone hope to have any success in business life.

Chaloner spent the last fifteen years of his life engaged in a variety of philanthropic and artistic projects and ventures—which is to say, he was no longer singularly pursuing financial success, for the most part having cast aside that ambition.[152] Toward the end of his life, Chaloner's exaltations of his X Faculty began to diminish as well, as he confessed that his relation to his prophetic aptitude had always been far more fraught than he had ever cared to admit. "Mr. Chaloner has a grudge against his subconsciousness," an article in the *New York Times* announced. "He thinks it pursues and tortures him. For a quarter of a century he has argued and wrangled and bickered with it every night and day about the severity of his lot on earth, which he does not think he deserves, he said."[153] It was a complicated affaire de coeur, to be sure, Chaloner admitted in another public lecture. "No lover," he proclaimed, "ever pursued a mistress with more determined—a more unrebuffable—passion than the passion I, for twenty five years, have laid at the stony altar of Science!"[154] In a few short years, the *New York Times* would reappraise Chaloner's X Faculty proclamations as little more than "'subconscious' ramblings." In a 1929 article that reads more like tabloid fodder than an example of journalism, Chaloner is reported to have received communications from George Washington and Abraham Lincoln, who allegedly informed him that "the cause of the world's unrest is that the millennium is coming and that Christ's second coming is near."[155]

"The Right of Life, Liberty, and Eccentricity"

By 1930, Chaloner had returned permanently to Virginia, where he spent the last few years of his life in disordered solitude, consuming what remained of his fortune and borrowing heavily from the siblings against whom he had once so adamantly warred. He drifted slowly, it seemed to others, into a state of wretchedness, submitting finally to whatever affliction he had managed throughout his adult life to keep at bay, at least to some degree. He delivered his final prophecy to the *Washington Post* in 1934 from his hospital bed, ten months before his death. The "one time millionaire," the paper called him, presaged, among other things, the end of all financial debt and international hostility. "There will be no debts and no politics," he prophesied.[156] He died of cancer at the beginning of June 1935.

Among Chaloner's posthumous papers, loosely filed, undated, and unsigned, is a short, single-paged, typed-sheet press release titled "Mr. Chaloner Vindicated."[157] It was published as an article with no byline and a slightly altered title in a local newspaper in mid-May 1919, shortly after the New York Supreme Court finally ruled in favor of Chaloner's mental competency.[158] "His own evidence shows him to be an extremely eccentric person," the press release read, "but the New York court, like those in the South, declines to consider eccentricity a proof of insanity." While there is evidence to indicate that Chaloner had once been averse to being called an "eccentric," the press release took an entirely different tone toward the category of eccentricity, one of embrace: "If [the court] had [considered eccentricity a proof of insanity], what would have become of hundreds of learned psychologists, poets, editors, statesmen and other persons of unusual intellectual pattern? In Chaloner's triumph they triumphed all."[159] Throughout the pages of his memoir recounting his institutionalization at Bloomingdale, Chaloner had decried the threat his incarceration posed to his "life, liberty, and property," reinforcing an American tendency to formulate civil freedoms in distinctly economic terms.[160]

The press release, however, was committed to a slightly different set of inalienable guarantees: "The right of life, liberty *and eccentricity* is now assured. A man may believe in ghosts, mediums, mascots and hoodoos without fear of being shut up in an asylum. This is something to be thankful for at a period which is producing so many mental suspects."[161] Chaloner was undoubtedly the author of the piece, though, from his standpoint, through mediumistic collaboration with his X Faculty. There ought to be, the press release suggested, a national organization for eccentrics, one that would doubtless be "powerful in numbers, if not in genius." Over and against the apprehensions of psychiatric practitioners, the press release praised both the virtue and the value of eccentricity. The statement was an unapologetic assertion that a true man of affairs could be nothing less than an eccentric, on the very edge of madness; for it was precisely the tinge of pathology that made the ordinary economic actor into a modern prophet, a diviner of markets, a heroic man of action, beaming with a Napoleonic grandeur and a daemonic (indeed Satanic) formidableness. This press release was one of many such performances in which Chaloner sought to take, at least to some degree, the authorial reins of his own eccentricity, to reclaim and stylize his idiosyncrasies and to cultivate them as a persona and form of life, and to do so precisely by saturating his prodigious and presumptive abnormality with a marked degree of economic value. Chaloner could not help but confirm that the most effective way to recoup the potential risk of eccentricity was to recast it as a value-generating resource, as the instrument of an enterpriser rather than the contrivance of a madman, to recognize that the only worth that eccentrics can possess is the value they produce.

John Chaloner's aspirational efforts to recast his own apparent eccentricities as an expression of anomalous virtuosity were facilitated by and grounded upon an underlying economic rationality—a style of reasoning that Chaloner was able to appropriate, cultivate, and direct inward as a lens by which to view himself, not as potentially insane but as latently extraordinary. Although his X Faculty adopted political, religious, and even messianic inflections—such as are culturally and conventionally associated with the ravings of madmen— his supposed oracular subconscious was, from the moment of its "discovery" and unfailingly thereafter, anchored to apparent successes in market speculations and business ventures. Chaloner tirelessly struggled to showcase his X Faculty as a redemptive source of financial value; it was the means by which he could stylize in economic terms his idiosyncrasies and likely pathologies both for himself and for the expert communities and lay publics who sought in different ways to valuate his social worth. He may not have been a literal prophet in the eyes of James, Jastrow, or his Roanoke business associates; but he was, at least for a period, revered as a prodigious and psychologically gifted enterpriser. And while such an appraisal might have been couched in the language and framework of psychical research and experimental psychology, it was ultimately validated by virtue of the X Faculty's ambitious and certainly fantastical promise of prognosticative profitability.

Through his constant self-reflection, Chaloner recognized that he was not altogether normal, that he experienced subjective states that would likely have been classified not only as abnormal but quite unsound. But immediately attached to that recognition was the conviction that his abnormality was not a liability but an economic resource. Chaloner displayed what I have called a pathological entitlement, the self-avowed privilege to view one's pathology as a lucrative resource rather than as a costly debt. That privilege was certainly reinforced by the apparent success of his investment in southern textiles. But what animated his pathological entitlement in the first place were the vectors of privilege that suffused his life from the very start: his inherited wealth, his social standing, his masculinity, and above all his racial self-identity. For, as we observed, Chaloner was not *passively* white, as it were. Much as his wealth and social position, he recognized his whiteness as a holding that he possessed, a trait that he could mobilize and accentuate in white supremacist terms, a form of property that in the United States was already laden with contingent economic value, contingent insofar as it was historically staked on the violent exclusion and subordination of Black and other nonwhite economic actors.

Chaloner's self-stylizations were emblematic of an unmistakable form of entrepreneurial self-fashioning, a manner of casting oneself as a wellspring of lucrative pathological value. Although he was undoubtedly a singular historical figure, Chaloner's self-performance was also a precursory pattern for a style of conduct that would eventually recede into the background of behavioral possibilities and ultimately saturate the most inimical self-aggrandizing features of the entrepreneurial ethos, particularly in the United States. Max Weber believed that the rationalizing spirit of modern industrial capitalism was most expressly embodied in the forms of life conduct observable in the United States, even as American "sociation" also provided important contrasts to the rationalizing imperatives of economic life.[162] We might add that the United States was not only where capitalist rationalism first truly flourished but also where entrepreneurial madness gained its earliest and most robust purchase as a normative style of conduct, accessible according to the constraints of class, gender, and racial privilege. As an entrepreneurial ethos disseminates today as a dominant and increasingly globalized style of self, its pathological underpinnings continue to take pride of place in the American landscape most especially. A distressing contemporary echo of it was evident in the behaviors of Donald Trump, who, against accusations of mental instability that emerged after the first year of his presidency, affirmed via social media that he was "a very stable genius," as nothing else explained how, in his words, "I went from VERY successful businessman, to top T.V. Star . . . to President of the United States (on my first try)."[163]

Chaloner was certainly not the progenitor of a style of conduct that Trump (and undoubtedly others) would later come to adopt; he was merely an early practitioner. For what this chapter above all illustrates is that the economic style of recasting pathology as value was not specific to the *institution* of psychiatry but to a diffuse system of thought whose deployments were not restricted to clinicians alone. The effort to mitigate and redeem pathologies and abnormalities by imbuing them with economic value was a veritable "technology of the self," a hermeneutic practice that could be individually appropriated and internalized. What this suggests is that in its detachment from the institution of psychiatry, in its individual appropriability, economic redemption was not restricted solely to psychopathologies and behavioral abnormalities. The social dissemination of an economic style of psychiatric reasoning, its diffusion as a hermeneutic practice of the self after the end of the nineteenth century, meant that it was possible for even the most morally pernicious behaviors—sociopolitical irrationalisms of the highest order—to be regarded as generative of pathological value and therefore something to be recuperated, if not revered.

The Economic Reason of Madness

"The Economic Reason of Madness" deliberately inverts a phrase that has otherwise been used to exemplify the irrational and contradictory nature of capitalism. For the Marxist geographer David Harvey, exposing "the madness of economic reason" is part of a long-standing tradition of revelatory critique dating back to Karl Marx's unfinished *Grundrisse* manuscript from the late 1850s. It was there that Marx first outlined his apprehensions about the money form, particularly its capacity to unbind itself from commodities and to enter into a permanent and interminable state of circulation, an endless accumulation of compound growth such that instead of *representing* value money becomes a form of value and something to be desired in itself. Marx referred to such a possibility as a "madness" (*Verrücktheit*) that was nevertheless "a moment of economics and . . . a determinant of the practical life of peoples."[1] The supposedly pathological nature of capitalist forms has only become more entrenched, Harvey explains, as the world continues increasingly to be "held hostage to the insanity of a bourgeois economic reason," which, above all, misrecognizes limitless accumulation for harmonious growth and well-being.[2]

Such invocations of economic "madness," however, are more than simply figurative exaggerations; they are sincere attempts to underscore both the structural disjointedness of capitalist forms and the pathogenic virulence of social relations that such forms beget. The Frankfurt School psychologist and psychoanalyst Erich Fromm felt that Marx in particular had recognized the psychiatric morbidity of the logic of capitalist relations as early as 1844, in *Economic and Philosophic Manuscripts*, when Marx first introduced the concept of alienation, the ontological schism that cleaved human beings from their authentic species-life.[3] Marx's early depiction of alienation was "the most fundamental expression of psychopathology," Fromm insisted, and therefore as much a medical as a political-economic critique. For Marx, alienation was the predominant "sickness of man," and it was precisely such an observation that linked Marx to Sigmund Freud, the two of whom were, according to Fromm, the preeminent diagnosticians of the pathologies of modern human life.[4] "Surveying the discussion of Freud's and Marx's respective views on mental illness," Fromm wrote, "it is obvious that Freud is primarily concerned with individual pathology, and Marx is concerned with the pathology common to a society and resulting from the particular system of that society."[5]

As Marx noted in *Manuscripts*, alienation was all but guaranteed by the earliest divisions of labor; it was, in other words, a historical inevitability, a conclusion that perhaps points to an unstated *psychiatric* legacy in Marx's writings, quite possibly derived (as were many of Marx's early concepts) from the philosophy of G. W. F. Hegel. In his *Encyclopedia of the Philosophical Sciences*, published and revised between 1817 and 1830, Hegel not only upended a long-standing silence in the history of philosophy on the topic of madness, but, by invoking the writings of Philippe Pinel, he posited "madness" (*Verrücktheit*) as a pivotal and necessary moment in the dialectical development of an objectively conscious form of subjectivity and, eventually, the very freedom of the mind.[6] Inverting the causality between health and illness, Hegel posited that madness was a metaphysical possibility that had to *precede* the emergence of health and rationality. We must be able to say, Hegel claimed, that a rational subject is capable in principle of becoming insane; the capacity to become insane, the "*privilege* of folly and madness" (*das Vorrecht der Narrheit und des Wahnsinns*), must be posited in advance of the actuality of health and reason, as their logical prerequisite.[7] The purely philosophical possibility of madness was a necessary precondition upon which rational subjectivity was grounded, even if very few of us in fact go mad. Anything, therefore, that was incapable in principle of genuine mental pathology—a computer, a non-human organism—could be called neither rational, healthy, nor free. (In the *Philosophy of Right*, Hegel added that alienation was not limited to madness but extended to legal, economic, and political deprivations as well.[8])

In the Marxist tradition, in other words, and on the strength of its Hegelian roots, the "madness of economic reason" was not a mere contingency but a historical certainty, a dialectical necessity on the path to emancipation, and a principle that extended beyond the remit of Marxian thought. For when Harvey deployed the phrase "madness of economic reason" to refer to the inherent irrationality of capitalism, he acknowledged that he derived the phrase not from Marx but from Jacques Derrida, who made use of it in a palpably non-Marxian sense. For Derrida, "the madness of economic reason" (*folie de la raison économique*) referred to Marcel Mauss's concept of the gift, an "aneconomic" origin and disruption to any and all economic systems.[9] The gift was a notion that Derrida employed to recast his famous "concept" of *différance* as the paradoxical condition of possibility and impossibility of any system, economic or otherwise—the gift being an inaugurating aporia or irresolvability, a "madness," at the heart of any intelligible system of exchange.[10]

According, therefore, to some of the most prominent philosophical and political-economic critiques, economic systems whether capitalist or "precapitalist" seem consistently and inescapably to yield one persistent feature about themselves: they are irrational—both mad and maddening. But over and against the madness of economic reason, this book has sought to elaborate an inverse possibility, namely, the *economic reason of madness*: the idea that while capitalism might indeed be irrational and pathogenic, madness nevertheless possesses, and is consonant with, an economic reason all its own. This is, after all, what I have meant all along by the notion of pathological value—the economic reason of madness—the idea that pathological states were transposed into economic terms and converted into value forms, which could then be appreciated or depreciated from a financial, rather than medical or moral, standpoint.

Over the course of this book, I have sought to recount the emergence of pathological value, to describe this "economization of madness" through a philosophically inflected history of psychiatric thought. The origins of that emergence, as I argued in the first two chapters, lay in a hermeneutic crisis that unfolded within psychiatric medicine around the problem of diagnosing the inscrutable and opaque social performances of borderland figures. While such figures served to substantiate psychiatry's professional appeal, they nevertheless introduced a diagnostic dilemma that conventional psychiatric reasoning about the health of the *mind* could not fully redress. Instead, psychiatric researchers and clinicians found a solution by turning to the status of conduct, by evaluating performances based on their social merit. Economic benchmarks, it turned out, facilitated determinations about the viability of social conduct far more readily than medical or moral standards, primarily because it was far easier to determine the economic advantage of a particular form of behavior than it was to adjudicate its medical normalcy or moral acceptability. Economic merit was simply subject to less interpretative uncertainty; the only question that mattered was whether wealth was maintained, gained, or lost.

This is precisely what an economic style of psychiatric reasoning entailed: adjudicating pathological conditions on the basis of their pecuniary translatability, while keeping open the possibility that states of madness could in many instances be compatible with sound economic action and, in some cases, even be a crucial ingredient for financial success. The economic style of psychiatric reasoning was, furthermore, as seemingly systematic as it was latent and diffuse, crossing not only national boundaries but institutional ones as well, since, as I endeavored to show in chapter 6, the economic style was capable of being adopted and deployed by lay observers as well as by clinical experts. What I tried most importantly to illustrate in the core chapters of the book, through diverse accounts of psychiatry's multifarious economic style of reasoning, was that when converted into an economic form, madness was adjudicated not in terms of its medical or moral implications but in terms of its economic promises and returns. As the different chapters illustrated, the economic value of madness, or its pathological value, could be either positive or negative, lucrative or costly. That difference was largely organized around a class boundary, with middle-class enterprisers and members of the propertied classes possessing more affordances to profitable pathological value than industrial laborers. The difference was also organized around racial and gender boundaries, as masculine white economic actors enjoyed far greater entitlements to lucrative forms of madness than their Black and nonwhite counterparts.

In this concluding chapter, I draw out what I take to be the culminating stakes of the story of the emergence of pathological value. More than anything else, these final remarks are intended to function as intellectual provocations, organized as variations on a crucial proposition that this book advances: that critical appraisals of capitalism must ultimately contend with the unsettling possibility that revealing capitalism's pathogenic irrationalism might not perform the castigatory check that it is otherwise supposed to accomplish. For madness does not signal the paradoxical outer limits of economic reason but, rather, a possibility quite compatible with it. Pathology and economic rationality are not locked in an agonistic tension; they are instead entwined like partners in a morbid dance, for despite its failings and incongruities, economic reason still functions as the great redeemer of madness.

In what follows, I draw out some of the implications of this proposition, and what it might mean not only with respect to the history and philosophy of psychiatry, but for social and critical theory as well. I end this chapter, and thus the book, on a relatively speculative though hopefully encouraging note, by considering what it might mean to think "beyond" the notion of pathological value. I do so by turning to the writings of Frantz Fanon, and his conviction that decolonial psychiatry could facilitate, at least in part, what he called the "reconversion" of colonized subjects, a prospect that I read somewhat liberally through Michel Foucault's notion of "spirituality," or the ethical practices of self-transformation. Fanonian psychiatry, when inflected through Foucault's writings, could be understood not merely as the medical effort to restore mental health but as a *spiritual* praxis of conversion. Reframing psychiatry in such a way, I conclude, as something akin to a "spiritual medicine," may very well hold the key to signaling what an "outside" to pathological value might possibly entail.

Psychiatry and the Social

Let me begin by asking: Is it justifiable to place so much stock in psychiatry as the key to unveiling the economic reason of madness? While economists have concerned themselves with the status of the irrational, the economic irrationalisms they invoke have tended simply to denote attitudes, dispositions, and forms of conduct that customarily fall outside the remit of a strict market rationality—emotions, moral and religious values, traditions, habits, implicit biases, poor judgments, and so on. Even the staunchest efforts to economize the gamut of human behaviors, including those behaviors that are "consistent with the emphasis on the subconscious in modern psychology and with the distinction between manifest and latent functions in sociology," have ultimately remained silent on the question of mental pathology as such.[11] Madness has been to the history of economic thought what it was to the history of early modern philosophy: a topic for the most part to remain silent about, to avoid, and essentially to disavow as ultimately irrelevant to the enterprise at hand. As Foucault writes with respect to Descartes's treatment of madness in the First Meditation, "Madness, quite simply, is no longer his concern."[12] Reason's other was not pathology but ignorance, foolishness, and perhaps absurdity. Insanity was a concept that did not need to be addressed, and that much seems largely true with respect to modern economic assessments as well.

If madness has exclusively been the domain of psychiatric inquiry, then its underlying economic rationality can only be wrested from the depths of psychiatric thought itself. It is for this reason that I have approached psychiatry as something akin to an *epistemic matrix* (in contrast to Thomas Kuhn's sociologically narrower "disciplinary matrix"[13]), as a diffuse system of thought rather than as a profession or institution, in which I have sought to unearth a latent style of reasoning. Styles of reasoning are not necessarily contained within national or even institutional boundaries—as I hope the previous chapters have sufficiently evinced—and what they reveal above all is that diagnostic practices are beset by deeply embedded norms that are not always, indeed rarely, manifest explicitly. The norms in question here were the set of economic vectors that infiltrated and eventually subsumed psychiatric valuations of mental illness. I have hesitated to call such norms an "ideology"—or have done so only guardedly—primarily because the term "ideology" denotes an occlusive function, one that would suppose that an economic style of reasoning covers over a more authentic material reality. I would insist, however, that an economic style of psychiatric reasoning does not conceal a more fundamental truth about madness but that, instead, economic valuations have recast the very ontology and reality of pathological states.

To suppose that some states of abnormality were adjudicated based on their ostensible economic worth introduces a number of scholarly implications with respect to the history and philosophy of psychiatric medicine. First, it makes a strong case for claiming that medical conceptions of mental health and unhealth have largely centered on the performative dimensions of social life—that, in the strongest terms possible, the propriety of conduct has always been, and likely will continue to be, the operative concern for psychiatric medicine. Hence, psychiatry has always been a form of practical reason, a moral discipline, the evidence of which lies in how readily we conflate moral judgments, such as good and evil, with medical appraisals about the status of person's psychological health. As I proposed in chapter 2, psychiatry's great contribution to the history of the human sciences was that it declared that there could be such a thing as a madness of conduct alone, and thus a medicine of the moral order whose function it was not simply to mend a broken mind but to ensure the organizational coherence of the entire social body. This is admittedly not a novel observation; Foucault asserted as much in *History of Madness* when he described how throughout the nineteenth century, madness "took on meaning in social morality," that it was "imprisoned in a moral world."[14] Foucault was, as is now well known, quite critical of psychiatry's self-purported function as a moral technology, in contrast, say, to his former mentor and eventual professional adversary, Henri Ey, who is generally counted, along with Jacques Lacan and Eugène Minkowski, as among the foremost psychiatric researchers of postwar France.[15]

For Ey, regulating the moral world was *precisely* psychiatry's function, its metaphysical imperative being entirely devoted to the "the *pathology of freedom*."[16] There was no better curator and technician of moral liberty than the psychiatrist. "From this point on," Ey wrote, "the function of psychiatry in the social sciences is to guarantee human values, insofar as it becomes aware that they are lacking in some people. It guarantees them by causing them to appear."[17] Ey's commitment to psychiatry's moral province made him especially antagonistic toward anything remotely smacking of an antipsychiatric attitude, which is why he saw in Foucault's *History of Madness* little more than a form of "psychiatricide."[18] Even today, psychiatry's penchant for moral regulation has hardly abated; in many cases the profession still considers itself a moral and even political panacea and social defense, as the entire field of political psychology from *The Authoritarian Personality* to recent efforts to pathologize the erratic behaviors of Donald Trump collectively attest, answering Freud's somewhat ambivalent conjecture at the end of *Civilization and Its Discontents* that "we may expect that one day someone will venture to embark upon a pathology of cultural communities."[19] More critical scholars of the history of psychiatry tend to agree that the field has come to possess a strongly moralizing function, though they view it far more unfavorably, as an apparatus for behavioral disciplining, a system of social control.[20] Either way, both advocates and critics of psychiatry's moral administration implicitly recognize that psychiatry has exceeded its remit as a medical field and become, or has at least endeavored to become, a veritable science of the social, a way of making sense of society's ailments and of offering normative pronouncements for possible and aspirational solutions.

By underscoring the performative dimensions of mental pathology, I too have sought in this book to emphasize psychiatry's function as a tacit theory of the social order, without, however, defending psychiatry's moral mandate or simply impugning it as a form of social engineering. Instead, I propose that psychiatry has remapped the very ontology of social relations such that our performances, dispositions, and even attitudes have become medically valuated in both moral *and economic* terms. Psychiatric thought has transformed into a diffuse hermeneutic by which we make sense of our behaviors as well as those of others. Styles of psychiatric reasoning—its economic style above all—have become among the most dominant hermeneutic frames by which social meaning is grasped and made sense of. Psychiatric styles of reasoning are crucial matrices of social intelligibility, and in that sense, psychiatry functions as a social science, a theoretical framework by which to decipher and medically recast social relations and modes of social legibility and self-expression.

But styles of reasoning are not effectuated by clinicians and medical experts alone, and so we are not simply the objects and targets of psychiatric systems of thought. Like John Chaloner (perhaps far less exaggeratedly) we are the perpetrators and perpetuators of psychiatric reasoning as well—a style of reasoning that has imbued medical and moral assessments of social action, particularly aberrant action, with an undeniable sense of economic value. In other words, we propagate styles of psychiatric reasoning without necessarily realizing that we do so, as latently perhaps as psychiatric clinicians themselves. There is a long-standing tradition in the history of social and sociological theory that has sought to corroborate Emily Martin's assertion that "in their everyday lives, most people have various degrees of awareness of reality and of the consequences of their actions, various degrees of 'reason' in their decisions and opinions."[21] The American sociologist Erving Goffman once claimed, "What an individual says he does, or what he likes that he does, has very little bearing very often on what he actually does."[22] The latency of styles of reasoning does not necessarily pertain simply to the covert intellectual labor of a community of medical and scientific experts. It can denote customary forms of social reasoning that operate at the level of what the sociologist Pierre Bourdieu called a "habitus"—that is, at a level of social embeddedness so entrenched and invisible as to remain quite automatic.[23] We can call these latent forms of customary reasoning the effects of a kind of social unconscious, but one that does not need to take on a psychoanalytic inflection. "We realize in everyday conduct and experience," wrote George Herbert Mead, "that an individual does not mean a great deal of what he is doing and saying."[24] These sorts of social incitements, what Émile Durkheim described as the coercive compulsions of social facts or what Gabriel Tarde called the laws of imitation (and for which Foucault gave us the almost always misunderstood term "power" to denote the dynamism at the heart of all social relations wherein one's performances simultaneously and inevitably affect and steer, however minimally, the conduct of others) all describe a similar phenomenon: that our conduct is often guided by vectors of force irreducible to conscious deliberation, biological impulses, internal drives, or repressed ideas.[25]

Pathological value is, therefore, the product of a social incitement to interpret aberrant conduct according to an entrenched valuational hermeneutic, an economic style of reasoning, the propagation of which we are all collectively complicit. Pathological value marks a significant transformation of the very ontology of social life such that our everyday performances have been modified according to a new reality wherein moral and medical appraisals are at once economic, and vice versa.

Acts of Valuation

Throughout this book, I have espoused a capacious construal of the concept of value and processes of valuation, in an effort to explore the overlaps and conflations of what are arguably the three most dominant forms of social valuation in the modern North Atlantic: the moral, medical, and economic. In contrast to the political domain, the sphere of the social is the context where norms, values, and conventions of customary reasoning circulate and converge; and within the social body, medicine, morality, and economic conduct typically exhibit the strongest forms of normative value. Therefore, acts of valuation must be understood as exceeding simple monetary appraisals, for they are the means by which social phenomena are assigned not only a sense of meaningfulness but also significance and worth in *every* normative sense.[26] Valuations are the processes by which an otherwise shared reality comes to *matter* for members of a social body in specific ways, value being what gives social experiences their coherence and sense of consequence.

The notion of *pathological* value, however, is a way of proposing that the moral, medical, and economic divisions of social value are not evenly partitioned and isolated but have become indistinguishably amalgamated. Pathological value introduces a particularly cynical outlook with respect to this amalgamation, by suggesting that economic appraisals are not only capable of subsuming other forms of social valuation but that they can reverse the conventional meanings of medical and moral worth. While mental illness, for example, is customarily assessed as a state of unhealth and often inflected, even today, with a stigmatizing sense of moral shortcoming, one often mitigated through the defusing language of medical risk, economic valuations nevertheless possess the power to reverse these customary moral-medical disapprobations by imbuing certain pathological states with a sense of fruitfulness and remunerative advantage. Pathological value signals the redemptive power of economic forms of valuation, that economic reason can transform madness into a vaunted and potentially desirable possession and, in so doing, represents something akin to an inversion of a conventional Weberian paradigm.

"The religious valuation of restless, continuous, systematic work in a worldly calling," Max Weber argued in *The Protestant Ethic and the Spirit of Capitalism*, "must have been the most powerful conceivable lever for the expansion of the attitude towards life which we have here called the spirit of capitalism."[27] If a Protestant moralism first animated a "capitalist way of life" and capitalist style of "life conduct" (*Lebensführung*), then we can say that pathological value represents the reverse scenario: that capitalist economic reasoning has come to permeate and reanimate noneconomic forms of moral judgment and conduct.[28] And it did so in large part thanks to the mediating role that psychiatry played in layering over the moral dimensions of social performances with a strong medical sense. One of the effects of the historical dominance of the human sciences in regulating and adjudicating moral life is that moral truth, in its attachment to medical reasoning, was ascribed a strong sense of scientific veracity. Medicine became the scaffolding through which conventions of moral value could be susceptible to an almost thoroughgoing economization. What was once judgable as unhealthy, unwholesome, deficient, and detrimental could be reappraised as advantageous, lucrative, and therefore worthwhile. The economization of madness must be

CONCLUSION

regarded as part and parcel of the economization of moral value in general. Any kind of social valuation can become thoroughly economized, and in such instances, objects and states that are conventionally devalued morally and medically can be overridden and redeemed as positive and productive—in a word, profitable—possibilities.

In his 1921 essay, "Critique of Violence," the cultural theorist Walter Benjamin proposed the unsettling conjecture that the entire sphere of moral relations had become subsumed by a juridical logic, a furtive legalism that, much like the law itself, successfully operated through the threat of violence. In the modern secular world, the moral order was essentially a proxy for the law, and moral propriety was achieved not through ambitions of virtuousness but in an effort to avoid castigation and punishment.[29] In a similar critical vein, I would add that moral value has become not only a juridical proxy but an economic one as well, that our moral economies have in many cases become more *economic* than ethical. Moral conduct can often be prompted not only by the desire to escape rebuke and retribution but also by the aspiration to accumulate moral value as a form of symbolic property that can be possessed in greater or lesser quantities. Does signaling virtue not at times serve as a proclamation of one's ethical net worth, the conspicuous consumption of one's purported moral capital?[30]

If pathological value means that even the most morbid, errant, and ignoble performances can be reappraised and redeemed, then we must conclude that the entire field of moral conduct is shot through with the possibility of near-total economization. In some ways, pathological value rehearses, in a modified form, Friedrich Nietzsche's bleak and infamous conjecture from *On the Genealogy of Morality* that modern moral propriety is animated by, among other sources, a fundamentally transactional logic, an abbreviation of the debtor-creditor relationship, and thus a powerful economic incitement.[31] To that end, pathological value introduces a troubling ambivalence with respect to the critique of political economy. Taking note of the *madness of economic reason*, or capitalism's irrational contradictions, may do little to mitigate the injuries of capitalist structures and relations. If madness possesses its own economic rationality, then accusations of madness do not quite carry the admonishing and nullifying force otherwise believed. If critiques of capital are executed in the name of moral right, then perhaps they do little but function as a means by which to augment one's moral net worth. Accordingly, we end up enjoying the symbolic returns on a form of economic reasoning at the very same moment that we strive to criticize it. It is one thing to critique and morally reject capitalism and quite another to live one's life in tenacious and open defiance of it.

Irrational Redemptions

The notion of pathological value, or the economic reason of madness, poses yet another challenge with respect to the theoretical aspirations of a rational society.[32] The great threat to the ideal social order is not represented by benign nonrational behaviors such as biases, emotions, traditions, and habits but by acute forms of sociopolitical irrationality typically cast in starkly pathological terms. Over the course of the twentieth century, for instance, the kernel of a potentially warped antidemocratic will and features of an authoritarian personality were increasingly believed to lie in incipient psychopathological disturbances.[33] Even forms of racism and racist violence have come to be framed as expressions of mental pathology.[34] And when not viewed as a threat to the social body, irrational conduct was otherwise construed through the lens of Cold War behavioral science, transposed into the rubric of computational processes and predictive algorithms in order to be filtered out as forms of noise and error that could be mitigated through calculative expertise.[35] With respect to the aims of a rational society, in other words, mental pathology has been viewed as either a germinal political risk or little more than informational residue—in either case, a spectral antipode against which to frame an imagined community where sociopolitical morbidities are either purged or populationally managed through an interminable dependency on pharmaceutical intervention.[36]

But if economic reason is the great redeemer of madness, what expiates pathological states and revalues them anew, then even the most perilous social and political irrationalisms will resist abolishment if they continue to confer pathological value. As I illustrated in chapters 3 and 6, even morbidly inflected forms of racialized hegemony and white supremacy thrived in a chiefly economic sense. The pathological value of racial capitalism unquestionably continues to grow today, and it is quite possible that the covert and genealogically inherited madness of the entrepreneurial form—especially given its normative appeal as an activity, attitude, and aspirational style of self—is partly what keeps the logic of racial capitalism alive and kicking. One reason why pathological value remains so resistant to moral-political admonishment is because both are animated by the same underlying economic rationality. Psychiatric moralism is especially compromised in that regard. Henri Ey's contention that psychiatry must function as the moral cure-all for modern society's ills is potentially shot through with an economic rationality all its own. In one of his final defenses of the profession, Ey contended that psychiatry's "mission" exceeded the simple remit of medical care. It was the duty of psychiatry, he wrote, to "enable and index . . . the maximum circulation of freedom [*maximum de circulation de liberté*] in society," which it ensured "not by the legislation of its mores but by the very organization of man."[37]

The "maximum circulation of freedom" is a striking way to describe psychiatry's extramedical ambitions, particularly given how closely it resembles the economist Milton Friedman's characterization of liberalism's political-economic aim, "to preserve the maximum degree of freedom for each individual."[38] Was Ey's conviction that psychiatry safeguard moral-political liberty simply an encodement of free market capitalism, that the "pathology of freedom" was ultimately staked on an *economic* liberty? If so, and in tacitly defending the position that freedom from mental illness was a guarantor of economic liberalism, Ey failed to recognize that madness can be just as remunerative as sanity, sometimes more so. If economic prosperity is the stated socio-moral ambition of psychiatric care, then it all but promises to fortify and reinforce the specter of pathological value.

Beyond Pathological Value

I end by asking: What would it mean to think *beyond* pathological value? As I described at the book's outset, freeing psychiatry from its capitalist imperatives may not actually purge psychiatry of the economic logic that underlies and animates many of its concepts. Consequently, thinking beyond pathological value may demand more than a modification to the economic structures and market imperatives of psychiatry as an institution and industry. I propose instead that positing a 'beyond' to pathological value means reimaging psychiatric thought and practice according to an entirely different ethical aim than the one that it has tacitly espoused since the moment of its professional formation. Psychiatry's most enduring ethical aspiration has been to ensure some manner or degree of freedom for the mentally ill—freedom, at least, from the subjugation of mental disease—and, therefore, the problem of freedom has long haunted the psychiatric imaginary. Freedom was the very event that marked the field's post-Revolutionary origin story, captured emblematically by Philippe Pinel's so-called humanitarian deliverance of the mentally ill patients at the Bicêtre asylum just south of Paris—an apocryphal fable, but one that nevertheless functioned as a powerful disciplinary founding myth.[39] Emil Kraepelin invoked the power of the narrative, that "on May 24th 1798, Pinel . . . removed the chains" from the patients at Bicêtre (even though Pinel had left Bicêtre for the Salpêtrière hospital in 1795). While Kraepelin conceded that the narrative was most likely an "invention," still "it reflects the general fact that it was the doctors who freed their patients from the chains."[40]

When Ey described madness as a "pathology of freedom," and psychiatry's function to ensure the "maximum circulation of freedom" in society, he too took the spirit (though likely not the facticity) of psychiatry's origin story at face value. But what Ey demonstrates above all is that psychiatry's deliverance has ultimately meant delivering both patients and nonpatients—that is, all users and subjects of psychiatric reasoning—into the structures and norms of modern political *and economic* liberalism. And so, the freedom that psychiatry typically ensures is one inevitably cast in a manifestly economic sense. In many respects, psychiatry continues to function even today, and not just in the North Atlantic but globally, according to an often-unstated appeal to freedom, to liberating patients from their suffering, re-conferring autonomy upon the alienated, and empowering them to partake in all the civil and economic liberties to which they are supposedly entitled. But must freedom continue to function as the ethical end of psychiatric thought and therapeutics? And if so, must it always be inflected in inescapably economic terms?

Ey was not alone in viewing mental illness as a "pathology of freedom." That very phrase was appropriated by Frantz Fanon, who early in his medical training was influenced by Ey's institutional authority in French psychiatric medicine.[41] Fanon accepted the formulation by Ey and his "French school" that "mental illness is presented as a veritable pathology of freedom" because it "situates the patient in a world in which his or her freedom, will and desires are constantly broken by obsessions, inhibitions, countermands, anxieties."[42] But where Ey's "pathology of freedom" gestured to a form of liberty commensurate with economic liberalism, for Fanon the freedom that madness forestalled was of an entirely different sort. Born in the French colony of Martinique, Fanon began his study of medicine, with a specialization in psychiatry, in the French city of Lyon. He quickly joined the psychiatric hospital of Blida-Joinville in Algeria in 1953—the epicenter of French colonial psychiatry—just a year prior to the onset of Algeria's war for independence from France, a political struggle with which Fanon's thought has become inextricably linked.[43] For Fanon the pathology of freedom was not counterposed to economic liberalism but to radical decolonial emancipation.

European colonialism represented for Fanon the enactment of a profoundly pathogenic form of political and psychical violence upon the minds, bodies, and social lives of indigenous populations. Consequently, Fanon's decolonial writings embodied the amalgamated attitude of both a physician and a political theorist, and from the standpoint of colonialism and its attendant forms of violence, the political realm and the psychiatric clinic were deeply interwoven spaces. To truly appreciate the annihilating devastation wrought by the colonizer, it was necessary to view political life through the lens of psychiatric health and unhealth and, conversely, to understand the psyche as an eminently political site.[44] The indigenous subject of colonial occupation was, as the social theorist Achille Mbembe writes, synonymous with "the figure of the patient."[45] This was a supposition that Fanon held in the strictest medical regard. "The human brain," he reportedly professed, "has enormous potentialities, but these potentialities must be able to develop in a coherent milieu." If any milieu, however, frustrates such a coherence, if it "does not authorize me to reply, I will atrophy, I will be halted."[46]

The political violence of the colonizer was always a form of the psychical destruction of the colonized subject; likewise, the therapeutic revitalization of the colonized psyche could only take place through the process of a decolonized form of political renewal. Psychiatry for Fanon, therefore, functioned as one component of a systematic practice of political emancipation, not only insofar as it was organized around novel forms of radical institutional psychotherapy, but because psychiatry could occasion a fervent self-transformation; and it did so precisely in its formal alliance to decolonial violence. For Fanon decolonial violence was a form of political insurgency that was both medically remedial and politically self-actualizing. In *The Wretched of the Earth*, Fanon would describe the practice of engaging in violent insurgency against colonial rule as a form of "absolute praxis" (*la praxis absolue*).[47]

Fanon's conception of violence drew strongly from Marx's early descriptions of labor as the practice through which individuals fulfilled the quintessential vital, social, and even ontological function of human nature: to create and produce. For it was through the laborious "creation" (*Erzeugung*) of the world and its history that human beings manifested "their own self-formation" (*seiner Geburt durch sich selbst*).[48] The Frankfurt School political philosopher Herbert Marcuse wrote that Marx's early conception of labor was "the activity through and in which man really first becomes what he is by his nature as man." Labor was an affirmational form of self-actualization, an act of self-creation, how "human existence is realized and confirmed."[49] Labor was therefore not only a manner of authentic self-creation but of revolutionary self-emancipation. "Man becomes free in his labor," Marcuse explained. "He freely realizes himself in the object of his labor."[50] Accordingly, for Fanon, it was imperative to conceive of decolonial violence as more than just a form of defense or rebellion. Violence was instead a self-creative and revolutionary *labor* in itself. "To work [*travailler*] means to work toward the death of the colonist," Fanon explained. "The colonized man liberates himself in and through violence."[51] The recovery of a fundamental and authentic form of labor, in the form of decolonial violence, was the means by which colonized subjects could attend to, and thereby reshape, their authentic human nature; it was the first instance of the reconstitution of oneself in light of the ravages of colonial brutality.[52]

At the same time, Fanon described decolonial violence as a "cleansing force" (*désintoxique*), a rehabilitative practice of "reintroducing man into the world" no longer as a fractured self but as a "total man."[53] In that sense, decolonial violence also possessed a powerful ethical dimension, even as it involved the likely bloody overthrow of colonizing forces. The ethical relation that decolonial violence occasioned, however, was not between the colonized and colonizer but between colonized subjects *and themselves*. As Fanon stressed, decolonial violence was a practice through which colonized subjects experienced a kind of conversion, a self-transformation, "the creation of a new man" (*création d'hommes nouveaux*).[54] The violence of decolonization was not simply an action directed at the forces of colonial rule, a means for "launching a new society." Decolonial violence was also something that colonized subjects performed upon themselves, a practice through which "a radical transformation [*bouleversement*] takes place within [the colonized] which makes any attempt to maintain the colonial system impossible and shocking." Violence was the very activity that made possible "the spiritual [*spirituelles*] and material conditions for the reconversion [*reconversion*] of man."[55]

In this sense, violent decolonial upheaval can be understood as a practical analogue to radical psychiatry, the two vectors of decolonial praxis. For while violence extricates colonized subjects from colonial subjugation, psychiatric therapeutics extricate colonized subjects from *themselves*, that is, from their prior identities as subjects of colonization. Militarized violence against the colonizer was structurally tethered to a psychiatric "violence" against one's colonized self, and while the former demanded insurgency and belligerent conflict, the latter demanded an analogously "violent" form of psychiatric self-transformation—violent in the sense that it demanded an upheaval *of oneself*. If decolonial violence was not just a tool for freedom, but freedom in practice, a revolutionary work in and of itself, then psychiatry for Fanon was similarly a practice of revolting against the pathological interiority that colonization forced colonized subjects to internalize. This was precisely what Fanon meant by the injunction that "psychiatry must be political."[56]

Fanon effectively postulated that colonization was too entrenched a logic to be undone simply at the level of institutions. The logic of colonization had to be abolished internally as well, and that abolishment had to take the form of a veritable annihilation of the colonized subject's former being in order to facilitate the emergence of a new salubrious self. It was necessary to recast psychiatry in precisely such a way—more than just a cure for the pernicious-ness of colonization but a weapon in the struggle to fashion oneself anew. Here there is a remarkable affinity between Fanonian psychiatry and what Foucault, toward the end of his life, dubbed "political spirituality," that is, the dominant practices, ethical tenets, and general belief systems "by which the individual is displaced, transformed, disrupted, to the point of renouncing their own individuality, their own subject position."[57] Foucault first coined the concept of political spirituality in 1978, in the context of his writings on the Iranian Revolution, where he believed one witnessed "the introduction of a spiritual dimension into political life," evident most especially in the writ-ings of the Iranian political theorist Ali Shariati, who was not only one of the conceptual architects of the Iranian Revolution but a reader and promulgator of Fanon's writings in Iran.[58] (Fanon is alleged to have corresponded directly with Shariati, who was an averred supporter of Algerian independence, and while Fanon was admittedly secular, he is nevertheless reported by Shariati—in what is likely a tactical confabulation—to have recognized, not unlike Fou-cault, the insurrectionary and anticolonial power of Islamic religiosity.[59])

But the affinity between Fanonian psychiatry and Foucault's notion of political spirituality is even stronger than their sympathies with respect to decolonial revolt. For Foucault, "spirituality," or the ethical traditions that displayed a commitment to self-transformation, while most dominant in Western antiquity through the Middle Ages, began to wane dramatically in the modern era. Still, Foucault noted that despite that decline vestiges of a spiritual ethics remained evident in two dominant modern doctrines, Marx-ism and psychoanalysis, which respectively harbored commitments to revo-lutionary and psychiatric styles of self-transformation, and were undeniably among the more important conceptual axes in Fanon's political and psychiat-ric thought.[60] In Foucault's rendering, spiritual self-transformation amounted to an extreme, even violent form of self-refashioning, a process that Fou-cault likened to a veritable uprising against oneself. "Rising up," Foucault ex-plained, "must be practiced, by which I mean one must practice rejecting the subject status in which one finds oneself, the rejection of one's identity, the rejection of one's own permanence, the rejection of what one is."[61] Radical resistance was not simply the struggle against political authority but the prac-tice of insurrection against the self who is formed within the structures and logics of that authority as well. "The political, ethical, social, philosophical problem of our day," Foucault wrote just two years before his death, "is not to try to liberate the individual from the state, and from the state's institutions, but to liberate us both from the state and from the type of individualization linked to the state. We have to promote new forms of subjectivity through the refusal of this kind of subjectivity that has been imposed on us for several centuries."[62] The insurgent refusal against oneself, the rising up and rejection of one's identity and subjectivity in order to "promote new forms of subjectivity" is precisely what I contend Fanon's radical psychiatry entailed and aspired to. And while Foucault was more inclined to view psychiatry as more disciplinary than emancipatory, Fanon saw in psychiatry an ethical possibility of tumultuous self-rebellion and self-upheaval. Fanonian psychiatry was passionately conversional in an almost religious sense, and though Foucault was likely more sympathetic to spirituality than Fanon, still we might be justified in calling Fanon's radical decolonial psychiatry a kind of *spiritual medicine*.[63]

Positing a "beyond" to the economic reason of madness, I conclude, would mean following this Fanonian-Foucauldian line of thought and admitting that pathological value is too entrenched a logic in the social body to be simply transformed at the level of institutions and infrastructures. Instead, it is necessary to imagine a psychiatry whose ethical ends exceed the return to health as a mere freedom from mental disease. This is not to deny the political significance of radical medical equity, of the need to enable full and absolute psychiatric access particularly for the most vulnerable and marginalized people and communities. But as I have said, this book has not been about the history of the political economy of psychiatric institutions, and the de-financialization of psychiatric care will likely not undo what the economization of madness has wrought—namely, that the very ontology of madness has been converted into an economic form. What is needed, I would propose, is not the de-financialization of psychiatry but an undoing—a *de-ontologization*, as it were—of madness's economic rationality. And the moral castigation of pathological value will unfortunately alone not suffice in such an endeavor, for, as I have suggested, the economization of madness has coincided with the economization of moral value itself. I contend that something more radical is required—something more provocative and in keeping with the fiercest traditions of radical psychiatry, which I take Fanon above all to represent. I propose that the economic reason of madness, the force of pathological value, can only dissipate if we imagine the ends of psychiatric thought and practice anew, not as a path to mental health alone but as the means by which we struggle to extricate ourselves from the vicious norms and valuational processes that we ourselves perpetrate and perpetuate, to unmoor ourselves from the covert economic valuations that suffuse our moral world, and to adopt styles of subjectivity that appear provocatively valueless, so to speak, and even disturbingly unprofitable. Overcoming the economic reason

of madness is first and foremost to abolish oneself as the vessel and propa-
gator of pathological value, to refashion oneself otherwise, and what such a
possibility calls for is a "psychiatry" that facilitates an ethics of radical self-
upheaval—a *spiritual medicine*, perhaps more spiritual than medical, more
sacred than secular.

Acknowledgments

It was at the University of Chicago, as a Harper-Schmidt Fellow in the Society of Fellows, that I first developed the core ideas of a new project that would eventually become *Madness and Enterprise*. There, I benefited from Chicago's deeply engrained culture of rigorous scholarly exchange. For their sustained conversations and thoughtful guidance over several years, I want to express my appreciation particularly to Marcello Barison, Alexis Becker, Mark Berger, Teri Chettiar, Joshua Craze, Isabel Gabel, Timothy Harrison, Jared Holley, Joel Isaac, Ashton Lazarus, Ainsley LeSure, Birte Loschenkohl, Pablo Maurette, Benjamin Morgan, Kareem Rabie, Eugene Raikhel, Robert Richards, Aviva Rothman, Emma Saunders-Hastings, David Sepkoski, David Simon, Simon Taylor, and Peter Wirzbicki.

My colleagues at Duke have been terrifically encouraging, both while I was an ACLS New Faculty Fellow and subsequently as an assistant professor. I am especially grateful to all my colleagues in the Program in Literature, particularly Rey Chow, Mark B. N. Hansen, and Fredric Jameson, as well as Roberto Dainotto, Anne Garreta, Markos Hadjioannou, Michael Hardt, Ranjana Khanna, Walter Mignolo, Toril Moi, Negar Mottahedeh, Luciana Parisi, Cate Reilly, Antonio Viego, and Robyn Wiegman. I've also benefited tremendously from dialogues with many other colleagues at Duke and neighboring institutions, including James Chappel, Malachi Hacohen, Deborah Jenson, Lloyd Kramer, Anthony La Vopa, Robert Mitchell, Noah Strote, K. Steven Vincent, and David Weinstein.

This book was profoundly enhanced by the comments and feedback that I received from friends and colleagues, in part through the John Hope Franklin Humanities Institute Faculty Manuscript Workshop at Duke, who generously read earlier and undoubtedly wanting drafts of the book, and whose input and reassurance were immeasurably helpful along the way. They include Jonny Bunning, James Chappel, Teri Chettiar, Rey Chow, Samuel Fury Childs

Daly, Stefani Engelstein, Stefanos Geroulanos, Malachi Hacohen, Mark B. N. Hansen, Evan Hepler-Smith, Ranjana Khanna, Satyel Larson, Camille Robcis, Gabriel Rosenberg, Harris Solomon, and Nasser Zakariya. Benjamin Aldes Wurgaft provided me with excellent editorial insights and suggestions at the last stages of manuscript preparation.

Like any other book that relies heavily on archival research, this book would simply not have been possible without the assistance of librarians at a number of key archives, including the Bibliothèque Charcot (particularly Guillaume Delaunay), the Bibliothèque médicale Henri Ey, the David M. Rubenstein Rare Book & Manuscript Library at Duke University (particularly Rachel Ingold and Brooke Guthrie), the Medical Center Archives of New York-Presbyterian/Weill Cornell Medicine, the Queen Square Archives, University College London Special Collections, University of Virginia Special Collections, and the Wellcome Library Special Collections. Stephen Casper, Emmanuel Delille, and Rhodri Hayward very generously provided me with some valuable research guidance along the way. I'm also very grateful to Karen Darling for her editorial patience and encouragement throughout the process, as well as to the rest of the editorial and production staff at the University of Chicago Press.

I would also like to express my gratitude to the people and institutions that so profoundly shaped me throughout my doctoral and early postdoctoral years. The University of California-Berkeley was an immensely formative place to develop as a scholar, and I will always remain grateful to the people I had the great fortune of encountering during my time there, particularly David Bates, Judith Butler, Marianne Constable, Whitney Davis, Catherine Malabou, Hélène Mialet, Kaja Silverman, Charis Thompson, Michael Wintroub, and the many graduate students whose counsel and comradery so enriched my graduate studies. At the Center for the Humanities at Wesleyan University, I was lucky to have been surrounded by a number of enormously thoughtful interlocutors, including Andrew Curran, Paul Erickson, Joseph Fitzpatrick, Matthew Garrett, Tushar Irani, Isaac Kamola, Jill Morawski, Joseph Rouse, Mary-Jane Rubenstein, Khachig Tölölyan, and Margot Weiss.

Finally, no words can truly express my veneration of and eternal indebtedness to Satyel Larson, whose unyielding support and devotion have, like a shining beacon, guided me through the most unnavigable of times. In the end, I dedicate this book to her.

Notes

Introduction

1. Kristian Bondo Hansen and Thomas Presskorn-Thygesen, "On Some Antecedents of Behavioral Economics," *History of the Human Sciences* 34, nos. 3–4 (2022): 58–83; Joseph Persky, "The Ethology of Homo Economicus," *Journal of Economic Perspectives* 9, no. 2 (1995): 221–31.

2. John Dewey, *Human Nature and Conduct: An Introduction to Social Psychology* (New York: Holt, 1922), 220.

3. Mark Blaug, "Was There a Marginal Revolution?," *History of Political Economy* 4, no. 2 (1972): 269–80; Donald Winch, "Marginalism and the Boundaries of Economic Science," *History of Political Economy* 4, no. 2 (1972): 325–43; Ronald L. Meek, "Marginalism and Marxism," *History of Political Economy* 4, no. 2 (1972): 499–511; R. D. Collison Black, "W. S. Jevons and the Foundation of Modern Economics," *History of Political Economy* 4, no. 2 (1972): 364–78; Janek Wasserman, *The Marginal Revolutionaries: How Austrian Economists Fought the War of Ideas* (New Haven: Yale University Press, 2019), 26–31.

4. John Forrester, "'A Sort of Devil' (Keynes on Freud, 1925): Reflections on a Century of Freud Criticism," *Österreichische Zeitschrift für Geschichtswissenschaften* 14, no. 2 (2003): 70–85; Rüdiger Graf, "Human Behavior as a Limit to and a Means of State Intervention: Günter Schmölders and Behavioral Economics," in *Nine Lives of Neoliberalism*, ed. Dieter Plehwe, Quinn Slobodian, and Philip Mirowski (New York: Verso Press, 2020); Günter Schmölders, "Finanzpsychologie," *FinanzArchiv/Public Finance Analysis*, n.s., 13, no. 1 (1951/52): 1–36, 4.

5. E. G. Winslow, "Keynes and Freud: Psychoanalysis and Keynes's Account of the 'Animal Spirits' of Capitalism," *Social Research* 53, no. 4 (1986): 549–78; Jonathan Levy, "Primal Capital," *Critical Historical Studies* (Fall 2019): 161–93; Geoff Mann, *In the Long Run We Are All Dead: Keynesianism, Political Economy, and Revolution* (New York: Verso, 2017), chap. 11.

6. John Maynard Keynes, "Economic Possibilities for Our Grandchildren (1930)," in *Essays in Persuasion* (London: Macmillan, 1931), 369.

7. John Maynard Keynes, *The General Theory of Employment, Interest, and Money* (London: Macmillan, 1936), 161–64; George A. Akerlof and Robert J. Shiller, *Animal Spirits: How Human Psychology Drives the Economy, and Why It Matters for Global Capitalism* (Princeton: Princeton University Press, 2009); D. E. Moggridge, "Correspondence," *Journal of Economic Perspectives* 6, no. 3 (1992): 207–12; R. C. O. Matthews, "Animal Spirits (Keynes Lecture in Economics)," *Pro-*

ceedings of the British Academy 70 (1984): 209229; Sonya Marie Scott, *Architectures of Economic Subjectivity: The Philosophical Foundations of the Subject in the History of Economic Thought* (London: Routledge, 2013), 169–80.

8. Floris Heukelom, *Behavioral Economics: A History* (Cambridge: Cambridge University Press, 2014); Nikolas Rose, *Our Psychiatric Future: The Politics of Mental Health* (Cambridge: Polity Press, 2019); Nikolas Rose and Joelle Abi-Rached, *Neuro: The New Brain Sciences and the Management of the Mind* (Princeton: Princeton University Press, 2013).

9. Robert J. Shiller, *Irrational Exuberance*, 3rd ed. (Princeton: Princeton University Press, 2015).

10. Elizabeth Lunbeck, *The Americanization of Narcissism* (Cambridge, MA: Harvard University Press, 2014), 18.

11. Emily Martin, *Bipolar Expeditions: Mania and Depression in American Culture* (Princeton: Princeton University Press, 2007), 53.

12. Alan Farnham, "Crazy and in Charge: Brilliant Tycoons Have Had a Tendency to Get Eccentric, or Worse," *Time*, December 7, 1998, 207–8.

13. Donald J. Trump with Tony Schwartz, *Trump: The Art of the Deal* (New York: Ballantine, 1987), 64.

14. Edward P. Lazear, "Economic Imperialism," *Quarterly Journal of Economics* 115, no. 1 (2000): 99–146, 103; Gary Becker, "The Economic Approach to Human Behavior," in *The Economic Approach to Human Behavior* (Chicago: University of Chicago Press, 1976), 7–8, 10.

15. Lukas Rieppel, William Deringer, and Eugenia Lean, eds., "Science and Capitalism: Entangled Histories," *Osiris* 33 (2018); Christy Ford Chapin, "What Historians of Medicine Can Learn from Historians of Capitalism," *Bulletin of the History of Medicine* 94, no. 3 (2020): 319–67.

16. I leave out of this discussion the question of aesthetic value, which I take to be irrevocably tied up with moral and chiefly economic forms of valuation. See, for example, Sianne Ngai, *Theory of the Gimmick: Aesthetic Judgement and Capitalist Form* (Cambridge, MA: Harvard University Press, 2020). See also Pierre Bourdieu, *Distinction: A Social Critique of the Judgment of Taste*, trans. Richard Nice (Cambridge, MA: Harvard University Press, 1984).

17. While I use the terms "economy" and "economic" throughout this book, primarily in reference to the forms and processes of valuation specific to industrial and market capitalism of turn-of-the-century Europe and the United States, I also accept William H. Sewell Jr.'s assertion that "capitalism is not reducible to an economic form [but is instead] a complex social whole, with specific political forms, psychologies, social relations, and cultural features." William H. Sewell Jr., "The Capitalist Epoch," *Social Science History* 28 (2014): 1–11, 2. See also Jonathan Levy, "Capital as Process and the History of Capitalism," *Business History Review* 91 (2017): 483–510.

18. Camille Robcis, *Disalienation: Politics, Philosophy, and Radical Psychiatry in Postwar France* (Chicago: University of Chicago Press, 2021).

19. Sigmund Freud, "Civilization and Its Discontents (1930)," in *The Standard Edition of the Complete Psychological Works of Sigmund Freud*, vol. 21, ed. James Strachey (London: Hogarth Press, 1961); Wilhelm Reich, *The Mass Psychology of Fascism*, trans. Vincent R. Carfagno (New York: Farrar, Straus & Giroux, 1970 [1933]); Theodor W. Adorno et al., *The Authoritarian Personality* (New York: Verso, 2019 [1950]); Robert Nye, *Crime, Madness, and Politics in Modern France: The Medical Concept of National Decline* (Princeton: Princeton University Press, 1984).

20. Michel Foucault, *History of Madness*, trans. Jonathan Murphy and Jean Khalfa (London: Routledge, 2006), 137; Foucault, *Histoire de la folie à l'âge classique* (Paris: Gallimard, 1972), 185; Foucault, "About the Concept of the 'Dangerous Individual' in 19th-Century Legal Psychiatry,"

trans. Alain Baudot and Jane Couchman, *International Journal of Law and Psychiatry* 1 (1978): 1–18.

21. Roger Smith, "What Is the History of the Human Sciences?," in *The Palgrave Handbook of the History of the Human Sciences*, ed. David McCallum (Basingstoke: Palgrave Macmillan, 2022).

22. Stephen Gaukroger, *Civilization and the Culture of Science: Science and the Shaping of Modernity, 1795–1935* (Oxford: Oxford University Press, 2020); Stephen Gaukroger, *The Natural and the Human: Science and the Shaping of Modernity, 1739–1841* (Oxford: Oxford University Press, 2016); Roger Smith, *Between Mind and Nature: A History of Psychology* (London: Reaktion Books, 2013); Smith, *Being Human: Historical Knowledge and the Creation of Human Nature* (Manchester: Manchester University Press, 2007); Smith, *The Norton History of the Human Sciences* (New York: Norton, 1997).

23. Dewey, *Human Nature*, 295–96.

24. Greg Eghigian, Andreas Killen, and Christine Leuenberger, "The Self as Project: Politics and the Human Sciences in the Twentieth Century," *Osiris* 22 (2007): 1–25.

25. Rhodri Hayward, *Resisting History: Religious Transcendence and the Invention of the Unconscious* (Manchester, UK: Manchester University Press, 2007), 64; Hayward, "Psychiatry and Religion," in *The Routledge History of Madness and Mental Health*, ed. Greg Eghigian (London: Routledge, 2017).

26. Raquel Weiss, "Durkheimian Revolution in Understanding Morality: Socially Created, Scientifically Grasped," in McCallum, *Palgrave Handbook of the History of the Human Sciences*, 140.

27. Talcott Parsons, "On Building Social System Theory: A Personal History," in *Social Systems and the Evolution of Action Theory* (New York: Free Press, 1977), 71–76; Parsons, "Belief, Unbelief, and Disbelief," in *Action Theory and the Human Condition* (New York: Free Press, 1978), 235–38; Howard Brick, "Talcott Parsons's 'Shift Away from Economics,' 1937–1946," *Journal of American History* 87, no. 2 (2000): 490–514.

28. Parsons, "Belief, Unbelief, and Disbelief," 235–38.

29. Max Weber, "Die protestantische Ethik und der 'Geist' des Kapitalismus," *Archiv für Sozialwissenschaft und Sozialpolitik* 20 (1905) 1–54; 21 (1905): 1–110; 20 (1905): 3. Max Weber, *The Protestant Ethic and the Spirit of Capitalism*, trans. Talcott Parsons (London: Routledge, 2010); Wilhelm Hennis, "Max Weber's 'Central Question,'" *Economy and Society* 12, no. 2 (1983): 135–80; Wolfgang Mommsen, "Personal Conduct and Societal Change: Towards a Reconstruction of Max Weber's Concept of History," trans. Rainhild Wells, in *Max Weber, Rationality, and Modernity*, ed. Sam Whimster and Scott Lash (London: Allen & Unwin, 1987).

30. Max Weber, "The Social Psychology of the World Religions," in *From Max Weber: Essays in Sociology*, trans., ed., and intro. H. H. Gerth and C. Wright Mills (New York: Oxford University Press, 1946), 270; Weber, "Die Wirtschaftsethik der Weltreligionen," in *Gesammelte Aufsätze zur Religionssoziologie* (Tübingen: J.C.B. Mohr, 1920), 19:241; David Chalcraft and Austin Harrington, eds., *The Protestant Ethic Debate: Weber's Replies to His Critics, 1907–1910* (Liverpool: Liverpool University Press, 2001), 74.

31. Simon Clarke, *Marx, Marginalism, and Modern Sociology* (London: Palgrave Macmillan, 1982), 204–7.

32. William Callison, "The Politics of Rationality in Early Neoliberalism: Max Weber, Ludwig von Mises, and the Socialist Calculation Debate," *Journal of the History of Ideas* 83, no. 2 (2022): 269–91.

33. Max Weber, *Economy and Society: An Outline of Interpretive Sociology*, ed. Guenther Roth and Claus Wittich (Berkeley: University of California Press, 1978), 21.

34. Paul Erickson et al., *How Reason Almost Lost Its Mind: The Strange Career of Cold War Rationality* (Chicago: University of Chicago Press, 2013), chaps. 2 and 6; S. M. Amadae, *Rationalizing Capitalist Democracy: The Cold War Origins of Rational Choice Liberalism* (Chicago: University of Chicago Press, 2003); William Derringer, "For What It's Worth: Historical Financial Bubbles and the Boundaries of Economic Rationality," *Isis* 106, no. 3 (2015): 646–56; Richard Samuels and Stephen P. Stich, "Rationality and Psychology," and Paul Weirich, "Economic Rationality," in *The Oxford Handbook of Rationality*, ed. Alfred R. Mele and Piers Rawling (New York: Oxford University Press, 2004); Herbert A. Simon, "A Behavioral Model of Rational Choice," *Quarterly Journal of Economics* 69, no. 1 (1955): 99–118. Cf. Kenneth J. Arrow, "Is Bounded Rationality Unboundedly Rational? Some Ruminations," in *Models of a Man: Essays in Memory of Herbert A. Simon*, ed. Mie Augier and James G. March (Cambridge, MA: MIT Press, 2004); Hunter Crowther-Heyck, *Herbert A. Simon: The Bounds of Reason in Modern America* (Baltimore: Johns Hopkins University Press, 2005); Gary S. Becker, "Irrational Behavior and Economic Theory," *Journal of Political Economy* 70, no. 1 (1962): 1–13, 5n7; Jon Elster, *Ulysses and the Sirens: Studies in Rationality and Irrationality* (Cambridge: Cambridge University Press, 1979); Stuart Sutherland, *Irrationality: Why We Don't Think Straight!* (New Brunswick, NJ: Rutgers University Press, 1994); Lisa Bortolotti, *Irrationality* (Cambridge: Polity Press, 2015).

35. Joseph A. Schumpeter, "The Meaning of Rationality in the Social Sciences," in *Rationality in the Social Sciences: The Schumpeter-Parsons Seminar 1939–1940 and Current Perspectives*, ed. Helmut Staubmann and Victor Lidz (Cham: Springer, 2018), 46; Joel Isaac, *Working Knowledge: Making the Human Sciences from Parsons to Kuhn* (Cambridge, MA: Harvard University Press, 2012), 63–91.

36. Jason A. Josephson-Storm, *The Myth of Disenchantment: Magic, Modernity, and the Birth of the Human Sciences* (Chicago: University of Chicago Press, 2018).

37. Sigmund Freud, "Character and Anal Eroticism (1908)," in *The Standard Edition of the Complete Works of Sigmund Freud*, vol. 9, ed. James Strachey (London: Hogarth Press, 1959); Silas L. Warner, "Sigmund Freud and Money," *Journal of the American Academy of Psychoanalysis* 17, no. 4 (1989): 609–22.

38. Herbert Marcuse, *One-Dimensional Man: Studies in the Ideology of Advanced Industrial Society* (London: Routledge, 2007 [1964]), 11, 230–31, 251; Theodor W. Adorno and Max Horkheimer, *Dialectic of Enlightenment*, trans. Edmund Jephcott (Stanford: Stanford University Press, 2002); Martin Jay, *The Dialectical Imagination: A History of the Frankfurt School and the Institute of Social Research* (Berkeley: University of California Press, 1996), 216–18; Frank Trentmann, *Empire of Things: How We Became a World of Consumers, from the Fifteenth Century to the Twenty-First* (New York: Harper, 2016); Lawrence B. Glickman, "From the History of Consumption to the History of Capitalism," *History Workshop Journal* 89 (2020): 271–82.

39. Georg Simmel, *The Philosophy of Money*, trans. Tom Bottomore and David Frisby (London: Routledge, 2011), 56; Simmel, *Philosophie des Geldes*, 2nd ed. (Leipzig: Verlag von Duncker & Humblot, 1907), viii.

40. Émile Durkheim, "Simmel (Georg)—Philosophie des Geldes (*Philosophie de l'argent*)," *L'année sociologique* 5 (1900–1901): 141–45.

41. Robert Castel, *From Manual Workers to Wage Laborers: Transformation of the Social Question*, trans. Richard Boyd (New Brunswick, NJ: Transaction, 2003), 121.

42. Hans Blumenberg, "Money or Life: Metaphors of Georg Simmel's Philosophy," *Theory, Culture & Society* 28, no. 7/8 (2012): 249–62; originally published as Hans Blumenberg, "Geld

oder Leben: Eine metaphorologische Studie zur Konsistenz der Philosophie Georg Simmels," in *Ästhetik und Soziologie um die Jahrhundertwende: Georg Simmel*, ed. Hannes Böhringer and Karlfried Gründer (Frankfurt am Main: Klostermann, 1976), 121–34.

43. Wilhelm Griesinger, "Vortrag zur Eröffnung der psychiatrischen Klinik zu Berlin," in *Gesammelte Abhandlungen* (Berlin: August Hirschwald 1872), 1:127–51, 150; Griesinger, "Introductory Lecture at the Reopening of the Psychiatric Clinic, at Berlin, May 2, 1867," *American Journal of Insanity* 24, no. 4 (1868): 393–417, 416 (trans. modified); Eric J. Engstrom, *Clinical Psychiatry in Imperial Germany: A History of Psychiatric Practice* (Ithaca: Cornell University Press, 2003), 51–87; Otto M. Marx, "Wilhelm Griesinger and the History of Psychiatry: A Reassessment," *Bulletin of the History of Medicine* 46, no. 6 (1972): 519–44.

44. Friedrich Nietzsche, *On the Genealogy of Morality*, trans. Carol Diethe (Cambridge: Cambridge University Press, 2007), 5; Eduard von Hartmann, *System der Philosophie im Grundriss*, vol. 5, *Grundriss de Axiologie oder Wertwägungslehre* (Bad Sachs im Harz: Hermann Haacke, 1908); Nicolae Râmbu, *The Axiology of Friedrich Nietzsche* (Frankfurt am Main: Peter Lang, 2016); Archie J. Bahm and Robert Ginsberg, eds., *Axiology: Science of Value* (Amsterdam: Rodopi, 1993).

45. Georges Canguilhem, *Knowledge of Life*, trans. Stefanos Geroulanos and Daniela Ginsburg (New York: Fordham University Press, 2008), 125.

46. Jonathan M. Metzl and Anna Kirkland, eds., *Against Health: How Heath Became the New Morality* (New York: New York University Press, 2010); Melanie A. Kiechle, "'Health Is Wealth': Valuing Health in the Nineteenth-Century United States," *Journal of Social History* 54, no. 3 (2021): 775–98.

47. Michel Foucault, "The Politics of Health in the Eighteenth Century," in *The Essential Works of Foucault, 1954–1984*, vol. 3, *Power*, ed. James D. Faubion (New York: New Press, 1994); Marion Fourcade, "Cents and Sensibility: Economic Valuation and the Nature of 'Nature,'" *American Journal of Sociology* 116, no. 6 (2011): 1721–77.

48. Georges Canguilhem, *Writings on Medicine*, trans. Stefanos Geroulanos and Todd Meyers (New York: Fordham University Press, 2012), 48, 52.

49. Canguilhem, *Writings on Medicine*, 35.

50. Don Kalb, "Elias Talks to Hayek (and Learns from Marx and Foucault): Reflections on Neoliberalism, Postsocialism, and Personhood," in *Neoliberalism, Personhood, and Postsocialism: Enterprising Selves in Changing Economies*, ed. Nicolette Makovicky (Farnham: Ashgate, 2014).

51. Marcel Gauchet and Gladys Swain, *La pratique de l'esprit humain: L'institution asilaire et la revolution démocratique* (Paris: Gallimard, 1980); Marcel Gauchet and Gladys Swain, *Madness and Democracy: The Modern Psychiatric Universe*, trans. Catherine Porter (Princeton: Princeton University Press, 1999); Klaus Dörner, *Bürger und Irre: Zur Sozialgeschichte und Wissenschaftssoziologie der Psychiatrie* (Frankfurt am Main: Europäische Verlagsanstalt, 1969); Klaus Dörner, *Madmen and the Bourgeoisie: A Social History of Insanity and Psychiatry*, trans. Joachim Neugroschel and Jean Steinberg (Oxford: Blackwell, 1981); Helmut Siefert, "Klaus Dörner," in *Enzyklopädie Medizingeschichte*, ed. Werner E. Gerabek et al. (Berlin: Walter de Gruyter, 2005), 319; Karl H. Beine, "Unterwegs zu einer anthropologischen Psychiatrie: Klaus Dörner zum 80. Geburtstag," *Psychiatrische Praxis* 40 (2013): 450–52. Cf. Doris Kaufmann, *Aufklärung, bürgerliche Selbsterfahrung und die "Erfindung" der Psychiatrie in Deutschland, 1770–1850* (Göttingen: Vandenhoeck & Ruprecht, 1995).

52. William H. Sewell Jr., "A Strange Career: The Historical Study of Economic Life," *History and Theory* 49 (2010): 146–66.

53. Andrew Scull, *The Most Solitary of Afflictions: Madness and Society in Britain, 1700–1900* (New Haven: Yale University Press, 1993), 29, 35, 38.

54. Arnold I. Davidson, "Ethics as Ascetics: Foucault, the History of Ethics, and Ancient Thought," in *Foucault and the Writing of History*, ed. Jan Goldstein (Oxford: Blackwell, 1994); Joelle M. Abi-Rached, *'Aṣfūriyyeh: A History of Madness, Modernity, and War in the Middle East* (Cambridge, MA: MIT Press, 2020), xxiii.

55. Michelle Murphy, *The Economization of Life* (Durham: Duke University Press, 2017).

56. Charly Coleman, *The Spirit of French Capitalism: Economic Theology in the Age of Enlightenment* (Stanford: Stanford University Press, 2021), 259; Arjun Appadurai, "Life after Debt," *differences: A Journal of Feminist Cultural Studies* 31, no. 3 (2020): 21–28.

57. Stanley Jeyaraja Tambiah, *Magic, Science, Religion, and the Scope of Rationality* (Cambridge: Cambridge University Press, 1990); Joshua Landy and Michael Saler, eds., *The Re-Enchantment of the World: Secular Magic in a Rational Age* (Stanford: Stanford University Press, 2009); David d'Avray, *Medieval Religious Rationalities: A Weberian Analysis* (Cambridge: Cambridge University Press, 2010); Richard Kieckhefer, "The Specific Rationality of Medieval Magic," *American Historical Review* 99, no. 3 (1994): 813–36.

58. George Miller Beard, *American Nervousness: Its Causes and Consequences* (New York: G. P. Putnam's Sons, 1881), 9–13; Lawrence Birken, "Freud's 'Economic Hypothesis': From Homo Oeconomicus to Homo Sexualis," *American Imago* 56, no. 4 (1999): 311–30; Richard T. Gray, "Economics as a Laughing Matter: Freud's Jokes and Their Relation to the Economic and Rhetorical Unconscious," *Germanic Review* 88 (2013): 97–120; John Forrester, *Truth Games: Lies, Money, and Psychoanalysis* (Cambridge, MA: Harvard University Press, 1997), 110–71.

59. Ian Hacking, "'Style' for Historians and Philosophers," in *Historical Ontology* (Cambridge, MA: Harvard University Press, 2002), 189; Luca Sciortino, "On Ian Hacking's Notion of Style of Reasoning," *Erkenntnis* 82 (2017): 243–64; Paul A. Roth, "Ways of Pastmaking," *History of the Human Sciences* 15, no. 4 (2002): 125–43; Martin Kusch, "Hacking's Historical Epistemology: A Critique of Styles of Reasoning," *Studies in History and Philosophy of Science* 41 (2010): 158–73.

60. Alistair Cameron Crombie, *Styles of Scientific Thinking in the European Tradition: The History of Argument and Explanation Especially in the Mathematical and Biomedical Sciences and Arts* (London: Duckworth, 1994). See also John V. Pickstone, *Ways of Knowing: A New History of Science, Technology, and Medicine* (Chicago: University of Chicago Press, 2000) and Chunglin Kwa, *Styles of Knowing: A New History of Science from Ancient Time to the Present*, trans. David McKay (Pittsburgh: University of Pittsburgh Press, 2011).

61. Michael Polanyi, *The Tacit Dimension* (Chicago: University of Chicago Press, 2009); Harry Collins, *Tacit and Explicit Knowledge* (Chicago: University of Chicago Press, 2010).

62. Luca Sciortino, "Styles of Reasoning, Human Forms of Life, and Relativism," *International Studies in the Philosophy of Science* 30, no. 2 (2016): 165–84.

63. Thomas Kuhn, *The Structure of Scientific Revolutions*, 4th ed. (Chicago: University of Chicago Press, 2012); Kuhn, "Second Thoughts on Paradigms," in *The Essential Tension* (Chicago: University of Chicago Press, 1977).

64. David Kaiser, "Thomas Kuhn and the Psychology of Scientific Revolutions," in *Kuhn's Structure of Scientific Revolutions at Fifty: Reflections on a Science Classic*, ed. Robert J. Richards and Lorraine Daston (Chicago: University of Chicago Press, 2016).

65. Ludwik Fleck, *Genesis and Development of a Scientific Fact*, trans. Fred Bradley and Thaddeus J. Trenn (Chicago: University of Chicago Press 1979), 41.

66. Émile Durkheim, *The Rules of Sociological Method and Selected Texts on Sociology and Its Methods*, trans. W. D. Halls (New York: Free Press, 2013), 43.

67. Fleck, *Genesis*, 64.

68. Fleck, *Genesis*, 99; Michael Hagner, "Perception, Knowledge, and Freedom in the Age of Extremes: On the Historical Epistemology of Ludwig Fleck and Michael Polanyi," *Studies in Eastern European Thought* 64 (2012): 107–20.

69. Lorraine Daston, "The Moral Economy of Science," *Osiris* 10 (1995): 2–24, 6; Steven Shapin, "Understanding the Merton Thesis," *Isis* 79, no. 4 (1988): 594–605, 598–99.

70. Michel Foucault, *The Archeology of Knowledge*, trans. A. M. Sheridan Smith (New York: Pantheon, 1972), 191; Foucault, "The Confession of the Flesh," in *Power/Knowledge: Selected Interviews and Other Writings, 1972–1977*, ed. Colin Gordon (New York: Pantheon, 1980), 197; Laura Stark, "Out of Their Depths: 'Moral Kinds' and the Interpretation of Evidence in Foucault's Modern Episteme," *History and Theory* 54 (2016): 131–47.

71. Foucault, *Archeology*, 60–62.

72. Dominique Lecourt, *L'épistémologie historique de Gaston Bachelard* (Paris: Vrin, 1969); Lecourt, *Marxism and Epistemology: Bachelard, Canguilhem, and Foucault*, trans. Ben Brewster (London: NLB, 1975); Mary Tiles, "Epistemological History: The Legacy of Bachelard and Canguilhem," in *Contemporary French Philosophy*, ed. A. Phillips Griffiths (Cambridge: Cambridge University Press, 1987); Georges Canguilhem, "Introduction: The Role of Epistemology in Contemporary History of Science," in *Ideology and Rationality in the History of the Life Sciences*, trans. Arthur Goldhammer (Cambridge, MA: MIT Press, 1988); François Dagognet, *Georges Canguilhem: Philosophie de la vie* (Le Plessis-Robinson, Essonne: Institut Synthélabo pour le progrès de la connaissance, 1997); Marjorie Grene, "The Philosophy of Science of Georges Canguilhem: A Transatlantic View," *Revue d'histoire des sciences* 53, no. 1 (2000): 47–63; Stuart Elden, *Canguilhem* (Cambridge: Cambridge University Press, 2019), 111–33.

73. Lorraine Daston, "Historical Epistemology," in *Questions of Evidence: Proof, Practice, and Persuasion across the Disciplines*, ed. James Chandler, Arnold I. Davidson, and Harry Harootunian (Chicago: University of Chicago Press, 1994), 254, 282–83; Ian Hacking, "Historical Meta-Epistemology," in *Wahrheit und Geschichte: Ein Kolloquium zu Ehren des 60. Geburtstages von Lorenz Krüger*, ed. Wolfgang Carl and Lorraine Daston (Göttingen: Vandenhoeck & Ruprecht, 1999); Uljana Feest and Thomas Sturm, "What (Good) Is Historical Epistemology? Editors' Introduction," *Erkenntnis* 75 (2011): 285–302.

74. Arnold I. Davidson, "Styles of Reasoning: From the History of Art to the Epistemology of Science" and "Foucault and the Analysis of Concepts," in *The Emergence of Sexuality: Historical Epistemology and the Formation of Concepts* (Cambridge, MA: Harvard University Press, 2001), 196.

75. John Forrester, "If *p*, Then What? Thinking in Cases," *History of the Human Sciences* 9, no. 3 (1996): 1–25; Forrester, *Thinking in Cases* (Cambridge: Polity Press, 2017), 2.

76. Robcis, *Disalienation*, 13, 147.

77. Joan W. Scott, "Histoire et psychanalyse," *Cliniques méditerranéennes* 95 (2017): 11–20; Joan W. Scott, "The Incommensurability of Psychoanalysis and History," *History and Theory* 51 (2012): 63–83.

78. For discussions and examples of the sociological movement of ideas, see James A. Secord, "Knowledge in Transit," *Isis* 95 (2004): 654–72; Theodore M. Porter, *Genetics in the Madhouse: The Unknown History of Human Heredity* (Princeton: Princeton University Press, 2018); John Harley Warner, *Against the Spirit of System: The French Impulse in Nineteenth-Century American Medicine* (Princeton: Princeton University Press, 1998); Daniel T. Rodgers, *Atlantic Crossings: Social Politics in a Progressive Age* (Cambridge, MA: Harvard University Press, 1989).

79. Elizabeth Popp Berman, *Thinking Like an Economist: How Efficiency Replaced Equality in U.S. Public Policy* (Princeton: Princeton University Press, 2022).

80. Peter E. Gordon, "Contextualism and Criticism in the History of Ideas," in *Rethinking Modern European Intellectual History*, ed. Darrin McMahon and Samuel Moyn (New York: Oxford University Press, 2014), 32, 34. Cf. Ian Hunter, "The Contest over Context in Intellectual History," *History and Theory* 58, no. 2 (2019): 185–209.

81. Eric Hobsbawm, *The Age of Capital: 1848–1875* (New York: Vintage, 1997 [1975]), 76.

82. E. P. Thompson, "The Moral Economy of the English Crowd in the Eighteenth Century," *Past and Present* 50 (1971): 76–136.

83. Martin, *Bipolar Expeditions*, 147–48; Judith Butler, *Gender Trouble: Feminism and the Subversion of Identity* (New York: Routledge, 2006); J. L. Austin, *How to Do Things with Words* (Cambridge, MA: Harvard University Press, 1975); Jacques Derrida, "Signature Event Context," in *Margins of Philosophy*, trans. Alan Bass (Chicago: University of Chicago Press, 1985).

84. Martin, *Bipolar Expeditions*, 83.

85. Elizabeth Lunbeck, *The Psychiatric Persuasion: Knowledge, Gender, and Power in Modern America* (Princeton: Princeton University Press, 1994), 21, 46.

86. Scull, *Most Solitary of Afflictions*, 391.

87. Martin, *Bipolar Expeditions*, 95.

88. Sara Ahmed, "A Phenomenology of Whiteness," *Feminist Theory* 8, no. 2 (2007): 149–68, 150.

89. Stuart Hall, "Race, Articulation, and Societies Structured in Dominance," in *Stuart Hall: Essential Essays*, vol. 1, ed., David Morley (Durham: Duke University Press, 2019), 216.

90. Abi-Rached, *'Asfûriyyeh*; Suman Seth, *Difference and Disease: Medicine, Race, and the Eighteenth-Century British Empire* (Cambridge: Cambridge University Press, 2018); Matthew Heaton, *Black Skin, White Coats: Nigerian Psychiatry, Decolonization, and the Globalization of Psychiatry* (Athens, OH: Ohio University Press, 2013); Richard C. Keller, *Colonial Madness: Psychiatry in French North Africa* (Chicago: University of Chicago Press, 2007); Warwick Anderson, *Colonial Pathologies: American Tropical Medicine, Race, and Hygiene in the Philippines* (Durham: Duke University Press, 2006); Jonathan Sadowsky, *Imperial Bedlam: Institutions of Madness in Colonial Southwest Nigeria* (Berkeley: University of California Press, 1999); Jock McCulloch, *Colonial Psychiatry and the "African Mind"* (Cambridge: Cambridge University Press, 1995); Megan Vaughan, *Curing Their Ills: Colonial Power and African Illness* (Stanford: Stanford University Press, 1991).

91. Frantz Fanon, *Black Skin, White Masks*, trans. Charles Lam Markmann (New York: Grove Press, 1967), 143.

92. Hacking, "Historical Ontology," in *Historical Ontology*, 2–4.

Chapter One

1. Isaac Ray, *A Treatise on the Medical Jurisprudence of Insanity* (Boston, 1838), 142–43.

2. James Cowles Prichard, *A Treatise on Insanity and Other Disorders Affecting the Mind* (London, 1835), 23, 383.

3. For example, John Timbs, *English Eccentrics and Eccentricities*, 2 vols. (London, 1866); Jules François Felix Fleury-Husson ("Champfleury"), *Les excentriques* (Paris, 1852); Firmin Boissin, *Excentriques disparus* (Paris: 1890); *Eccentric Biography; or, Memories of Remarkable Female Characters, Ancient and Modern* (London, 1803); Miranda Gill, *Eccentricity and the Cultural Imagination in Nineteenth-Century Paris* (Oxford: Oxford University Press, 2009).

4. John Stuart Mill, *On Liberty* (London, 1859), 120–23.

5. Joseph Williams, *Insanity: Its Causes, Prevention, and Cure; including Apoplexy, Epilepsy, and Congestion of the Brain*, 2nd ed. (London, 1852 [1848]), 8–9.

6. Daniel Noble, *Elements of Psychological Medicine: An Introduction to the Practical Study of Insanity* (London, 1853), 104; see also 216–20.

7. Wilhelm Griesinger, *Die Pathologie und Therapie der psychischen Krankheiten, für Ärtze und Studierende*, 2nd ed. (Stuttgart, 1861 [1845]), 125–26; Griesinger, *Mental Pathology and Therapeutics*, trans. C. Lockhart Robertson and James Rutherford (London, 1867), 86 (trans. modified).

8. Bénédict Auguste Morel, *Traité des maladies mentales* (Paris, 1860), 258–59, 546–47; Gill, *Eccentricity*, 240–45; Jacques-Joseph Moreau, *La psychologie morbide dans ses rapports avec la philosophie de l'histoire ou de l'influence des névropathies sur le dynamisme intellectuel* (Paris, 1859), 188, 385; Jules Falret, "Folie raisonnante ou folie morale, premier discours," in *Études cliniques sur les maladies mentales et nerveuses* (Paris, 1890), 475–525, 519; originally published as Jules Falret, "Discussion sur la folie raisonnante," *Annales médico-psychologiques* 7 (1866): 382–426; Paul Moreau de Tour, *Les excentriques: Étude psychologique et anecdotique* (Paris, 1894), 7–9; Miranda Gill, "A Little Bit Mad/Almost Mad/Not Quite Mad? Eccentricity and the Framing of Mental Illness in Nineteenth-Century French Culture," in *Framing and Imagining Disease in Cultural History*, ed. George Sebastian Rousseau (New York: Palgrave Macmillan, 2003), 153–72.

9. Joseph Mortimer Granville, "Eccentricities of the Mentally Affected," in Andrew Wynter, *The Borderlands of Insanity and Other Papers*, ed. J. Mortimer Granville (London, 1877), 195.

10. George H. Savage, "Moral Insanity," *Journal of Mental Science* 27 (July 1881): 147–55, 148.

11. Richard von Krafft-Ebing, *Lehrbuch der Psychiatrie auf klinischer Grundlage für praktische Ärzte und Studirende* (Stuttgart, 1879), 1:195, 136–37.

12. William F. Bynum, "Tuke's *Dictionary* and Psychiatry at the Turn of the Century," in *150 Years of British Psychiatry, 1841–1991*, ed. German E. Berrios and Hugh Freeman (London: Royal College of Psychiatrists, 1991), 163–79.

13. Daniel Hack Tuke, "Eccentricity," in *A Dictionary of Psychological Medicine*, ed. Daniel Hack Tuke (London, 1892), 1:419–20.

14. Michel Foucault, *Psychiatric Power: Lectures at the Collège de France, 1973–1974*, trans. Graham Burchell (New York: Palgrave Macmillan, 2006), 39–58; Elizabeth Lunbeck, *The Psychiatric Persuasion: Knowledge, Gender, and Power in Modern America* (Princeton: Princeton University Press, 1994), 46–77.

15. Daniel Hack Tuke, *Insanity in Ancient and Modern Life, with Chapters on Its Prevention* (London, 1878), 124, 126 (emphasis added).

16. Norbert Elias, *The Civilizing Process: Sociogenetic and Psychogenetic Investigations*, trans., Edmund Jephcott (Oxford: Blackwell, 2000), 398–99.

17. Charles Taylor, "Interpretation and the Sciences of Man," *Review of Metaphysics* 25, no. 1 (1971): 3–51, 10.

18. Benjamin Ball, "Les frontières de la folie," *Revue scientifique de la France et de l'étranger* 31, 3rd series (Jan. 1883): 1–5, 1; Ball, "Les frontières de la folie," *L'encéphale* (Jan. 1883): 1–20; reprinted as Benjamin Ball, *La morphinomanie, les frontières de la folie, le dualisme cérébral, les rêves prolongés, la folie gémellaire ou aliéntation mentale chez les jumeaux* (Paris, 1885); Denis Tiberghien, "The Chair of Mental and Brain Disease: Charcot's Pupils—Benjamin Ball, Alix Joffroy, and Gilbert Ballet," in *Following Charcot: A Forgotten History of Neurology and Psychiatry*, ed. Julien Bogousslavsky (Basel: Karger, 2011); Jan Goldstein, *Console and Classify: The*

French Psychiatric Profession in the Nineteenth Century (Chicago: University of Chicago Press, 2001), 333.

19. Alexandre Cullerre, *Les frontières de la folie* (Paris, 1888), 24.

20. Henry Maudsley, *Responsibility in Mental Disease* (London, 1874), 40.

21. Krafft-Ebing, *Lehrbuch der Psychiatrie*, 1:28–29, 32; Harry Oosterhuis, *Stepchildren of Nature: Krafft-Ebing, Psychiatry, and the Making of Sexual Identity* (Chicago: University of Chicago Press, 2000), 87; Paul Hoff, "Emil Kraepelin and Forensic Psychiatry," *International Journal of Law and Psychiatry* 21, no. 4 (1998): 343–53.

22. Andrew Wynter, *The Borderlands of Insanity and Other Allied Papers* (London, 1875), 1; Elaine Showalter, *The Female Malady: Women, Madness, and English Culture, 1830–1908* (New York: Pantheon, 1985), 105–10.

23. Peter Cryle and Elizabeth Stephens, *Normality: A Genealogy* (Chicago: University of Chicago Press, 2017), 261–91.

24. G. W. Balfour, "The Borderland," *Edinburgh Medical Journal*, n.s., 9 (1901): 1–20, 109–23, 1.

25. Sigmund Freud, "The Psychopathology of Everyday Life (1901)," in *The Standard Edition of the Complete Works of Sigmund Freud*, vol. 6, ed. James Strachey (London: Hogarth Press, 1960), 278; Katja Guenther, "Psychoanalysis: Freud and Beyond," in *Cambridge History of Modern European Thought*, ed. Warren Breckman and Peter E. Gordon (Cambridge: Cambridge University Press, 2019), 2:49.

26. Henry Maudsley, *The Pathology of Mind: A Study of Its Distempers, Deformities, and Disorders* (London, 1895), 1–3.

27. Maudsley, *Pathology of Mind*, 4.

28. Lunbeck, *Psychiatric Persuasion*, 20.

29. Marcel Gauchet, *L'inconscient cérébral* (Paris: Éditions du Seuil, 1992), 32; Rhodri Hayward, *The Transformation of the Psyche in British Primary Care, 1870–1970* (London: Bloomsbury, 2014); Matt Ffytche, *The Foundation of the Unconscious: Schelling, Freud, and the Birth of the Modern Psyche* (Cambridge: Cambridge University Press, 2012); Angus Nicholls and Martin Liebscher, eds., *Thinking the Unconscious: Nineteenth-Century German Thought* (Cambridge: Cambridge University Press, 2010); Henri F. Ellenberger, *The Discovery of the Unconscious: The History and Evolutionary of Dynamic Psychiatry* (New York: Basic Books, 1970).

30. Daniel Hack Tuke, "Automatism (Mental)," in Tuke, *Dictionary of Psychological Medicine*, 1:115–16.

31. Richard Noakes, *Physics and Psychics: The Occult and the Sciences in Modern Britain* (Cambridge: Cambridge University Press, 2019); Alexandra Bacopoulos-Viau, "Automatism, Surrealism, and the Making of French Psychopathology: The Case of Pierre Janet," *History of Psychiatry* 23, no. 3 (2012): 259–76; Andreas Sommer, "Psychical Research and the Origins of American Psychology: Hugo Münsterberg, William James, and Eusapia Palladion," *History of the Human Sciences* 25, no. 2 (2012): 23–44; Sofie Lachapelle, *Investigating the Supernatural: From Spiritism and Occultism to Psychical Research and Metaphysics in France, 1853–1931* (Baltimore: Johns Hopkins University Press, 2011); John Warne Monroe, *Laboratories of Faith: Mesmerism, Spiritism, and Occultism in Modern France* (Ithaca: Cornell University Press, 2008); Edward Bizub, *Proust et le moi divisé* (Geneva: Libraire Droz, 2006), 63–83; Alison Winter, *Mesmerized: Powers of Mind in Victorian Britain* (Chicago: University of Chicago Press, 1998); Janet Oppenheim, *The Other World: Spiritualism and Psychical Research in England, 1850–1914* (Cambridge: Cambridge University Press, 1985).

32. Jean-Martin Charcot and Paul Richet, "Note on Certain Facts of Cerebral Automatism Observed in Hysteria during the Cataleptic Period of Hypnotism," *Journal of Nervous and Men-*

tal Disease 10, no. 1 (1883): 1–13; Théodule Ribot, *Les maladies de la volonté* (Paris, 1883), 89–91; Pierre Janet, *L'automatisme psychologique: Essai de psychologie expérimentale sur les forme inférieures de l'activité humaine* (Paris, 1889); Gaëtan Gatian de Clérambault, *Oeuvre psychiatrique*, ed. Jean Fretet (Paris: Presses universitaires de France, 1942), 2:455–654.

33. Henri Ey, "La notion d'automatisme en psychiatrie," *L'évolution psychiatrique* 2 série, no. 3 (1932): 9–35, 12.

34. Henri Ey, *Hallucinations et délire: Les formes hallucinatoires de l'automatisme verbal* (Paris: Félix Lacan, 1934; L'Harmattan, 1999 [Félix Lacan, 1934]), 2–3.

35. Ey, *Hallucinations et délire*, 12.

36. François-Pierre-Gontier de Biran, *Influence de l'habitude sur la faculté de penser* (1802), *Oeuvres de Maine de Biran*, ed. Pierre Tisserand (Paris: F. Alcan, 1922), 2:104.

37. Joseph John Murphy, *Habit and Intelligence: A Series of Essays on the Laws of Life and Mind* (London, 1869), 87, 407–8.

38. Cesare Lombroso, *Criminal Man*, trans. Mary Gibson and Nicole Hahn Rafter (Durham: Duke University Press, 2006), 295; Lombroso, *L'uomo delinquente*, 4th ed. (Turin, 1889), 2:431.

39. Georges Canguilhem, *La formation du concept de réflexes aux XVIIe et XVIIIe siècles* (Paris: Presses universitaires de France, 1955); Franklin Fearing, *Reflex Action: A Study in the History of Physiological Psychology* (Cambridge, MA: MIT Press, 1970); Edwin Clarke and L. S. Jacyna, *Nineteenth-Century Origins of Neuroscientific Concepts* (Berkeley: University of California Press, 1992), chap. 4; Roger Smith, *Inhibition: History and Meaning in the Sciences of Mind and Behavior* (Berkeley: University of California Press, 1992), 66–112; Katja Guenther, *Localization and Its Discontents: A Genealogy of Psychoanalysis and the Neuro Disciplines* (Chicago: University of Chicago Press, 2015), chap. 1; Albrecht Hirschmüller, "The Development of Psychiatry and Neurology in the Nineteenth Century," trans. Magda Whitrow, *History of Psychiatry* 10 (1999): 395–423; L. S. Jacyna, "Somatic Theories of Mind and the Interests of Medicine in Britain, 1850–1879," *Medical History* 26 (1982): 233–58.

40. Johannes Müller, *Handbuch der Physiologie des Menschen*, 4th ed. (Coblenz, 1844), 1:732–41; Wilhelm Griesinger, "Ueber psychische Reflexactionen: Mit einem Blick auf das Wesen der psychischen Krankheiten," *Archiv für physiologische Heilkunde* 2. Jahrgang (Stuttgart, 1843), 76–113, 112.

41. Clarke and Jacyna, *Nineteenth-Century Origins*, 145; L. S. Jacyna, "The Physiology of Mind, the Unity of Nature, and the Moral Order of Victorian Thought," *British Journal for the History of Science* 14, no. 2 (1981): 109–32.

42. Thomas Laycock, "Analytic Essay on Irregular and Aggravated Forms of Hysteria," *Edinburgh Medical and Surgical Journal* 52 (1839): 43–87, 50, 59–60; Laycock, *Treatise on the Nervous Diseases of Women; Comprising an Inquiry into the Nature, Causes, and Treatment of Spinal and Hysterical Disorders* (London, 1840), 105–8; Laycock, "On the Reflex Function of the Brain," *British and Foreign Medical Review* 19 (1845): 298–311, 310–11; Laycock, "Further Researches into the Function of the Brain," *British and Foreign Medico-Chirurgical Review* 16 (1855): 120–44, 127; Laycock, *Mind and Brain: or, the Correlations of Consciousness and Organization* (Edinburgh and London, 1860), 2:39–40, 54–59; Alex Leff, "Thomas Laycock and the Cerebral Reflex: A Function Arising from and Pointing to the Unity of Nature," *History of Psychiatry* 2 (1991): 385–407; F. E. James, "Thomas Laycock, Psychiatry, and Neurology," *History of Psychiatry* 9 (1998): 491–502.

43. Thomas Laycock, "On the Reflex Function of the Brain: A Correction of Dates," *British Medical Journal* 1, no. 700 (May 30, 1874): 705–6, 705; Laycock, "Reflex, Automatic, and Unconscious Cerebration: A History and a Criticism," *Journal of Mental Science* 21, no. 96 (1876):

477–98; no. 97 (1876): 1–17, 489; William Carpenter, "On the Voluntary and Instinctive Actions of Living Beings," *Edinburgh Medical and Surgical Journal* 48 (1837): 22–44, 25–29; Carpenter, *Principles of Human Physiology* (London, 1842), 54; Carpenter, "On the Influence of Suggestion in Modifying and Directing Muscular Movement, Independent of Volition" (1852), in *Nature and Man: Essays Scientific and Philosophical* (London, 1888), 169–72, 170.

44. William Carpenter, "The Automatic Execution of Voluntary Movements" (1850), in *Nature and Man*, 165 (emphasis in original).

45. William Carpenter, *Principles of Human Physiology*, 4th ed. (London, 1853), 165–66, 843–44; William James, *Principles of Psychology* (New York, 1890), 1:104; David E. Leary, *The Routledge Guidebook to James' Principles of Psychology* (New York: Routledge, 2018), chap. 6.

46. UCL Special Collections College Archives Fees Books (1849–50 through 1853–54) [PROF 17–20, UCL Special Collections]; for UCL academic calendar, see The University College, London, Calendar for the Session 1853–4 through Calendar for Session 1858–9 [UCL Special Collections].

47. Carpenter, *Human Physiology*, 4th ed., vii, xi, 672, 811–20.

48. William Carpenter, *Principles of Mental Physiology*, 2nd ed. (London, 1875), 280.

49. Carpenter, *Human Physiology*, 4th ed., 800; Carpenter, *Unconscious Action of the Brain* and *Epidemic Delusions* (Boston: Estes & Lauriat, 1872), 214–15, 221.

50. Carpenter, *Principles of Mental Physiology*, 4th ed., 658; Carpenter, *Principles of Human Physiology*, 5th ed. (London, 1855), 653.

51. Laycock, *Mind and Brain*, 1:97; see also 2:82. Cf. Laycock, "Reflex, Automatic, and Unconscious Cerebration," 2; Kurt Danziger, "Mid-Nineteenth-Century British Psycho-Physiology: A Neglected Chapter in the History of Psychology," in *The Problematic Science: Psychology in Nineteenth-Century Thought*, ed. William R. Woodward and Mitchel G. Ash (New York: Praeger, 1982).

52. Jules Baillarger, "Application de la physiologie des hallucinations à la physiologie du délire considéré d'une manière générale: Théorie de l'automatisme (1845)," in *Recherches sur les maladies mentales* (Paris, 1890), 1:494–500; Baillarger, "La théorie de l'automatisme: Étudié dans le manuscrit d'un monomaniaque (1856)," in *Recherches sur les maladies mentales*, 1:563–75; originally published in *Annales médico-psychologiques*, no. 2 (1856): 57–70; Baillarger, "Recherches sur la structure de la couche corticale des circonvolutions du cerveau," *Mémoires de l'Académie royal de médecine* 8 (1840): 149–83; Baillarger, "Mode de formation des centres nerveux," *Annales médico-psychologiques* 2 (1843): 343–57; Jules Luys, *Recherches sur le système nerveux cérébro-spinal, sa structure, ses fonctions et ses maladies* (Paris, 1865), 271; H. Maurel, "Approche historique de la notion d'automatisme en psychiatrie," *Annales médico-psychologiques* 147 (1989): 946–50; A. Gorceix, "La notion d'automatisme psychologique d'Aristote à Janet," *Annales médico-psychologiques* 147 (1989): 944–45; Goldstein, *Console and Classify*, 262–63; T. Haustgen, "Jules Baillarger (1809–1890)," *Dictionnaire biographique de psychiatrie par des membres de la Société Médico-Psychologique/Annales médico-psychologiques* 162 (2004): 237–39; Ian R. Dowbiggin, *Inheriting Madness: Professionalization and Psychiatric Knowledge in Nineteenth-Century France* (Berkeley: University of California Press, 1991), 32–37, 76–79; Martin Parent and André Parent, "Jules Bernard Luys in Charcot's Penumbra," in Bogousslavsky, *Following Charcot*, 125–36.

53. Luys, *Recherches sur le système nerveux cérébro-spinal*, 361–62; Luys, "Des obsessions pathologique: Dans leurs rapports avec l'activité automatique des éléments nerveux," *L'encéphale* 3, no. 1 (1883): 20–61, 20–22.

54. Jules Luys, *Le cerveau et ses fonctions* (Paris, 1876), 162; Luys, *The Brain and Its Functions* (New York, 1890), 205.

55. Jules Luys, *De l'automatisme dans les opérations de l'activité mentale* (Paris, 1890), 12.

56. Théodule Ribot, *La psychologie anglaise contemporaine* (Paris, 1870); Ribot, *L'hérédité: Étude psychologique sur ses phénomènes, ses lois, ses causes, ses conséquences* (Paris, 1873), 307, 313; Ribot to Espinas, "Vesoul, 5 juillet 1868," in Raymond Lenoir, "Lettre de Théodule Ribot à Espinas," *Revue philosophique de la France et de l'étranger* 147 (1957): 1–14, 3 ; Jacqueline Carroy and Régine Plas, "The Origins of French Experimental Psychology: Experiment and Experimentalism," *History of the Human Sciences* 9, no. 1 (1996): 73–84; D. P. Faber, "Théodule Ribot and the Reception of Evolutionary Ideas in France," *History of Psychiatry* 8 (1997): 445–58; John I. Brooks III, "Philosophy and Psychology at the Sorbonne, 1885–1913," *Journal of the History of the Behavioral Sciences* 29 (1993): 123–45; Brooks, *The Eclectic Legacy: Academic Philosophy and the Human Sciences in Nineteenth-Century France* (Newark: University of Delaware Press), 72–75, 83–87.

57. Prosper Despine, *Étude scientifique sur l'somnambulisme, sur les phénomènes qu'il présente et sur son action thérapeutique dans certaines maladies nerveuses du rôle important qu'il joue dans l'épilepsie, dans l'hystérie, et dans les névroses dites extraordinaires* (Paris, 1880), 15.

58. Resident physicians at the National were required to hold a second appointment in general medicine at other institutions. Gordon Holmes, *The National Hospital, Queen Square, 1860–1948* (Edinburgh/London: Livingstone, 1954).

59. Henry Monro, *Remarks on Insanity: Its Nature and Treatment* (London, 1851); Francis Anstie, *Stimulants and Narcotics, Their Mutual Relations: With Special Researches on the Action of Alcohol, Ether, and Chloroform on the Vital Organism* (London, 1864); Claude Bernard, *Leçons de pathologie expérimentale* (Paris, 1872), 205–6; John Hughlings Jackson, "Remarks on the Double Condition of Loss of Consciousness and Mental Automatism Following Certain Epileptic Seizures," *Medical Times Gazette*, July 19, 1873, 63–64, 64. Jackson was *not* drawing from John Russell Reynolds (*Epilepsy: Its Symptoms, Treatment, and Relations to Other Chronic Convulsive Diseases* [London, 1861], 9) as it has been otherwise suggested by German E. Berrios in "Positive and Negative Symptoms and Jackson," *Archives of General Psychiatry* 42 (1985): 95–97; and in "French Views on Positive and Negative Symptoms: A Conceptual History," *Comprehensive Psychiatry* 32, no. 5 (1991): 395–403.

60. John Hughlings Jackson, *Neurological Fragments* (London: Oxford University Press, 1925), 28.

61. John Hughlings Jackson, "On Temporary Mental Disorders after Epileptic Paroxysms," in *Selected Writings of John Hughlings Jackson*, ed. James Taylor (New York: Basic Books, 1958), 1:123; "Obituary, John Hughlings Jackson," *British Medical Journal*, October 14, 1911, 950–54; Macdonald Critchley and Eileen A. Critchley, *John Hughlings Jackson: Father of English Neurology* (New York: Oxford University Press, 1998), 53–60; Martin N. Raitiere, *The Complicity of Friends: How George Eliot, G. H. Lewes, and John Hughlings Jackson Encoded Herbert Spencer's Secret* (Lewisburg: Bucknell University Press, 2012).

62. Théodule Ribot, "Jackson—Les localisations du mouvement dans le cerveau," *Revue philosophique de la France et de l'étranger* 1 (1876): 214–16; Eduard Hitzig, "Hughlings Jackson and the Cortical Motor Centers in the Light of Physiological Research," *Brain* 23 (1900): 545–81, 547; Constantin von Monakow and Raoul Mourgue, *Introduction biologique à l'étude de la neurologie et de la psychopathologie: Intégration et désintégration de la fonction* (Paris: Félix Alcan, 1928); Jean Delay, "Ribot et le Jacksonisme," *Dialectica* 5, no. 3/4 (1951): 413–44; Delay, "Jacksonism

and the Works of Ribot," *AMA Archives of Neurology and Psychiatry* 78 (1957): 505–15; Antoine Porot, "Jacksonisme et Néo-Jacksonisme-Dissolution," in *Manuel alphabétique de psychiatrie clinique et thérapeutique*, ed. Antoine Porot (Paris: Presses universitaires de France, 1960), 319–21; Kenneth Dewhurst, *Hughlings Jackson on Psychiatry* (Oxford: Sandford Publications, 1982), 103–36; S. P. Fullinwider, "Sigmund Freud, John Hughlings Jackson, and Speech," *Journal of the History of Ideas* 44, no. 1 (1983): 151–58; Anne Harrington, *Medicine, Mind, and the Double Brain* (Princeton: Princeton University Press, 1987), chap. 8; Benedetto Farina et al., "Henri Ey's Neo-jacksonism and the Psychopathology of Disintegrated Mind," *Psychopathology* 38 (2005): 285–90; Stephen Jacyna, "The Contested Jacksonian Legacy," *Journal of the History of the Neurosciences* 16 (2007): 307–17; Serge Nicolas, "'A Big Piece of News': Théodule Ribot and the Founding of the *Revue Philosophique de la France et de l'Etranger*," *Journal of the History of the Behavioral Sciences* 49, no. 1 (2013): 1–17; Nima Bassiri, "Freud and the Matter of the Brain: On the Rearrangements of Neuropsychoanalysis," *Critical Inquiry* 40, no. 1 (2013): 83–108.

63. John Hughlings Jackson, "Remarks on Evolution and Dissolution in Nervous Maladies (with Special Reference to Double Mental Conditions in Paroxysms of Uncinate Epilepsies and to So-Called 'Procursive Epilepsy')," 2–3 (Queen Square Archives, File E11). The document is undated but probably written sometime around 1890.

64. Michael J. Clark, "The Rejection of Psychological Approaches to Mental Disorder in Late Nineteenth-Century British Psychiatry," in *Madhouses, Mad-Doctors, and Madmen: The Social History of Psychiatry in the Victorian Era*, ed. Andrew Scull (Philadelphia: University of Pennsylvania Press, 1981); Smith, *Inhibition*, 40, 162–78.

65. Cesare Lombroso, "Atavism and Evolution," *Contemporary Review* 68 (1895): 42–49, in *The Criminal Anthropological Writings of Cesare Lombroso Published in English-Language Periodical Literature During the Late 19th and Early 20th Centuries*, ed. David M. Horton and Katherine E. Rich, Criminology Studies 22 (Lewiston, NY: Edwin Mellen, 2004), 53–62.

66. Jackson, "Remarks on the Double Condition," 63–64, 63n (b). This is not at all to suggest that Jackson always avoided neurologically based civilization comparisons. See Jackson, "Physiological Aspects of Education," in Taylor, *Selected Writings*, 2:265–69.

67. Herbert Spencer, *First Principles* (London, 1862), 221; 2nd ed. (London, 1867), 352, 519–41; Herbert Spencer, *Principles of Psychology*, 2nd ed. (London, 1870), 105–8, 116; C. U. M. Smith, "Evolution and the Problem of Mind: Part I, Herbert Spencer," *Journal of the History of Biology* 15, no. 1 (1982): 55–88; Smith, "Evolution and the Problem of Mind: Part II, John Hughlings Jackson," *Journal of the History of Biology* 15, no. 2 (1982): 241–62; Janet Browne, *Charles Darwin: A Biography*, vol. 2, *The Power of Place* (Princeton: Princeton University Press, 2003), 183–89. See also J. W. Burrow, *The Crisis of Reason: European Thought, 1848–1914* (New Haven: Yale University Press, 2000), chap. 1; Allan Young, "Bodily Memory and Traumatic Memory," in *Tense Past: Cultural Essays in Trauma and Memory*, ed. Paul Antze and Michael Lambek (New York: Routledge, 1996).

68. John Hughlings Jackson, "Remarks on Evolution and Dissolution of the Nervous System," in Taylor, *Selected Writings*, 2:76–118.

69. John Hughlings Jackson, "Some Implications of Dissolution of the Nervous System" (1882), in Taylor, *Selected Writings*, 2:41–42.

70. John Hughlings Jackson, "Remarks on the Hierarchy of the Nervous System" (Queen Square Archives, C28, c. 1899), 4 (emphasis in original).

71. John Hughlings Jackson, "Notes on the Physiology and Pathology of Language" (1866), in Taylor, *Selected Writings*, 2:121–28.

72. John Hughlings Jackson, "Observations on the Physiology of Language" (Queen Square Archives, A20, [QSA/839], 1868), 34.

73. Jackson, "Remarks on the Double Condition," 63.

74. "Hors de combat" was a turn of phrase that Jackson repeated in many instances. See, for example, John Hughlings Jackson, "On Affectations of Speech from Disease of the Brain," in Taylor, *Selected Writings*, 2:203-4.

75. Smith, *Inhibition*.

76. Jackson, *Neurological Fragments*, 144.

77. Jackson, "On Temporary Mental Disorders," 122.

78. Jackson, "Some Implications," in Taylor, *Selected Writings*, 2:30; John Hughlings Jackson, "Remarks on Dissolution of the Nervous System as Exemplified by Certain Post-Epileptic Conditions" (1881), in Taylor, *Selected Writings*, 2:6.

79. Jackson, *Neurological Fragments*, 49.

80. Jackson viewed the right brain hemisphere as related to automatic functions. See Harrington, *Medicine, Mind, and the Double Brain*, chap. 7.

81. Jackson, "On Temporary Mental Disorders," 123.

82. John Hughlings Jackson, "On Normal Nervous Discharges" (Queen Square Archives, C30, c. 1898), 12.

83. Jackson, "On Normal Nervous Discharges," 17.

84. John Hughlings Jackson, George H. Savage, Charles Mercier, and John Milne Bradwell, "On Imperative Ideas: Being a Discussion on Dr. Hack Tuke's Paper," *Brain* 18 (1895): 318-51, 320.

85. John Hughlings Jackson, "On Post-Epileptic States," in Taylor, *Selected Writings*, 1:382.

86. Jackson, "On Post-Epileptic States," 383.

87. John Hughlings Jackson, "The Factors of Insanities," in Taylor, *Selected Writings*, 2:415-16.

88. Jackson, *Neurological Fragments*, 48.

89. Jackson, "An Address on the Psychology of Joking," in Taylor, *Selected Writings*, 2:359.

90. Jackson, "An Address on the Psychology of Joking," 362.

91. Henri Ey and Julien Rouart, *Essai d'application des principes de Jackson à une conception dynamique de la neuro-psychiatrique* (Paris: G. Doin, 1938); Henri Ey, *Des idées de Jackson à un modèle organo-dynamique en psychiatrie* (Toulouse: Privat, 1975); Philip Evans, "Henri Ey's Concepts of the Organization of Consciousness and Its Disorganization: An Extension of Jacksonian Theory," *Brain* 95 (1972): 413-40; Benedetto Farina et al., "Henri Ey's Neojacksonism and the Psychopathology of Disintegrated Mind," *Psychopathology* 38 (2005): 285-90; G. E. Berrios, "Henri Ey, Jackson et les idées obsédantes," *L'évolution psychiatrique* 42 (1977): 685-99.

92. Roger Smith, "Mental Disorder, Criminal Responsibility, and the Social History of Theories of Volition," *Psychological Medicine* 9 (1979): 13-19, 15; G. E. Berrios and M. Gili, "Will and Its Disorders: A Conceptual History," *History of Psychiatry* 6 (1995): 87-104; Kathleen Haack, Ekkehardt Kumbier, and Sabine C. Herpertz, "Illness of the Will in 'Pre-Psychiatric' Times," *History of Psychiatry* 21, no. 3 (2010): 261-77; Roger Smith, *Free Will and the Human Sciences in Britain, 1870-1910* (London: Pickering, 2013); Smith, *Inhibition*, 167; Lorraine J. Daston, "The Theory of Will versus the Science of Mind," in Woodward and Ash, *Problematic Science*, 100-103; Ian Hacking, "Nineteenth-Century Cracks in the Concept of Determinism," *Journal of the History of Ideas* 44, no. 3 (1983): 455-75; Jonathan Miller, "Going Unconscious," in *Hidden Histories of Science*, ed. Robert B. Silvers (New York: New York Review of Books, 1995); Jonathan Miller, "Mental Action and the Threat of Automaticity," in *Decomposing the Will*, ed. Andy Clark, Julian Kiverstein, and Tillmann Vierkant (Oxford: Oxford University Press, 2013).

93. Leon Solomons and Gertrude Stein, "Normal Motor Automatism," *Psychological Review* 3, no. 5 (1896): 492–512, 509 (emphasis in original); Janet Hobhouse, *Everybody Who Was Anybody: A Biography of Gertrude Stein* (New York: Putnam, 1975), 13–18; Michael J. Hoffman, "Gertrude Stein in the Psychology Laboratory," *American Quarterly* 17, no. 1 (1965): 127–32.

94. Gertrude Stein, "Cultivated Motor Automatism: A Study of Character in Its Relation to Attention," *Psychological Review* 5, no. 3 (1898): 295–306, 298.

95. Pierre Janet, "Les actes inconscients et la mémoire: Pendant le somnambulisme," *Revue philosophique de la France et de l'étranger* 25 (1888): 238–79, 252.

96. F. W. H. Myers, "French Experiments on the Strata of Personality," *Proceedings of the Society for Psychical Research* 5 (1888–1889): 374–97, 386.

97. Luys, *De l'automatisme*, 23.

98. Ernest Mesnet, *De l'automatisme de la mémoire et du souvenir dans le somnambulisme pathologique: Considérations médico-légale* (Paris, 1874), 9, 14, 29–30. The case was reported for British readers as "Automatism," *Journal of Mental Science* 21, no. 95 (1875): 415–21.

99. Pierre Janet, *L'automatisme psychologique: Essai de psychologie expérimentale sur les forms inférieures de l'activité humaine* (Paris, 1889), 470.

100. Janet, *L'automatisme*, 460.

101. Maxime Laignel-Lavastine, *Automatisme mental et organicité*, extrait de *La Presse Médicale* 4, Mai 1923 (Paris: Masson, 1928), 5–6; Alain Segal and Alain Lellouch, "Maxime Laignel-Lavastine (1875–1953)," *Histoire des sciences médicales* 27, no. 3 (1993): 201–6; Julien Rouart, "Janet et Jackson," *L'évolution psychiatrique* 3 (1950): 485–96; P. Estingoy, "Le concept d'automatisme à l'épreuve de la filiation: De Charles Richet à Pierre Janet," *Annales médico-psychologiques* 166 (2008): 177–84.

102. Janet, *L'automatisme*, 444–45; Jan Goldstein, "Neutralizing Freud: The Lycée Philosophy Class and the Problem of the Reception of Psychoanalysis in France," *Critical Inquiry* 40, no. 1 (2013): 40–82, 65–66.

103. Hippolyte Bernheim, *De la suggestion dans l'état hypnotique et dans l'état de veille* (Paris, 1884), 87.

104. Hippolyte Bernheim, *Automatisme et suggestion* (Paris, 1917), 12.

105. Auguste Forel, *Der Hypnotismus: Seine Bedeutung und seine Handhabung* (Stuttgart, 1889), 49.

106. Andreas Mayer, *Sites of the Unconscious: Hypnosis and the Emergence of the Psychoanalytic Setting* (Chicago: University of Chicago Press, 2013); Elisabeth Roudinesco, *Histoire de la psychanalyse en France*, vol. 1, *1885–1939* (Paris: Fayard, 1994), 51–84.

107. Charcot and Richet, "Note on Certain Facts of Cerebral Automatism," 12–13.

108. Eduard Hitzig, *Welt und Gehirn: Ein essay* (Berlin: Hirschwald, 1905), 45; Eduard Hitzig, "The World and Brain," *International Quarterly* 10 (Oct. 1904–Jan. 1905): 165–80, 319–48, 332; Mayer, *Sites of the Unconscious*, chaps. 1 and 2; Jacqueline Carroy and Régine Plas, "The Origins of French Experimental Psychology: Experiment and Experimentalism," *History of the Human Sciences* 9, no. 1 (1996): 73–84.

109. Archibald Stodart Walker, "On Hypnotism and Crime," *Edinburgh Medical Journal*, n.s., 3 (1898): 65–68, 68; William Julius Mickle, "General Paralysis," in Tuke, *Dictionary of Psychological Medicine*, 1:529; Mickle, *General Paralysis of the Insane* (London, 1888); Nye, *Crime, Madness, and Politics in Modern France*; Susanna Barrows, *Distorting Mirrors: Visions of the Crowd in Late Nineteenth-Century France* (New Haven: Yale University Press, 1981); Ruth Harris, "Murder under Hypnosis in the Case of Gabrielle Bompard: Psychiatry in the Courtroom in Belle Époque Paris," in *The Anatomy of Madness*, vol. 3, *The Asylum and Its Psychiatry*, ed. W. F. Bynum, Roy

Porter, and Michael Shepherd (London: Routledge, 1988), 197–241; Emese Lafferton, "Murder by Hypnosis? Altered States and the Mental Geography of Science," in *Medicine, Madness, and Social History: Essays in Honour of Roy Porter*, ed. Roberta Bivins and John V. Pickstone (New York: Palgrave Macmillan, 2007), 182–96.

110. Jules Luys, *Questions médico-légales afférentes à l'hypnotisme* (Clermont, 1891), 3, 11.

111. Ernest Dupré and Georges Rocher, *L'hypnotisme devant la loi, XIIIe Congrès International de Médecine, Section de Médecine Légale, Paris 1900* (Clermont, 1901), 14.

112. Nima Bassiri, "Epileptic Insanity and Personal Identity: John Hughlings Jackson and the Formations of the Neuropathic Self," in *Plasticity and Pathology: On the Formations of the Neural Subject*, ed. David Bates and Nima Bassiri (New York: Fordham University Press, 2016).

113. Fulgence Raymond, "Les délires ambulatoires ou les fugues," *Gazette des hôpitaux* 79 (July 9, 1895): 787–93, 793.

114. Adrien Proust, "Un cas curieux d'automatisme ambulatoire chez un hystérique," *Tribune médicale* 27 (March 27, 1890): 202–3, 203.

115. Luys, *Questions médico-légales*, 7. A holographic will is a will and testament handwritten and signed by a testator in the absence of any corroborating witnesses.

116. Albert Moll, *Der Hypnotismus*, 2nd expanded ed. (Berlin, 1890), 284.

117. "Ambulatory Automatism," *The Lancet*, April 20, 1889, 807; Ian Hacking, *Mad Travelers: Reflections on the Reality of Transient Mental Illnesses* (Cambridge, MA: Harvard University Press, 1998), 57.

118. Jean-Martin Charcot, *Leçons du mardi à la Salpêtrière, Policliniques, 1887–1888, Notes de Cours* (Paris, 1887), 155–70; Charcot, *Leçons du mardi à la Salpêtrière, Policlinique, 1887–1888*, vol. 1, 2nd ed. (Paris, 1892), 112–25; Charcot, *Leçons du mardi à la Salpêtrière, Policlinique, 1888–1889* (Paris, 1889), 203–15; Charcot, *Charcot, the Clinician: The Tuesday Lessons; Excerpts from Nine Case Presentations on General Neurology Delivered at the Salpêtrière Hospital in 1887–88*, trans. Christopher Goetz (New York: Raven Press, 1987), 26–46; Georges Sous, *De l'automatisme comitial ambulatoire* (Paris, 1890), 28–35; Christopher G. Goetz and Emmanuel Drouin, "Two Faces of the Teacher: Comparing Editions of Charcot's *Leçons du mardi*," *Journal of the History of the Neurosciences*, 2022, 1–12.

119. Ian Hacking, "Les aliénés voyageurs: How Fugue Became a Medical Entity," *History of Psychiatry* 7 (1996): 425–49, 435.

120. Charcot, *Leçons* (1889), 303.

121. Charcot, *Leçons* (1889), 320.

122. Philippe Auguste Tissié, *Les aliénés voyageurs: Essai médico-psychologique* (Bordeaux, 1887); Ian Hacking, "*Automatisme ambulatoire*: Fugue, Hysteria, and Gender at the Turn of the Century," *Modernism/modernity* 3, no. 2 (1996): 31–41.

123. Hacking, "Les aliénés voyageurs," 430; Charcot, *Leçons* (1889), 307–8.

124. Charcot, *Leçons* (1889), 316.

125. Daniel Hack Tuke, *Sleep-Walking and Hypnotism* (London, 1884), 8; Charcot, *Leçons du mardi à la Salpêtrière, 1888–1889*, 307–8.

126. Charcot, *Leçons* (1887), 161; (1892), 117.

127. Charcot, *Leçons* (1887), 156; (1892), 113.

128. Charcot, *Leçons* (1887), 166; (1892), 121; (1889), 310.

129. Charcot, *Leçons* (1887), 158; (1892), 115.

130. Charcot, *Leçons* (1887), 158; (1892), 115.

131. Charcot, *Leçons* (1889), 311.

132. Charcot, *Leçons* (1887), 160; (1892), 116.

133. Charcot, *Leçons* (1889), 305.

134. Charcot, *Leçons* (1887), 164; (1892), 119.

135. Charcot, *Leçons* (1887), 169; (1892), 124.

136. Charcot, *Leçons* (1889), 317–25.

137. Bassiri, "Epileptic Insanity and Personal Identity," 80–82.

138. Jackson, "On Temporary Mental Disorders," 125–26 (emphasis in original, which Jackson notes represents the sections underlined by the patient); John Hughlings Jackson, "Les troubles intellectuels momentanés qui suivent les accès épileptiques," *Revue philosophique de la France et de l'étranger* 34, 2nd series, 5th year (Feb. 19, 1876): 169–78.

139. Jackson, "On Temporary Mental Disorders," 126.

140. Charcot, *Leçons* (1887), 167.

141. Charcot, *Leçons* (1889), 316; (1892), 123.

142. Charcot, *Leçons* (1889), 316.

143. Jackson, "On Temporary Mental Disorders," 22.

144. Erika Lorraine Milam and Robert A. Nye, "An Introduction to *Scientific Masculinities*," *Osiris* 30 (2015): 1–14.

145. Jessie Hewitt, *Institutionalizing Gender: Madness, the Family, and Psychiatric Power in Nineteenth-Century France* (Ithaca: Cornell University Press, 2020).

Chapter Two

1. Charles L. Dana, "The Partial Passing of Neurasthenia," *Boston Medical and Surgical Journal* 150, no. 13 (1904): 339–44, 339–40; M. Dominic Beer, "The Dichotomies: Psychosis/Neurosis and Functional/Organic; A Historical Perspective," *History of Psychiatry* 7 (1996): 231–55.

2. Freud, "The Psychopathology of Everyday Life," 278.

3. Daniel Hack Tuke, "Imperative Ideas," in *Dictionary of Psychological Medicine*, 2:678–81; Tuke, "Imperative Ideas," *Brain* 17 (1894): 179–97; Michael J. Clark, "'Morbid Introspection,' Unsoundness of Mind, and British Psychological Medicine, c. 1830–c. 1900," in Bynum, Porter, and Shepherd, *Anatomy of Madness*, 3:77–80.

4. Tuke, "Imperative Ideas," *Brain*, 179.

5. Tuke, "Imperative Ideas," *Brain*, 182–83.

6. Tuke, "Imperative Ideas," *Brain*, 181.

7. Tuke, "Imperative Ideas," *Brain*, 194.

8. Jackson et al., "On Imperative Ideas," 323.

9. Jackson et al., "On Imperative Ideas," 328.

10. Alfred Binet, "Hughlings-Jackson, Savage, Mercier, Milne-Bramwell—On Imperative Ideas (Sur les idées impératives), Brain été 1895, 318–51," *L'année psychologique* 2 (1895): 889–90.

11. Pierre Janet, "Pierre Janet," trans. Dorothy Olson, in *A History of Psychology in Autobiography*, ed. Carl Murchison (Worcester, MA: Clarke University Press, 1930), 1:130–33; Pierre Janet, *L'amour et la haine: Notes de cours recueillies et rédigées par M. Miron Epstein* (Paris: Tessier, 1932), 117–32.

12. René Descartes, "Meditations on First Philosophy," in *The Philosophical Writings of Descartes*, ed. John Cottingham, Robert Stoothoff, and Dugald Murdoch (Cambridge: Cambridge University Press, 1985), 2:21; C. Adam and P. Tannery, eds., *Oeuvres de Descartes* (Paris: J. Vrin, 1974–1982), 7:32.

13. Janet, "Pierre Janet," 131.

14. John B. Watson, *Behavior: An Introduction to Comparative Psychology* (New York: Holt, 1914), 27.

15. Janet, "Pierre Janet," 131.

16. Mark Addison Amos, "'For Manners Make Man': Bourdieu, de Certeau, and the Common Appropriation of Noble Manners in the *Book of Courtesy*," Anna Dronzek, "Gendered Theories of Education in Fifteenth-Century Conduct Books," and Jennifer Fisk Rondeau, "Conducting Gender: Theories and Practices in Italian Confraternity Literature," in *Medieval Conduct*, ed. Kathleen Ashley and Robert L. A. Clark (Minneapolis: University of Minnesota Press, 2001).

17. Aristotle, *Nicomachean Ethics*, trans. Roger Crisp (Cambridge: Cambridge University Press, 2014).

18. Roger Smith, "What Is the History of the Human Science?," in McCallum, *Palgrave Handbook of the History of the Human Sciences*; Smith, *Being Human*; Smith, *Norton History of the Human Sciences*.

19. Wilhelm Dilthey, *Introduction to the Human Sciences: An Attempt to Lay a Foundation for the Study of Society and History*, trans. Ramon J. Betanzos (Detroit: Wayne State University Press, 1988), 83–88; Dilthey, *Einleitung in die Geisteswissenschaften* (Leipzig, 1883), 17–26; Rudolf A. Makkreel, *Dilthey: Philosopher of the Human Sciences*, 2nd ed. (Princeton: Princeton University Press, 1992), 35–73; Rudolf A. Makkreel, "The Emergence of the Human Sciences from the Moral Sciences," in *The Cambridge History of Philosophy in the Nineteenth Century (1790–1870)*, ed. Allen W. Wood (Cambridge: Cambridge University Press, 2012).

20. Canguilhem, "The Normal and the Pathological," in *Knowledge of Life*, 125.

21. Gaukroger, *The Natural and the Human*, 5.

22. Michel Foucault, *Les mots et les choses* (Paris: Gallimard, 1966), 365; Foucault, *The Order of Things* (New York: Vintage, 1994), 354 (trans. modified).

23. Paul Veyne, *Comment on écrit l'histoire: Essai d'épistémologie* (Paris: Editions du Seuil, 1971), 290; Veyne, *Writing History: Essay on Epistemology*, trans. Mina Moore-Rinvolucri (Middletown, CT: Wesleyan University Press, 1984), 246.

24. Helen E. Longino, *Studying Human Behavior: How Scientists Investigate Aggression and Sexuality* (Chicago: University of Chicago Press, 2013), 166.

25. Arnold I. Davidson, "In Praise of Counter-Conduct," *History of the Human Sciences* 24, no. 4 (2011): 25–41, 31; Liz McFall, Paul du Gay, and Simon Carter, eds., *Conduct: Sociology and Social Worlds* (Manchester: Manchester University Press, 2008).

26. Helmut Lethen, *Cool Conduct: The Culture of Distance in Weimar Germany*, trans. Don Reneau (Berkeley: University of California Press, 2002), 18.

27. Max Weber, *Wirtschaft und Gesellschaft*, in *Grundrisse der Sozialökonomik*, pt. 3 (Tübingen: J.C.B. Mohr (P. Siebeck), 1922), 2; Weber, *Economy and Society*, 4 (trans. modified).

28. Durkheim, *Rules of Sociological Method*, 15.

29. Gustave Le Bon, *Psychologie des foules* (Paris, 1895); Le Bon, *The Crowd: A Study of the Popular Mind* (London and New York, 1896); Georg Simmel, "The Metropolis and Mental Life (1903)," in *George Simmel: On Individuality and Social Forms*, ed. Donald N. Levine (Chicago: University of Chicago Press, 1971); George Herbert Mead, *Mind, Self, and Society* (Chicago: University of Chicago Press, 1934) 68; Marcel Mauss, "Techniques of the Body (1934)," *Economy and Society* 2, no. 1 (1973): 70–88; Jef C. Verhoeven, "An Interview with Erving Goffman, 1980," in *Erving Goffman*, ed. Gary Alan Fine and Gregory W. H. Smith (London: Sage, 2000), 1:217–18; Anne Warfield Rawls, "The Interaction Order Sui Generis: Goffman's Contribution to Social

Theory," in Fine and Smith, *Erving Goffman*, vol. 2; Richard W. Burkhardt Jr., *Patterns of Behavior: Konrad Lorenz, Niko Tinbergen, and the Founding of Ethology* (Chicago: University of Chicago Press, 2005).

30. Freud, "Psychopathology of Everyday Life," 157–58n1.

31. Joseph De Rivera, ed., *Field Theory as Human-Science: Contributions of Lewin's Berlin Group* (New York: Gardner Press, 1976); Elias, *The Civilizing Process*, 408–9; Norbert Elias, *Über den Prozess der Zivilisation: Soziogenetische und psychogenetische Untersuchungen*, 2nd ed. (Bern: Franke Verlag, 1969 [1939]), 2:388–89; Bernard Lahire, "Elias, Freud, and the Human Science," in *Norbert Elias and Social Theory*, ed. François Dépelteau and Tatiana Savoia Landini (New York: Palgrave Macmillan, 2013); Johan Goudsblom, "The Sociology of Norbert Elias: Its Resonance and Significance," *Theory, Culture & Society* 4 (1987): 323–37; Richard Kilminster, "The Dawn of Detachment: Norbert Elias and Sociology's Two Tracks," *History of the Human Sciences* 27, no. 3 (2014): 96–115; Pierre Bourdieu, *Outline of a Theory of Practice*, trans. Richard Nice (Cambridge: Cambridge University Press, 1977); Bourdieu, *The Logic of Practice*, trans. Richard Nice (Stanford: Stanford University Press, 1992); Bourdieu, "Habitus," in *Habitus: A Sense of Place*, ed. Jean Hillier and Emma Rooksby, 2nd ed. (Burlington, VT: Ashgate, 2005); Craig Calhoun, Edward LiPuma, and Moishe Postone, eds., *Bourdieu: Critical Perspectives* (Chicago: University of Chicago Press, 1993); Richard Shusterman, ed., *Bourdieu: A Critical Reader* (Oxford: Blackwell, 1999); Rob Boddice, *The History of Emotions* (Manchester: Manchester University Press, 2018), 194–204.

32. John Dewey, "The New Psychology," *Andover Review* 2 (September 1884): 278–90, 279. Cf. Edwin Boring, *A History of Experimental Psychology* (New York: Appleton-Century, 1929), 377; Gary Hatfield, "Psychology: Old and New," in *The Cambridge History of Philosophy, 1870–1945*, ed. Thomas Baldwin (Cambridge: Cambridge University Press, 2003).

33. Woodward and Ash, *Problematic Science*; Mitchell G. Ash and William R. Woodward, eds., *Psychology in Twentieth-Century Thought and Society* (Cambridge: Cambridge University Press, 1987); Ellenberger, *Discovery of the Unconscious*; Jacqueline Carroy and Régine Plas, "The Origins of French Experimental Psychology: Experiment and Experimentalism," *History of the Human Sciences* 9, no. 1 (1996): 73–84; Kurt Danziger, *Constructing the Subject: Historical Origins of Psychological Research* (Cambridge: Cambridge University Press, 1990); Martin Kusch, *Psychological Knowledge: A Social History and Philosophy* (New York: Routledge, 1999). See also Eric Shiraev, "Psychology and Psychiatry," in *The Fin-de-Siècle World*, ed. Michael Saler (New York: Routledge, 2015); George H. Savage, *The Harveian Oration on Experimental Psychology and Hypnotism: Delivered Before the Royal College of Physicians of London, October 18, 1909* (London: Henry Frowde, 1909).

34. David W. Jones, "Moral Insanity and Psychological Disorder: The Hybrid Roots of Psychiatry," *History of Psychiatry* 28, no. 3 (2017): 263–79; Jan Verplaetse, *Localizing the Moral Sense: Neuroscience and the Search for the Cerebral Seat of Morality, 1800–1930* (New York: Springer, 2009); Eric Carlson and Norman Dain, "The Meaning of Moral Insanity," *Bulletin of the History of Medicine* 36 (1962): 130–40; Henri Ey, "La notion de 'maladie morale' et de 'traitement moral' dans la psychiatrie française et allemande du début du XIXe siècle," *Perspectives psychiatriques* 65 (1978): 12–36.

35. Immanuel Kant, "Anthropology from a Pragmatic Point of View," in Günter Zöller and Robert B. Louden, *Anthropology, History, and Education* (Cambridge: Cambridge University Press, 2007), 324; William Battie, *A Treatise on Madness* (London, 1758), 5–6; Thomas Arnold, *Observations on the Nature, Kinds, Causes, and Prevention of Insanity and Lunacy or Madness*

(Leicester, 1782), 1:80; Anne C. Vila, *Enlightenment and Pathology: Sensibility in the Literature and Medicine of Eighteenth-Century France* (Baltimore: Johns Hopkins University Press, 1998), chap. 3; Roger Smith, "The Law and Insanity in Great Britain, with Comments on Continental Europe," in *Ziek of schuldig? Twee eeuwen forensische psychiatrie en psychologie*, ed. F. Koenraadt (Arnhem: Gouda Quint, 1991); Andrew Scull, "Moral Treatment Reconsidered: Some Sociological Comments on an Episode in the History of British Psychiatry," in Scull, *Madhouses, Mad-Doctors, and Madmen*.

36. Gauchet and Swain, *Madness and Democracy*, 158–59.

37. Philippe Pinel, *Traité médico-philosophique sur l'aliénation mentale*, 2nd ed. (Paris, 1809), 93, 102. The second edition represents a very slight modification to an equivalent passage from the first edition. See Philippe Pinel, *Traité médico-philosophique sur l'aliénation mentale ou la manie* (Paris, 1801), 23, 81; Goldstein, *Console and Classify*, chap. 5.

38. Samuel Tuke, *Description of the Retreat, an Institution near York, for Insane Persons of the Society of Friends, Containing an Account of Its Origin and Progress, the Modes of Treatment, and a Statement of Cases* (York, 1813), 133–34, 139–42; Anne Digby, *Madness, Morality, and Medicine: A Study of the York Retreat, 1896–1914* (Cambridge: Cambridge University Press, 1985), 93–100; Benjamin Rush, *Medical Inquiries and Observations upon the Diseases of the Mind* (Philadelphia, 1812), 263; Nicole Rafter, "The Unrepentant Horse-Slasher: Moral Insanity and the Origins of Criminological Thought," *Criminology* 42, no. 4 (2004): 979–1008.

39. John Conolly, *An Inquiry Concerning the Indications of Insanity, with Suggestion for the Better Protection and Care of the Insane* (London, 1830), 300; Andrew Scull, *Social Order/Mental Disorder: Anglo-American Psychiatry in Historical Perspective* (Berkeley: University of California Press, 1989), chap. 7.

40. Johann Gaspar Spurzheim, *Observations on the Deranged Manifestations of the Mind, or Insanity* (London, 1817), 68.

41. Johann Gaspar Spurzheim, *Observations on the Deranged Manifestations of the Mind, or Insanity*, 2nd American ed. (Boston, 1835), 54, 58–59.

42. Nietzsche, *On the Genealogy of Morality*, 37–38.

43. Johann Christian Reil, "Ueber den Begriff der Medicin und ihre Verzweigungen, besonders in Beziehung auf die Berichtigung der Topik der Psychiatrie," *Beyträge zur Beförderung einer Kurmethode auf psychischem Wege* 1 (1808): 161–279; Max Neuburger, "British and German Psychiatry in the Second Half of the Eighteenth and Early Nineteenth Century," *Bulletin of the History of Medicine* 18, no. 2 (1945): 121–45.

44. Johann Christian Reil, *Rhapsodieen über die Anwendung der psychischen Curmethode auf Geisteszerrüttungen*, 2nd ed. (Halle, 1818), 361–73; Johann Christian Heinroth, *Lehrbuch der Störungen des Seelenlebens oder der Seelenstörungen und ihre Behandlung* (Leipzig, 1818), 1:36; Heinroth, *System der psychisch-gerichtlichen Medizin* (Leipzig, 1825), 177; Otto M. Marx, "German Romantic Psychiatry, Part 1," *History of Psychiatry* 1 (1990): 351–81; Marx, "German Romantic Psychiatry, Part 2," *History of Psychiatry* 2 (1991): 1–25; Theodore Ziolkowski, *German Romanticism and Its Institutions* (Princeton: Princeton University Press, 1990), 138–217; Robert J. Richards, *The Romantic Conception of Life: Science and Philosophy in the Age of Goethe* (Chicago: University of Chicago Press, 2002), 254–88; Robert J. Richards, "Rhapsodies on a Cat-Piano, or Johann Christian Reil and the Foundations of Romantic Psychiatry," *Critical Inquiry* 24, no. 3 (1998): 700–736; LeeAnn Hansen, "Metaphors of Mind and Society: The Origins of German Psychiatry in the Revolutionary Era," *Isis* 89, no. 3 (1998): 387–409; Otto M. Marx, "J. C. A. Heinroth (1773–1843) on Psychiatry and the Law," *Journal of the History of the Behavioral Sciences* 4, no.

2 (1968): 163–79; Marx, "The Beginning of Psychiatric Historiography in Nineteenth-Century Germany," in *Discovering the History of Psychiatry*, ed. Mark S. Micale and Roy Porter (New York: Oxford University Press, 1994), 39–52; Gerlof Verwey, *Psychiatry in an Anthropological and Biomedical Context: Philosophical Presuppositions and Implications of Germany Psychiatry, 1820–1870* (Dordrecht: D. Reidel, 1985), 15–20; Holger Steinberg, "The Sin in the Aetiological Concept of Johann Christian August Heinroth (1773–1843), Part 1: Between Theology and Psychiatry; Heinroth's Concept of 'Whole Being,' 'Freedom,' 'Reason,' and 'Disturbance of the Soul,'" *History of Psychiatry* 15, no. 3 (2004): 329–44.

45. Johann Christoph Hoffbauer, *Die Psychologie in ihren Hauptanwendungen auf die Rechtspflege nach den allgemeinen Gesichtspunkten der Gesetzgebung*, 2nd ed. (Halle, 1823 [1808]), 11n.

46. Johan Christoph Hoffbauer, *Médecine légale relative aux aliénés et aux sourds-muets, ou les lois appliquées aux désordres de l'intelligence*, trans. A.-M. Chambeyron (Paris, 1827); Goldstein, *Console and Classify*, 128–47, 152–89; Raymond de Saussure, "The Influence of the Concept of Monomania on French Medico-Legal Psychiatry (from 1825 to 1840)," *Journal of the History of Medicine and Allied Sciences* 1, no. 3 (1946): 365–97.

47. Yohan Trichet and Agnès Lacroix, "Esquirol's Change of View towards Pinel's Mania without Delusion," *History of Psychiatry* 27, no. 4 (2016): 443–57. See also Kathleen Haack, Ekkehardt Kumbier, and Sabine C. Herpertz, "Illnesses of the Will in 'Pre-psychiatric' Times," *History of Psychiatry* 21, no. 3 (2010): 261–77.

48. Hoffbauer, *Médecine légale*, 311–12.

49. Étienne Esquirol, *Note sur la monomanie-homicide* (Paris, 1827), 51; Esquirol, *Des maladies mental considérée sous les rapports medical, hygiénique et medico-légal* (Paris, 1838), 2:360.

50. Étienne Georget, *Discussion médico-légale sur la folie ou aliénation mentale* (Paris 1826), 70; Goldstein, *Console and Classify*, 174; Étienne Georget, *Examen médical des procès criminels des nommés Léger, Feldtmann, Lecouffe, Jean-Pierre et Papavoine, dans lesquels l'aliénation mentale a été alléguée comme moyen de défense; suivi de quelques considérations médico-légales sur la liberté morale* (Paris, 1825), 67–68; Ulysse Trélat, *Recherches historiques sur la folie* (Paris, 1839), 123.

51. Prichard, *A Treatise on Insanity*, 3; Verplaetse, *Localizing the Moral Sense*, 194–96; Hannah Franziska Augstein, "JC Prichard's Concept of Moral Insanity—a Medical Theory of the Corruption of Human Nature," *Medical History* 40 (1996): 311–43.

52. Prichard, *Treatise on Insanity*, 6.

53. Prichard, *Treatise on Insanity*, 14.

54. Prichard, *Treatise on Insanity*, 4.

55. Esquirol *Des maladies mentales*, 1:364–65; 2:5; Jules Falret, "On Moral Insanity," *American Journal of Insanity* 23–24 (1867): 407–24, 516–46, 52–63.

56. Alexandre Brière de Boismont, *Observations médico-légales sur la monomanie homicide* (Paris, 1827), 36–37.

57. Ray, *Treatise on the Medical Jurisprudence of Insanity*, 48–49; S. P. Fullinwider, "Insanity as the Loss of Self: The Moral Insanity Controversy Revisited," *Bulletin of the History of Medicine* 49, no. 1 (1975): 87–101.

58. Joel Peter Eigen, "Diagnosing Homicidal Mania: Forensic Psychiatry and the Purposeless Murder," *Medical History* 54 (2010): 433–56.

59. Isaac Ray, *Mental Hygiene* (Boston, 1863), 35–36.

60. Ray, *Mental Hygiene*, 62.

61. Forbes Winslow, "On Medico-Legal Evidence in Cases of Insanity," *American Journal of Insanity* 9 (1853): 365–84, 383–84. Cf. David W. Jones, "Moral Insanity and Psychological Dis-

order: The Hybrid Roots of Psychiatry," *History of Psychiatry* 28, no. 3 (2017): 263-79. See also David Skae, "Remarks on that Form of Moral Insanity Called Dipsomania, and the Legality of Its Treatment by Isolation," *Edinburgh Medical Journal* 3 (1858): 769-83, 771; Alexandre Brière de Boismont, *De la folie raisonnante et de l'importance du délire des actes pour le diagnostic et la médecine légale* (Paris, 1867), 94.

62. H. Manning, "Moral Insanity. Case of Homicidal Insanity," *Journal of Mental Science* 28 (1882): 369-72, 369-70 (emphasis in original).

63. Edward Jarvis, "Criminal Insane: Insane Transgressors and Insane Convicts," *American Journal of Insanity* 13, no. 3 (1857): 195-231, 195, 199; James R. King, "The Mysterious Case of the 'Mad' Rector of Bletchingdon: The Treatment of Mentally Ill Clergy in Late Thirteenth-Century England," in *Madness in Medieval Law and Custom*, ed. Wendy J. Turner (Leiden: Brill, 2010), 57-80.

64. John P. Gray, "Mental and Physical Characteristics of Pauperism," *American Journal of Insanity* 13 (1857): 309-20, 310-12.

65. Thomas Laycock, "Suggestions for Rendering Medico-Mental Science Available to the Better Administration of Justice and the More Effectual Prevention of Lunacy and Crime," *Journal of Mental Science* 14 (Oct. 1868): 334-45, 342.

66. J. Bruce Thompson, "Criminal Lunacy in Scotland for Quarter of a Century, viz., from 1846 to 1870, Both Inclusive," *Edinburgh Medical Journal* 17 (1871): 21-29, 21; Smith, *Free Will and the Human Sciences*; Ian Hacking, "Nineteenth-Century Cracks in the Concept of Determinism," *Journal of the History of Ideas* 44, no. 3 (1983): 455-75; Lorraine Daston, "British Responses to Psycho-Physiology, 1860-1900," *Isis* 69, no. 2 (1978): 192-208.

67. Verplaetse, *Localizing Moral Sense*, 198.

68. Bénédict Augustin Morel, *Traité des maladies mentales* (Paris, 1860), 545.

69. Morel, *Traité des maladies mentales*, 258-59.

70. Valentin Magnan, *Leçons cliniques sur les maladies mentales*, 2nd ed. (Paris, 1893), 296-97.

71. Henry Maudsley, *Body and Mind: An Inquiry into Their Connection and Mutual Influence, Especially in Reference to Mental Disorders* (London, 1873), 136.

72. Henry Maudsley, *The Physiology and Pathology of the Mind* (London, 1867), 312.

73. Ulysse Trélat, *La folie lucide: Étudiée et considérée au point de vue de la famille et de la société* (Paris, 1861), 7-12, 317.

74. Griesinger, *Die Pathologie und Therapie*, 71; Forbes Winslow, *On the Incubation of Insanity* (London, 1846), 14.

75. Falret, "Folie raisonnante," 475-525, 485-89; Maudlsey, *Physiology and Pathology*, 322; Maudsley, *Responsibility in Mental Disease*, 75; Jan Goldstein, "Professional Knowledge and Professional Self-Interest: The Rise and Fall of Monomania in 19th-Century Europe," *International Journal of Law and Psychiatry* 21, no. 4 (1998): 385-96.

76. Jules Falret, *Des aliénés dangereux et des asiles spéciaux pour les aliénés dits criminels* (Paris, 1869), 25.

77. Richard von Krafft-Ebing, *Lehrbuch der gerichtlichen Psychopathologie* (Stuttgart, 1875), 155-59, 162-63; Krafft-Ebing, *Lehrbuch der Psychiatrie*,1: 49; Krafft-Ebing, *Beiträge zur Erkennung und richtigen forensischen Beurtheilung krankhafter Gemüthszustände für Aerzte, Richter und Vertheidiger* (Erlangen, 1867), 19.

78. Richard von Krafft-Ebing, *Grundzüge der Kriminalpsychologie auf Grundlage des Strafgesetzbuchs des deutschen Reichs für Aerzte und Juristen* (Erlangen, 1872), 28.

79. Krafft-Ebing, *Grundzüge der Kriminalpsychologie*, 98. See Verplaetse, *Localizing the*

Moral Sense, 200–203; William Johnston, *The Austrian Mind: An Intellectual and Social History, 1848–1938* (Berkeley: University of California Press, 1972), 232–35.

80. Eugen Bleuler, *Der geborene Verbrecher: Eine kritische Studie* (Munich, 1896), 21, 26, 47.

81. Alexander Holländer, "Zur Lehre von der Moral Insanity," *Jahrbücher für Psychiatrie* 4 (1883): 1–18, 2, 13.

82. Emil Kraepelin, *Psychiatrie: Eine kurzes Lehrbuch für Studirende und Aerzte*, 2nd ed. (Leipzig, 1887), 55, 523–25.

83. Erdmann Müller, "Ueber 'Moral Insanity,'" *Archiv für Psychiatrie und Nervenkrankheiten* 31 (1899): 325–77, 335, 369.

84. Johannes Longard, "Ueber 'Moral Insanity,'" *Archiv für Psychiatrie und Nerven-krankheiten* 43 (1907): 135–233, 231; Arnold Pick, "Ueber Krankheitsbewusstsein in psychischen Krankheiten," *Archiv für Psychiatrie und Nervenkrankheiten* 13 (1908): 518–81, 533. See also Julius Ludwig August Koch, *Die psychopathischen Minderwertigkeiten* (Ravensburg, 1891–1893); James Horley, "The Emergence and Development of Psychopathy," *History of the Human Sciences* 27, no. 5 (2014): 91–110; Greg Eghigian, "A Drifting Concept for an Unruly Menace: A History of Psychopathy in Germany," *Isis* 106, no. 2 (2015): 283–309; Richard Noll, "Psychosis," in Eghigian, *Routledge History of Madness and Mental Health*.

85. Daniel Hack Tuke, "Moral or Emotional Insanity," *Journal of Mental Science* 31 (1885): 174–90, 183, 189; Tuke, "Moral Insanity," in Tuke, *Dictionary of Psychological Medicine*, 2:813–16, 813.

86. Foucault, "About the Concept of the 'Dangerous Individual,'" 7.

87. Tuke, "Moral Insanity," 814.

88. Charles Mercier, "Vice, Crime, and Insanity," in *A System of Medicine by Many Writers*, vol. 8, *Diseases of the Brain and Mental Diseases*, ed. Thomas Clifford Allbutt and Humphry Davy Rolleston (London: Macmillan, 1911), 842–74, 851.

89. Charles Mercier, "The Study of Insanity," *Journal of Mental Science* 27 (July 1881): 166–77; (Jan. 1882): 512–27, 527n.

90. Charles Arthur Mercier, "Recollections," in Jackson, *Neurological Fragments*, 40–46.

91. Charles Mercier, *A New Logic* (London: W. Heinemann, 1912).

92. Charles Arthur Mercier, *Conduct and Its Disorders: Biologically Considered* (London: Macmillan, 1911), vii.

93. Mercier, *Conduct and Its Disorders*, viii, xi.

94. Mead, *Mind, Self, and Society*, 41; Sam Parkovnick, "The Behaviorism of George Herbert Mead," *American Sociologist* 46 (2015): 288–93; John D. Baldwin, "George Herbert Mead and Modern Behaviorism," *Pacific Sociological Review* 24, no. 4 (1981): 411–40.

95. Mercier, *Conduct and Its Disorders*, 23–26.

96. Mercier, *Conduct and Its Disorders*, 53–59.

97. Charles Arthur Mercier, "Who Is Insane?," *The Hospital*, April 15, 1916, 49; Paul Bowden, "Pioneers in Forensic Psychiatry: Charles Arthur Mercier (1852–1919); Wit without Understanding," *Journal of Forensic Psychiatry* 5, no. 2 (1994): 321–53; "Medico-Psychological Association of Great Britain and Ireland," *Journal of Mental Science* 54, no. 227 (1908): 780–98; "Charles Arthur Mercier," *British Medical Journal* 2, no. 3063 (September 13, 1919): 363–65, 364; George H. Savage, "Obituary," *Journal of Nervous and Mental Disease* 52, no. 2 (1920): 189–92.

98. Mercier, "Who Is Insane?," 49.

99. H. Bryan Donkin, "Charles Arthur Mercier," *Journal of Mental Science* 66, no. 272 (1920): 1–10, 5. See also Charles Mercier, "The Study of Conduct," *British Medical Journal*, Jan. 27, 1912, 220; H. Hayes Newington, "The Study of Conduct," *British Medical Journal*, Feb. 3, 1912, 272.

100. Mercier, "Study of Insanity," 166–67.

101. Mercier, "Study of Insanity," 173.

102. Mercier, "Study of Insanity," 515–18.

103. Mercier, "Study of Insanity," 522; also 527n.

104. Charles Mercier, *Sanity and Insanity* (London, 1890), 99.

105. Mercier, *Sanity and Insanity*, 100.

106. John Conolly, *An Inquiry Concerning the Indications of Insanity, with Suggestions for the Better Protection and Care of the Insane* (London, 1830), chap. 5.

107. Mercier, *Sanity and Insanity*, 102.

108. George H. Savage and Charles Mercier, "Insanity of Conduct," *Journal of Mental Science* 42 (Jan. 1896): 1–17, 12.

109. Mercier, "Study of Insanity," 519.

110. Charles Mercier, *The Nervous System and the Mind: A Treatise on the Dynamics of the Human Organism* (London, 1888), 159.

111. Savage and Mercier, "Insanity of Conduct," 2; George H. Savage, "Moral Insanity," *Journal of Mental Science* 27 (July 1881): 147–55, 147–48.

112. Savage and Mercier, "Insanity of Conduct," 8.

113. Charles Mercier, *A Text-Book of Insanity* (London: Swan Sonnenschein, 1902), 96–98; 2nd ed. (London: George Allen and Unwin, 1914), 132.

114. Mercier, *Text-Book*, 12.

115. Charles Mercier, "Conduct," in Tuke, *Dictionary of Psychological Medicine*, 1:242.

116. Mercier, "Conduct," 242.

117. Savage, "Obituary," 192.

118. Mercier, *Conduct*, xi.

119. Ludwig von Mises, *Human Action* (New Haven: Yale University Press, 1949); Jean J. Ostrowski, "Notes biographiques et bibliographiques sur Alfred Espinas," *Revue philosophique de la France et de l'étranger* 157 (1967): 385–91.

120. Blaug, "Was There a Marginal Revolution?"; Winch, "Marginalism and the Boundaries of Economic Science"; Meek, "Marginalism and Marxism"; Black, "W. S. Jevons and the Foundation of Modern Economics"; Wasserman, *Marginal Revolutionaries*, 26–31.

121. J. M. Clark, "Economics and Modern Psychology: I," *Journal of Political Economy* 26, no. 1 (1918): 1–30, 4. Cf. Frank H. Knight, "Ethics and Economic Interpretation," *Quarterly Journal of Economics* 36, no. 3 (1922): 454–81, 474; Angus Burgin, "The Radical Conservatism of Frank H. Knight," *Modern Intellectual History* 6, no. 3 (2009): 513–38.

122. Francesco Boldizzoni, "Capital, Class, and Empire: Nineteenth-Century Political Economy and Its Imaginary," in Breckman and Gordon, *Cambridge History of Modern European Thought*, vol. 1; Bert Mosselmans, *Marginalism* (Newcastle upon Tyne: Agenda, 2018), 1–29.

123. Charles D. Salley, "Gustav Schmoller, Wilhelm Dilthey, and the German Rejection of Positivism in Economics," *History of Economic Ideas* 1, no. 3 (1993): 81–91.

124. Savage and Mercier, "Insanity of Conduct," 12.

125. Freud, "Psychopathology of Everyday Life," 157–58n1.

Chapter Three

1. William Saunders Hallaran, "An Inquiry into the Causes Producing the Extraordinary Addition to the Number of Insane . . . ," in *Three Hundred Years of Psychiatry, 1523–1860: A*

History Presented in Selected English Texts, ed. Richard Hunter and Ida Macalpine (New York: Oxford University Press, 1963), 648–55, 652.

2. Ray, *Mental Hygiene*, 245; Silas Weir Mitchell, *Wear and Tear, or Hints for the Overworked* (Philadelphia, 1871), 38–39.

3. Ernst Waltraud, "Therapy and Empowerment, Coercion and Punishment: Historical and Contemporary Perspectives of Work, Psychiatry, and Society," in *Work, Psychiatry, and Society, c. 1750–2015*, ed. Ernst Waltraud (Manchester: Manchester University Press, 2016).

4. Henry Maudsley, "Middle-Class Hospitals for the Insane," *Journal of Mental Science* 8 (1862): 356–63; Janet Oppenheim, *"Shattered Nerves": Doctors, Patients, and Depression in Victorian England* (New York: Oxford University Press, 1991), 85; Scull, *Social Order/Mental Disorder*; Scull, *Most Solitary of Afflictions*; Leonard D. Smith, "The County Asylum in the Mixed Economy of Care, 1808–1845"; and Peter Bartlett, "The Asylum and Poor Law: The Productive Alliance," in *Insanity, Institutions, and Society, 1800–1914: A Social History of Madness in Comparative Perspective*, ed. Joseph Melling and Bill Forsythe (New York: Routledge, 1999); J. K. Walton, "Casting Out and Bringing Back in Victorian England: Pauper Lunatics, 1840–70," in *The Anatomy of Madness: Essays in the History of Psychiatry*, vol. 2, *Institutions and Society*, ed. W. F. Bynum, Roy Porter, and Michael Shepherd (London: Routledge, 1985), 132–46.

5. Tuke, *Insanity in Ancient and Modern Life*, 104.

6. William Gowers, *Lectures on the Diagnosis of Diseases of the Brain, Delivered at University College London* (London, 1885), 119.

7. Jerrold Seigel, *Modernity and Bourgeois Life: Society, Politics, and Culture in England, France, and Germany since 1750* (Cambridge: Cambridge University Press, 2012), 168–69; Peter T. Cominos, "Late-Victorian Sexual Respectability and the Social System," *International Review of Social History* 8 (1963): 18–48, 216–50.

8. Étienne Esquirol, "Rapport statistique sur la maison royale de Charenton, pendant les années 1826, 1827, et 1828," *Annales d'hygiène publique et de médecine légale* 1 (1829): 101–51, 119–22.

9. François-Emmanuel Fodéré, *Traité du délire, appliqué à la médecine, à la morale, et à la législation* (Paris, 1817), 2:57–58.

10. Joseph Williams, *Insanity: Its Causes, Prevention, and Cure; including Apoplexy, Epilepsy, and Congestion of the Brain*, 2nd ed. (London, 1852 [1848]), 30.

11. Henry Maudsley, *The Physiology and Pathology of the Mind* (London, 1867), 205.

12. George H. Savage, "Mental Diseases: Introduction," in Allbutt and Rolleston, *A System of Medicine by Many Writers*, 842–74, 834–35.

13. Christof Dejung, David Motadel, and Jürgen Osterhammel, eds., *The Global Bourgeoisie: The Rise of the Middle Classes in the Age of Empire* (Princeton: Princeton University Press, 2019); Michael Zakim, *Accounting for Capitalism: The World the Clerk Made* (Chicago: University of Chicago Press, 2018); A. Ricardo López and Barbara Weinstein, eds., *The Making of the Middle Class: Toward a Transnational History* (Durham: Duke University Press, 2012); Geoffrey Crossick, "The Emergence of the Lower Middle Class in Britain: A Discussion," in *The Lower Middle Class in Britain, 1870–1914*, ed. Geoffrey Crossick (London: Croom Helm, 1977); Arno J. Mayer, "The Lower Middle Class as Historical Problem," *Journal of Modern History* 47, no. 3 (1975): 409–36. Cf. Dror Wahrman, *Imagining the Middle Class: The Political Representation of Class in Britain, c. 1780–1840* (Cambridge: Cambridge University Press, 1995); and Sarah Maza, *The Myth of the French Bourgeoisie: An Essay on the Social Imaginary, 1750–1850* (Cambridge, MA: Harvard University Press, 2003).

14. Weber, *Economy and Society*, 302–7, 927.

15. Ray, *Treatise on the Medical Jurisprudence of Insanity*, 11, 13–14; James Moran, "Travails of Madness," in Waltraud, *Work, Psychiatry, and Society*.

16. Goldstein, *Console and Classify*, 166–69.

17. Ray, *Treatise on the Medical Jurisprudence of Insanity*, 240.

18. Ray, *Treatise on the Medical Jurisprudence of Insanity*, 455.

19. Ray, *Treatise on the Medical Jurisprudence of Insanity*, 466–67.

20. Goldstein, *Console and Classify*, 285–87.

21. Charles-Bonaventure-Marie Toullier, *Le droit civil français, suivant l'ordre du Code Napoléon* (Rennes, 1811), 524.

22. Georget, *Examen médical des procès criminels*, 103.

23. Étienne Esquirol, *Aliénation mentale; des illusions chez les aliénés; question médico-légale sur l'isolement des aliénés* (Paris, 1832), 35.

24. Esquirol, *Des maladies mentales*, 1:48.

25. Prichard, *Treatise on Insanity*, 48–49, 352, 371.

26. Alexandre Brière de Boismont, *Études médico-légales sur la perversion des facultés morales et affectives dans la période prodromique de la paralysie générale* (Paris, 1860), 8–9.

27. Williams, *Insanity*, 17, 75.

28. Williams, *Insanity*, 101.

29. Daniel Paul Schreber, *Memoirs of My Nervous Illness*, trans. Ida Macalpine and Richard A. Hunter (New York: New York Review Books, 2000); Schreber, *Denkwürdigkeiten eines Nervenkranken nebst Nachträgen, und einem Anhang über die Frage: "Unter welchen Voraussetzungen darf eine für geisteskrank erachtete Person gegen ihren erklärten Willen in einer Heilanstalt festgehalten werden?"* (Leipzig: O. Mutze, 1903); Sigmund Freud, "Psychoanalytische Bemerkungen über einen autobiographisch beschriebenen Fall von Paranoia (Dementia Paranoides)," *Jahrbuch für psychoanalytische und psychopathologische Forschungen* 3 (1911): 9–68.

30. Schreber, *Memoirs*, Addendum C, 356, 382–84; Eric L. Santner, *My Own Private Germany: Daniel Paul Schreber's Secret History of Modernity* (Princeton: Princeton University Press, 1996), 81–82; Peter Goodrich, *Schreber's Law: Jurisprudence and Judgment in Transition* (Edinburgh: Edinburgh University Press, 2018); Vincent Crapanzano, "'Lacking Now Is Only the Leading Idea, That Is—We, the Rays, Have No Thoughts': Interlocutory Collapse in Daniel Paul Schreber's *Memoirs of My Nervous Illness*," *Critical Inquiry* 24, no. 3 (1998): 737–67.

31. Noble, *Elements of Psychological Medicine*, 136.

32. Henry Maudsley, *The Pathology of Mind* (London, 1880), 248.

33. Maudsley, *Responsibility in Mental Disease*, 295.

34. Charles Mercier, "Insanity as Disorder of Conduct," *Journal of Mental Science* 56 (1910): 405–18, 410–11.

35. Jennifer Karns Alexander, "Efficiency and Pathology: Mechanical Discipline and Efficient Worker Seating in Germany, 1929–1932," *Technology and Culture* 47, no. 2 (2006): 286–310; Maarten Derksen, "Turning Men into Machines? Scientific Management, Industrial Psychology, and the 'Human Factor,'" *Journal of the History of the Behavioral Sciences* 50, no. 2 (2014): 148–65.

36. Thorstein Veblen, "The Preconceptions of Economic Science," *Quarterly Journal of Economics* 14, no. 2 (1900): 240–69, 244; Veblen, "The Limitations of Marginal Utility," *Journal of Political Economy* 17, no. 9 (1909): 620–36, 622–23, 626; William Stanley Jevons, *The Theory of Political Economy*, 4th ed. (London, 1911 [1871]), 39; David Wilson and William Dixon, *A History of Homo Economicus: The Nature of the Moral in Economic Theory* (New York: Routledge, 2012).

37. Mercier, *Sanity and Insanity*, 104–5.

38. Mercier, *Sanity and Insanity*, 113–14.

39. Mercier, *The Nervous System and the Mind*, 161.

40. Mercier, *The Nervous System and the Mind*, 167–68.

41. Mercier, *The Nervous System and the Mind*, 185.

42. Mercier, *The Nervous System and the Mind*, 186.

43. Mercier, *The Nervous System and the Mind*, 186.

44. Mercier, *The Nervous System and the Mind*, 188.

45. Mercier, *The Nervous System and the Mind*, 189.

46. Mercier, *The Nervous System and the Mind*, 189.

47. Mercier, *The Nervous System and the Mind*, 189.

48. Mercier, *The Nervous System and the Mind*, 189–90.

49. Mercier, *The Nervous System and the Mind*, 190–91.

50. The entries appear as anonymously authored articles. They are all titled "Human Temperaments," and numerically ordered from I to XIII. The first entry, with Roman numeral I, is titled "Insanity and Temperament." The dates of the publication are May 6, 13, 27; June 10, 24; July 8, 22; August 5, 19; September 9, 23; and October 7, all in *The Hospital* 60–61 (1916).

51. Charles Mercier, *Human Temperaments: Studies in Character* (London: Scientific Press, 1916), 87.

52. Charles Mercier, *Human Temperaments: Studies in Character*, 3rd ed. (1918), 101–2.

53. Mercier, *Human Temperaments*, 3rd ed., 117.

54. Mercier, *Human Temperaments*, 3rd ed., 119.

55. Joseph Schumpeter, *Theorie der wirtschaftlichen Entwicklung* (Leipzig: Verlag von Duncker & Humblot, 1911), 525. Chapters 2 and 7 from the 1st edition of *Theorie* were removed from later editions and the English translation. Those excluded chapters are translated in Markus C. Becker, Thorbjørn Knudsen, and Richard Swedberg, eds., *The Entrepreneur: Classic Texts by Joseph A. Schumpeter* (Stanford: Stanford University Press, 2011). Much of the translation of the second chapter was published earlier as Joseph Schumpeter, Markus C. Becker, and Thorbjørn Knudsen, "New Translations: Theorie der wirtschaftlichen Entwicklung," *American Journal of Economics and Sociology* 61, no. 2 (2002): 405–37.

56. Robert F. Hébert and Albert N. Link, *The Entrepreneur: Mainstream Views and Radical Critiques* (New York: Praeger, 1988); Robert F. Hébert and Albert N. Link, *A History of Entrepreneurship* (London: Routledge, 2009); Joseph Schumpeter, "Economic Theory and Entrepreneurial History" (1949)," in *Explorations in Enterprise*, ed. Hugh G. J. Aitken (Cambridge, MA: Harvard University Press, 1965), 45–64.

57. Thorstein Veblen, *The Theory of Business Enterprise* (New York: Augustus M. Kelley, 1965 [1904]), chap. 3; Robert Fredona and Sophus A. Reinert, "The Harvard Research Center in Entrepreneurial History and the Daimonic Entrepreneur," *History of Political Economy* 49, no. 2 (2017): 267–314, 294.

58. Richard Swedberg, *Schumpeter, A Biography* (Princeton: Princeton University Press, 1991), 15; Thomas K. McCraw, *Prophet of Innovation: Joseph Schumpeter and Creative Destruction* (Cambridge, MA: Harvard University Press, 2007), 57–60; Richard Swedberg, "Rebuilding Schumpeter's Theory of Entrepreneurship," in *Marshall and Schumpeter on Evolution: Economic Sociology of Capitalist Development*, ed. Yuichi Shionoya and Tamotsu Nishizawa (Cheltenham: Edward Elgar, 2008), 188–203; Erwin Dekker, "Schumpeter: Theorist of the Avant-Garde; The Embrace of the New in Schumpeter's Original Theory of Economic Conduct," *Review of Austrian Economics* 31 (2018): 177–94.

59. Joseph Schumpeter, *Das Wesen und der Hauptinhalt der theoretischen Nationalökonomie* (Leipzig: Duncker & Humblot, 1908); Erich W. Streissler, "The Influence of German and

Austrian Economics on Joseph A. Schumpeter," in *Schumpeter in the History of Ideas*, ed. Yuichi Shionoya and Mark Perlman (Ann Arbor: University of Michigan Press, 1994), 33; Hébert and Link, *Entrepreneur*, 152–53; Thomas K. McCraw, "Schumpeter Ascending," *American Scholar* 60, no. 3 (1991): 371–92.

60. Schumpeter, *Theorie*, 542–43.

61. Schumpeter, *Theorie*, 128.

62. Schumpeter, *Theorie*, 123.

63. Knight, "Ethics and Economic Interpretation," 474.

64. Schumpeter, *Theorie*, 127–28.

65. Schumpeter, *Theorie*, 130.

66. Schumpeter, *Theorie*, 104.

67. Schumpeter, "Economic Theory and Entrepreneurial History," 51; Esben Sloth Andersen, *Joseph A. Schumpeter: A Theory of Social and Economic Evolution* (New York: Palgrave Macmillan, 2011), 62–63.

68. Schumpeter, *Theorie*, 542–43.

69. Schumpeter, *Theorie*, 133.

70. Schumpeter, *Theorie*, 147, 157; Yuichi Shionoya, "Schumpeter and Evolution: An Ontological Exploration," in Shionoya and Nishizawa, *Marshall and Schumpeter on Evolution*, 15–35.

71. Schumpeter, *Theorie*, 171, 158.

72. Joseph A. Schumpeter, *Business Cycles: A Theoretical, Historical, and Statistical Analysis of the Capitalist Process* (New York: McGraw-Hill, 1923), 1:87–109; Joseph Schumpeter, *Capitalism, Socialism, and Democracy*, 2nd ed. (New York: George Allen & Unwin, 1947 [1942]), 132; Harald Hagemann, "Capitalist Development, Innovations, Business Cycles, and Unemployment: Joseph Alois Schumpeter and Emil Hans Lederer," *Journal of Evolutionary Economics* 25 (2015): 117–31.

73. Schumpeter, *Theorie*, 172, 177.

74. Schumpeter, *Theorie*, 162–63.

75. Schumpeter, *Theorie*, 130.

76. Schumpeter, *Theorie*, 145; Joseph Schumpeter, *The Theory of Economic Development*, trans. Redvers Opie (New Brunswick, NJ: Transaction, 1983), 91–92. The original English edition of *Theorie* was translated from the 3rd German edition, which was simply a reprint of the 2nd edition, which appeared in 1926.

77. Swedberg, *Schumpeter*, 173; Nicholas W. Balabkins, "Adaptation without Attribution? The Genesis of Schumpeter's Innovator," Helge Peukert, "The Missing Chapter in Schumpeter's *The Theory of Economic Development*," and Marcel A. G. van Meerhaeghe, "The Lost Chapter of Schumpeter's 'Economic Development,'" in *Joseph Alois Schumpeter: Entrepreneurship, Style and Vision*, ed. Jürgen Backhaus (Dordrecht: Kluwer, 2003).

78. Joseph Schumpeter, "Entrepreneur" (1928), in Becker, Knudsen, and Swedberg, *Entrepreneur*, 227–60, 241–42.

79. Gustav de Molinari, *La morale économique* (Paris, 1888), 17, 29, 86, 136, 267, 269; Hébert and Link, *Entrepreneur* and *History of Entrepreneurship*. Cf. Philippe Fontaine, "The Capitalist Entrepreneur in Eighteenth-Century Economic Literature," *Journal of the History of Economic Thought* 15, no. 1 (1993): 72–89.

80. Schumpeter, *Capitalism, Socialism, and Democracy*, 132 (emphasis added).

81. Schumpeter, "Economic Theory and Entrepreneurial History," 62. Fredona and Reinert, "Harvard Research Center," 295–99; Christopher Adair-Toteff, "Max Weber's Charisma," *Journal of Classical Sociology* 5, no. 2 (2005): 189–204.

82. Frank H. Knight, *Risk, Uncertainty, and Profit* (Boston: Houghton Mifflin, 1921), 163n.

83. Dewey, *Human Nature and Conduct*, 146.

84. Joseph Schumpeter, "Letter to Irving Fisher, 19 March 1936," in Joseph Alois Schumpeter, *Briefe/Letters*, ed. Ulirch Hedtke and Richard Swedborg (Tübingen: Mohr Siebeck, 2000), 282.

85. Irving Fisher and Eugene Lyman Fisk, *How to Live: Rules for Healthful Living Based on Modern Science* (New York: Funk & Wagnalls, 1915), 323–24.

86. Fisher and Fisk, *How to Live*, 300–301; Francis Galton, *Hereditary Genius*, 2nd ed. (London, 1889 [1869]); Janet Browne, "Inspiration to Perspiration: Francis Galton's *Hereditary Genius* in Victorian Context," in *Genealogies of Genius*, ed. Joyce E. Chaplin and Darrin M. McMahon (New York: Palgrave Macmillan, 2016), 77–95.

87. Lennard J. Davis, "Genius and Obsession: Do You Have to Be Mad to Be Smart?," in Chaplin and McMahon, *Genealogies of Genius*, 63–75; Susan A. Ashley, *"Misfits" in Fin-de-Siècle France and Italy: Anatomies of Difference* (London: Bloomsbury, 2017), 29–53.

88. Maudsley, *Pathology of Mind*, 297.

89. Maudsley, *Pathology of Mind*, 298.

90. Maudsley, *Pathology of Mind*, 299, 302–3.

91. Maudsley, *Pathology of Mind*, 303 (emphasis added).

92. Maudsley, *Pathology of Mind*, 303–4.

93. Henry Maudsley, *Life in Mind and Conduct: Studies of Organic in Human Nature* (London: Macmillan, 1902), 358.

94. Cesare Lombroso, *The Man of Genius* (London, 1891 [1889]), 359.

95. Lombroso, *Man of Genius*, 361; Tobias Dahlkvist, "The Epileptic Genius: The Use of Dostoevsky as Example in the Medical Debate over the Pathology of Genius," *Journal of the History of Ideas* 76, no. 4 (2015): 587–608.

96. J. F. Nisbet, *The Insanity of Genius and the General Inequality of Human Faculty, Physiologically Considered* (London, 1891), xii; Anne Stiles, "Literature in *Mind*: H. G. Wells and the Evolution of the Mad Scientist," *Journal of the History of Ideas* 70, no. 2 (2009): 317–39.

97. Nisbet, *Insanity of Genius*, 56.

98. Nisbet, *Insanity of Genius*, 263 (emphasis added).

99. Nisbet, *Insanity of Genius*, 239.

100. Havelock Ellis, "A Study of British Genius," *Current Literature* 31, no. 4 (1901): 422–24, 423; Ellis, *A Study of British Genius* (London: Hurst & Blackett, 1904).

101. Max Nordau, *Entartung*, 2nd ed. (Berlin, 1893 [1892]), 23–24; Nordau, *Degeneration* (New York, 1895).

102. William James, "Degeneration and Genius," *Psychological Review* 2 (1895): 287–94, 294.

103. William James, *The Varieties of Religious Experience: A Study of Human Nature, being the Clifford Lectures on Natural Religion Delivered at Edinburgh in 1901–1902* (New York: Longmans, Green, 1922), 17.

104. A. C. Taymans, "Tarde and Schumpeter: A Similar Vision," *Quarterly Journal of Economics* 64, no. 4 (1950): 611–22; Luc Marco, "Entrepreneur et innovation: Les sources français de Joseph Schumpeter," *Économies et sociétés* 10 (1985): 89–106; Faridah Djellal and Faïz Gallouj, "Les lois de l'imitation et de l'invention: Gabriel Tarde et l'économie évolutionniste de l'innovation," *Revue économique* 68 (2017): 643–71.

105. Gabriel Tarde, *The Laws of Imitation*, trans. Elsie Clews Parsons (New York: Henry Holt and Company, 1903), 144.

106. Cited in Taymans, "Tarde and Schumpeter," 615; Gabriel Tarde, "La psychologie en économie politique," *Revue philosophique de la France et de l'étranger* 12 (1881): 232–50, 401–18.

107. Schumpeter, "Letter to Edward S. Mason, mid- to late 1949," in *Briefe*, 381; Panayotis G.

Michaelides and Theofanis Papageorgiou, "Joseph Schumpeter and the Origins of His Thought," in McCallum, *Palgrave Handbook of the History of the Human Sciences*, 1569–75.

108. Gabriele Tarde, *La logique sociale* (Paris, 1895), 76–77.

109. Esquirol, *Des maladies mentales*, 1:333.

110. Trélat, *La folie lucide*, 91.

111. Trélat, *La folie lucide*, 117–18.

112. Tuke, *Insanity in Ancient and Modern Life*, 38.

113. John Reid, *Essays on Insanity, Hypochondriasis, and Other Nervous Affections* (London, 1816), 71–72.

114. Moses Alen Starr, *Organic and Functional Nervous Diseases*, 2nd ed. (New York: Baillière & Co., 1907), 548. The first edition was titled simply *Organic Nervous Disease* (New York: Lea Brothers & Co., 1903).

115. Joseph Schumpeter, "Comments on a Plan for the Study of Entrepreneurship," in *The Economics and Sociology of Capitalism*, ed. Richard Swedberg (Princeton: Princeton University Press, 1991), 412–13; Swedberg, *Schumpeter*, 34.

116. Weber, *Economy and Society*, 321.

117. Arnold, *Observations on the Nature*, 1:21–22, 25–27.

118. Reid, *Essays*, 108.

119. Ray, *Mental Hygiene*, 246–47.

120. Mitchell, *Wear and Tear*, 47; Mitchell, *Wear and Tear*, 5th ed. (Philadelphia, 1891), 29–30.

121. Gerhard Masur, *Imperial Berlin* (New York: Basic Books, 1970), 74. For a discussion of the Americanization of Germany in the early to mid-twentieth century, see Mary Nolan, *Visions of Modernity: American Business and the Modernization of Germany* (New York: Oxford University Press, 1994).

122. Richard von Krafft-Ebing, *Ueber gesunde und kranke Nerven*, 5th ed. (Tübingen: Laupp, 1903 [1883]), 114.

123. Weber, *Protestant Ethic*, 14; Chalcraft and Harrington, *Protestant Ethic Debate*, 31.

124. George Beard, *American Nervousness: Its Causes and Consequences* (New York, 1881), 340–42.

125. Benjamin Pailhas, "Des idées de richesses et de grandeur chez les émigrés aliénés," *Annales médico-psychologiques* 5 (1897): 50–54.

126. Beard, *American Nervousness*, 13; Silas Weir Mitchell, *Lectures on Diseases of the Nervous System, Especially in Women* (Philadelphia, 1881), 201–2. See also Tom Lutz, *American Nervousness, 1903: An Anecdotal History* (Ithaca: Cornell University Press, 1991).

127. Andreas Killen, *Berlin Electropolis: Shock, Nerves, and German Modernity* (Berkeley: University of California Press, 2006), chap. 2.

128. Charles L. Dana, "The Partial Passing of Neurasthenia," *Boston Medical and Surgical Journal* 150, no. 13 (1904): 339–44, 341.

129. Beard, *American Nervousness*, 9–10.

130. Beard, *American Nervousness*, 102.

131. C. H. Hughes, "Brain Bankruptcy of Business Men: A Comment on Psychical Sanitation," *Alienist and Neurologist* 20, no. 3 (1899): 463–69, 466; recounted in *Annales médico-psychologiques* 17 (1903): 289–90.

132. Beard, *American Nervousness*, 122.

133. Tuke, *Insanity in Ancient and Modern Life*, 197.

134. Beard, *American Nervousness*, 159.

135. Frank B. Wilderson III, "Afropessimism and the Ruse of Analogy: Violence, Freedom

Struggles, and the Death of Black Desire," in *Antiblackness*, ed. Moon-Kie Jung and João H. Costa Vargas (Durham: Duke University Press, 2021).

136. Angus McLaren, *The Trials of Masculinity: Policing Sexual Boundaries, 1870–1930* (Chicago: University of Chicago Press, 1997); John Pettegrew, *Brutes in Suits: Male Sensibility in America, 1890–1920* (Baltimore: Johns Hopkins University Press, 2007).

137. Mary Gluck, "Decadence and the 'Second Modernity,'" in Breckman and Gordon, *Cambridge History of Modern European Thought*, 1:353–71; Gerald N. Izenberg, *Modernism and Masculinity: Mann, Wedekind, Kandinsky through World War I* (Chicago: University of Chicago Press, 2000), 4–9.

138. Michael S. Kimmel, *Manhood in America: A Cultural History*, 2nd ed. (New York: Oxford University Press, 2006); Leonore Davidoff and Catherine Hall, *Family Fortunes: Men and Women of the English Middle Class, 1780–1850* (Chicago: University of Chicago Press, 1987), 229–71.

139. Zakim, *Accounting for Capitalism*; Carole Srole, *Transcribing Class and Gender: Masculinity and Femininity in Nineteenth-Century Courts and Offices* (Ann Arbor: University of Michigan Press, 2019); Leslie Salzinger, "Sexing Homo Oeconomicus: Finding Masculinity at Work," in *Mutant Neoliberalism: Market Rule and Political Rupture*, ed. William Callison and Zachary Manfredi (New York: Fordham University Press, 2019); Leslie Salzinger, "Re-Marking Men: Masculinity as a Terrain of the Neoliberal Economy," *Critical Historical Studies* 7 (2020): 1–25; Miranda Joseph, "Gender, Entrepreneurial Subjectivity, and Pathologies of Personal Finance," *Social Politics* 20, no. 2 (2013): 242–73.

140. Paul Lerner, *Hysterical Men: War, Psychiatry, and the Politics of Trauma in Germany, 1890–1930* (Ithaca: Cornell University Press, 2003), 8, 34.

141. Marieke de Goede, *Virtue, Fortune, and Faith: A Genealogy of Finance* (Minneapolis: University of Minnesota Press, 2005), 41–46; E. Anthony Rotundo, *American Manhood: Transformations in Masculinity from the Revolution to the Modern Era* (New York: Basic Books, 1993).

142. Lunbeck, *Psychiatric Persuasion*, 236–55.

143. Jack Halberstam, *Female Masculinity* (Durham: Duke University Press, 2018 [1998]).

144. Cheryl I. Harris, "Whiteness as Property," in *Critical Race Theory: The Key Writings That Formed the Movement*, ed. Kimberlé Crenshaw, Neil Gotanda, Gary Peller, and Kendall Thomas (New York: New Press, 1995); originally published as Cheryl I. Harris, "Whiteness as Property," *Harvard Law Review* 106, no. 8 (1993): 1707–1791. See also Nancy Leong, "Racial Capitalism," *Harvard Law Review* 126, no. 8 (2013): 2151–81.

145. See Martin Summers, *Manliness and Its Discontents: The Black Middle Class and the Transformation of Masculinity, 1900–1930* (Chapel Hill: University of North Carolina Press, 2004). Cf. Gail Bederman, *Manliness and Civilization: A Cultural History of Gender and Race in the United States, 1880–1917* (Chicago: University of Chicago Press, 1995); Melissa N. Stein, *Measuring Manhood: Race and the Science of Masculinity, 1830–1934* (Minneapolis: University of Minnesota Press, 2015).

146. Leslie Brown, *Upbuilding Black Durham: Gender, Class, and Black Community Development in the Jim Crow South* (Chapel Hill: University of North Carolina Press, 2008), 19, 320.

147. E. Franklin Frazier, "Durham: Capital of the Black Middle Class," in *The New Negro: An Interpretation*, ed. Alain Locke (New York: Albert and Charles Boni, 1925), 333–34.

148. Frazier, "Durham," 338.

149. E. Franklin Frazier, *Black Bourgeoisie* (Glencoe, IL: Free Press, 1957 [1955]), 85, 230.

150. W. E. Burghardt Du Bois, *Black Reconstruction in America: An Essay Toward a History of the Part Which Black Folk Played in the Attempt to Reconstruct Democracy in America, 1860–1880* (Cleveland: World Pub., 1962 [1935]), 15.

151. Cedric J. Robinson, *Black Marxism: The Making of the Black Radical Tradition* (Chapel Hill: University of North Carolina Press, 1983), 206.

152. Edward Franklin Frazier, "The Pathology of Race Prejudice," *Forum* 77, no. 6 (June 1927): 856–62, 856.

153. Frazier, "The Pathology of Race Prejudice," 857.

154. Robinson, *Black Marxism*, 9–28; Justin Leroy and Destin Jenkins, eds., *Histories of Racial Capitalism* (New York: Columbia University Press, 2021); Walter Johnson, "To Remake the World: Slavery, Racial Capitalism, and Justice," *Boston Review*, February 20, 2018, http://boston review.net/forum/walter-johnson-to-remake-the-world; Michael Ralph and Maya Singhal, "Racial Capitalism," *Theory and Society* 48 (2019): 851–81.

155. Wendy Brown, *Undoing the Demos: Neoliberalism's Stealth Revolution* (New York: Zone Books, 2015), 36; Michel Foucault, *Birth of Biopolitics: Lectures at the Collège de France, 1978–79*, trans. Graham Burchell (New York: Palgrave Macmillan, 2008), 226–30; Plehwe et al., *Nine Lives of Neoliberalism*; Dieter Plehwe and Philip Mirowski, eds., *The Road from Mont Pèlerin: The Making of the Neoliberal Thought Collective* (Cambridge: MA, Harvard University Press, 2015); Philip Mirowski, *Never Let a Serious Crisis Go to Waste: How Neoliberalism Survived the Financial Meltdown* (New York: Verso, 2013).

Chapter Four

1. Armand Trousseau, *Clinique médicale de l'Hôtel-Dieu de Paris*, 3rd ed. (Paris, 1868 [1861]), 2:653–54. While the first edition appeared in 1861, the chapter on aphasia only appeared for the first time in the second edition. J. M. S. Pearce, "Armand Trousseau—Some of His Contributions to Neurology," *Journal of the History of the Neurosciences* 11, no. 2 (2002): 125–35.

2. William Henry Broadbent, "On the Cerebral Mechanisms of Speech and Thought," *Medico-Chirurgical Transactions* 55, no. 1 (1872): 145–94, 147–48; Mervyn Eadie, "William Henry Broadbent (1835–1907) as a Neurologist," *Journal of the History of the Neurosciences* 24 (2015): 137–47.

3. E.g., Adolf Kussmaul, *Die Störungen der Sprache: Versuch einer Pathologie der Sprache* (Leipzig, 1877), 22, 170.

4. John Hughlings Jackson, "On Affections of Speech from Disease of the Brain," *Brain* 1 (1878): 304–30, 315.

5. Marjorie Perlman Lorch, "Speaking for Yourself: The Medico-Legal Aspects of Aphasia in Nineteenth-Century Britain," in *The Neurological Patient in History*, ed. L. Stephen Jacyna and Stephen T. Casper (Rochester: University of Rochester Press, 2012).

6. "The Pathology of Speech," *British Medical Journal*, September 1879, 378–81, 380.

7. Frank Warren Langdon, *The Aphasias and Their Medico-Legal Relations* (Norwalk, OH: Laning, 1898), 28.

8. Joseph Collins, *The Genesis and Dissolution of the Faculty of Speech* (New York, 1898), 407, 412.

9. Susanna L. Blumenthal, *Law and the Modern Mind: Consciousness and Responsibility in American Legal Culture* (Cambridge, MA: Harvard University Press, 2016), chaps. 3 and 4.

10. Jens Beckert, *Inherited Wealth*, trans. Thomas Dunlap (Princeton: Princeton University Press, 2004), 15; Thomas Piketty, *Capital in the Twenty-First Century*, trans., Arthur Goldhammer (Cambridge, MA: Harvard University Press, 2017), 301–3.

11. Michel Foucault, *Foucault Live: Interviews, 1961–1984*, ed. Sylvère Lotringer (New York: Semiotext(e), 1996), 197; Goldstein, *Console and Classify*, 407.

12. Weber, *Economy and Society*, 319.

13. For an extended discussion of the "legal coding of capital," see Katharina Pistor, *The Code of Capital: How the Law Creates Wealth and Inequality* (Princeton: Princeton University Press, 2019).

14. Stefanos Geroulanos and Todd Meyers, *The Human Body in the Age of Catastrophe: Brittleness, Integration, Science, and the Great War* (Chicago: University of Chicago Press, 2018).

15. Paul Broca, "Remarques sur le siège de la faculté du langage articulé suivies d'une observation d'aphémie (perte de la parole)," *Bulletins de la Société anatomique de Paris*, 2e série, t. 6 (1861): 330–57; L. S. Jacyna, "The 1875 Aphasia Debate in the *Berliner Gesellschaft für Anthropologie*," *Brain and Language* 69, no. 1 (1999): 5–15.

16. L. S. Jacyna, *Lost Words: Narratives of Language and the Brain, 1825–1926* (Princeton: Princeton University Press, 2000), chap. 2; Jules Falret, "Des troubles du langage et de la mémoire des mots dans les affections cérébrales (aphémie, aphasie, alalie, amnésie verbale)," *Archives générales de médecine* 1, 6e série, t. 3 (1864): 336–54, 338; reprinted in Falret, *Études cliniques sur les maladies mentales et nerveuses*, 410–74.

17. Guenther, *Localization and Its Discontents*; Robert M. Young, *Mind, Brain, and Adaptation in the Nineteenth Century: Cerebral Localization and Its Biological Context from Gall to Ferrier* (Oxford: Clarendon, 1970).

18. Jacyna, *Lost Words*, 64–87; Michael Hagner, *Homo cerebralis: Der Wandel vom Seelenorgan zum Gehirn* (Berlin: Berlin Verlag, 1997).

19. Jules Falret, "Aphasie," in *Dictionnaire encyclopédique des sciences médicales*, ed. A. Dechambre (Paris, 1876), 624.

20. Broca, "Remarques," 345–46.

21. Trousseau, *Clinique médicale de l'Hôtel-Dieu de Paris*, 2:676 (emphasis in original).

22. Frederick Bateman, *On Aphasia, or Loss of Speech in Cerebral Disease*, 4 pts. (London, 1868–1869), pt. 4, p. 5.

23. Henry Charlton Bastian, "On the Various Forms of Loss of Speech in Cerebral Disease," *British and Foreign Medical Chirurgical Review* 43, no. 85 (1869): 209–36; no. 86 (1869): 470–92; 218, 225.

24. J. Fayrer, "A Case of Aphasia," *Edinburgh Medical Journal* 15 (1870): 700–708, 707.

25. Falret, "Aphasie," 621–22.

26. Kussmaul, *Die Störungen der Sprache*, 16–17.

27. Kussmaul, *Die Störungen der Sprache*, 154. Kussmaul's text was translated as "Disturbances of Speech," in *Cyclopedia of the Practice of Medicine*, ed. Hugo von Ziemssen, trans. E. Buchanan Baxter et al. (New York, 1877), 14:580–875.

28. Henri Legrand du Saulle, *Étude médico-légale sur les testaments contestés pour cause de folie* (Paris, 1879), 215.

29. Legrand du Saulle, *Étude médico-légale*, 240.

30. John Wyllie, "The Disorders of Speech," *Edinburgh Medical Journal* 37 and 38 (1892–1894): 585–604, 777–93, 897–907, 289–307, 401–21, 696–715, 785–802; 696–97.

31. Wyllie, "Disorders of Speech," 698.

32. F. Page Atkinson, "A Few Ideas Regarding the Subject of Aphasia," *Edinburgh Medical Journal* 16 (1870): 311–13, 312; Adrien Proust, "De l'aphasie," *Archives générale de médecine* 1 (6e série, t. 19) (1872): 147–66, 303–18, 653–85; 151, 303; William Broadbent, *Selections from the Writings Medical and Neurological of Sir William Broadbent* (Oxford: H. Frowde, 1908), 338; Jean-Martin Charcot, *Clinical Lectures on Diseases of the Nervous System, Delivered at the Infirmary of La Salpêtrière*, trans. Thomas Savill (London, 1889), 3:365.

33. Gilbert Ballet, *Le langage intérieur et les diverses formes de l'aphasie* (Paris, 1886), 13.

34. Henri Legrand du Saulle, "L'aphasie et les aphasiques," *Gazette des hopitaux* 37 (1882): 289–92; 40 (1882): 313–16; 45 (1882): 353–55; 48 (1882): 377–79; 52 (1882): 409–11; 54 (1882): 425–27; 57 (1882): 449–51; 65 (1882): 513–15; 68 (1882): 537–38; 71 (1882): 561–64; 74 (1882): 585–87; 77 (1882): 609–11; 450–51, 538.

35. John Hughlings Jackson, Untitled analysis of a patient with attacks of unconsciousness, or aphasia, or affection of the right arm and leg. First-person account of attacks of aphasia (QSA/855, Queen Square Archives, Folder C36) (emphasis added).

36. Jules Baillarger, "De l'aphasie au point de vue psychologique" (1865), in *Recherches sur les maladies mentales* (Paris, 1890), 1:584–601. See Th. Alajouanine, "Baillarger and Jackson: The Principle of Baillarger-Jackson in Aphasia," *Journal of Neurology, Neurosurgery, and Psychiatry* 23, no. 3 (1960): 191–93.

37. Marjorie Perlman Lorch, "The Unknown Source of John Hughlings Jackson's Early Interest in Aphasia and Epilepsy," *Cognitive Behavioral Neurology* 17, no. 3 (2004): 124–32; Samuel H. Greenblatt, "Hughlings Jackson's First Encounter with the Work of Paul Broca: The Physiological and Philosophical Background," *Bulletin of the History of Medicine* 44 (1970): 555–70; John Forrester, *Language and the Origins of Psychoanalysis* (New York: Columbia University Press, 1980), 29.

38. Marjorie Perlman Lorch, "The Merest *Logomachy*: The 1868 Norwich Discussion of Aphasia by Hughlings Jackson and Broca," *Brain* 131, no. 6 (2008): 1658–1670, 1670. Cf. John Hughlings Jackson, "On the Nature of the Duality of the Brain" (1874), in Taylor, *Selected Writings*; John Hughlings Jackson, "On the Physiology of Language," *Medical Times and Gazette* 2 (September 5, 1868), 2; reprinted in *Brain* 38 (1915): 59–64, 60.

39. Jackson, "On the Physiology of Language," 2; John Hughlings Jackson, "Notes on the Physiology and Pathology of Language" (1866), in Taylor, *Selected Writings*, vol. 2; Jackson, "Observations on the Physiology of Language" (Queen Square Archives, A20, [QSA/839], 1868), 24.

40. Jackson, "On the Physiology of Language"; Jackson, "Notes on the Physiology and Pathology of Language," 227–28; Jackson, "Observations on the Physiology of Language," 5.

41. John Hughlings Jackson, "Middle Level of the Cerebral Sub-System," unpublished manuscript (Queen Square Archives, C40), 10–11.

42. Jackson, *Neurological Fragments*, 141.

43. Jackson, "On Affections of Speech," 190, 192.

44. Lorch, "Merest *Logomachy*," 1667.

45. Jackson, "Observations on the Physiology of Language," 4.

46. John Hughlings Jackson, "On Affections of Speech from Disease of the Brain" (1878), in Taylor, *Selected Writings*, 2:159. See Critchley and Critchley, *John Hughlings Jackson*, chaps. 12–14.

47. Jackson cites an earlier 1868 article in "On the Nature of the Duality of the Brain," 130; Walther Riese, "The Sources of Hughlings Jackson's View on Aphasia," *Brain* 88 (1965): 811–22.

48. Jackson, "On Affections," in Taylor, *Selected Writings*, 2:159.

49. Roman Jakobson, *Studies on Child Language and Aphasia* (The Hague: Mouton, 1971), 114.

50. John Hughlings Jackson, "Swearing," in *The London Hospital Reports*, vol. 1, 1864, 453 (Queen Square Archives, QSA 834).

51. Jackson, "On the Nature of the Duality of the Brain," 137.

52. Jackson, "On the Nature of the Duality of the Brain," 141.

53. Forrester, *Language and the Origins of Psychoanalysis*, 19.

54. Jackson, "On the Physiology of Language," 2.

55. Jackson, "Observations on the Physiology of Language," 25.

56. Edward Coke, *The First Part of the Institutes of the Lawes of England, or, A Commentarie Upon Littleton* (London: 1628), book 3, section 405, 246–47.

57. John Locke, *Two Treatises of Government* (London, 1690), II, sec. 58, sec. 60; John Brydall, *Non Compos Mentis: or, the Law Relating to Natural Fools, Mad-Folks, and Lunatick Persons, Inquisited, and Explained, for Common Benefit* (London, 1700); Lloyd Bonfield, *Devising, Dying, and Dispute: Probate Litigation in Early Modern England* (Burlington, VT: Ashgate, 2012); James Tully, *A Discourse on Property: John Locke and His Adversaries* (Cambridge: Cambridge University Press, 1980).

58. Smith, "The Law and Insanity in Great Britain, with Comments on Continental Europe," 249.

59. William Blackstone, *Abridgment of Blackstone's Commentaries*, ed. William C. Sprague, 2nd ed. (Detroit, 1893), 261.

60. Theodric Beck, *Elements of Medical Jurisprudence*, 2nd ed. (London, 1825 [1823]), 255.

61. Blackstone, *Abridgment*, 261.

62. Georget, *Examen médical des procès criminels*, 130.

63. Harvey Prindle Peet, "On the Legal Rights and Responsibilities of the Deaf and Dumb," *American Journal of Insanity* 13, no. 2 (1856): 97–171, 119, 169.

64. Isaac Ray, "The Angell Will Case," *American Journal of Insanity* 20 (1863): 145–86, 145.

65. Isaac Redfield, *The Law of Will: Embracing the Jurisprudence of Insanity; the Making and Construction of Wills; and the Effect of Extrinsic Evidence upon Such Construction*, 3rd ed. (Boston, 1869), 1:62–63.

66. "The Maclean Medical Case," *Edinburgh Medical Journal* 7, no. 3 (1861): 298–306, 304. See also W. T. Gairdner, "Remarks on Certain Medico-Legal Aspects of the Maclean Will Case," *Edinburgh Medical Journal* 7 (1862): 797–808, 804.

67. Edmund Wetmore, "Mental Unsoundness as Affecting Testamentary Capacity," *American Journal of Insanity* 20 (1864): 314–35, 320.

68. Alexandre Brière de Boismont, *Des hallucinations ou histoire raisonnée des apparitions, des visions, des songes, de l'extase, des rêves, du magnétisme et du somnambulisme* 3rd ed. (Paris, 1862 [1845]), 708.

69. Krafft-Ebing, *Lehrbuch der gerichtlichen Psychopathologie*, 371.

70. Douglas Maclagan, "On Civil Incapacity," *Edinburgh Medical Journal* 10 (1865): 865–82, 877.

71. Arthur Mitchell, "Some of the Medico-Legal Relations of Insanity to Will-Making," *Edinburgh Medical Journal* 17 (1872): 673–89, 865–76, 1057–69; 1060, 687.

72. Henry C. Chapman, *A Manual of Medical Jurisprudence: Insanity and Toxicology* (Philadelphia, 1893), 165.

73. Alexandra Braun, "Testamentary Freedom and Its Restrictions in French and Italian Law: Trends and Shifts," and Elizabeth Cooke, "Testamentary Freedom: A Study of Choice and Obligation in England and Wales," in *Freedom of Testation/Testierfreiheit*, ed. Reinhard Zimmermann (Tübingen: Mohr Siebeck, 2012); Beckert, *Inherited Wealth*, 69.

74. T. C. Fry, "The Ethics of Wills," *Economic Review*, April 1893, 190–200; Roger Kerridge, "Testamentary Formalities in England and Wales," in *Comparative Succession Law*, vol. 1, *Testamentary Formalities*, ed. Kenneth G. C. Reid, Marius J. de Waal, and Reinhard Zimmerman (Oxford: Oxford University Press, 2011), 305–28, 313.

75. Henry Sumner Maine, *Ancient Law: Its Connection with the Early History of Society, and Its Relation to Modern Ideas* (London, 1861), 172, 177.

76. George Allnutt, *The Practice of Wills and Administrations*, 3rd ed. (London, 1852); Catherine O. Frank, *Law, Literature, and the Transmission of Culture in England, 1837–1925* (Burlington, VT: Ashgate, 2010).

77. Orrin K. McMurray, "Liberty of Testation and Some Modern Limitations Thereon," *Illinois Law Review* 14 (1919): 96–123, 115.

78. Braun, "Testamentary Freedom," 62–64.

79. Beckert, *Inherited Wealth*, 40–48.

80. Reinhard Zimmerman, "Testamentary Formalities in Germany," in Reid, Waal, and Zimmerman, *Comparative Succession Law*, 182–86; Beckert, *Inherited Wealth*, 57.

81. Beckert, *Inherited Wealth*, 276–84.

82. Isaac Ray, "Testamentary Capacity," *Alienist and Neurologist* 5, no. 1 (1884): 102–23.

83. Collins, *Genesis and Dissolution*, 407.

84. Thomas Piketty and Gabriel Zucman, "Wealth and Inheritance in the Long Run," in *Handbook of Income Distribution*, ed. Anthony B. Atkinson and François Bourguignon (New York: Elsevier, 2015), 2:1303–68; Thomas Piketty, "On the Long-Run Evolution of Inheritance: France 1820–2050," *Quarterly Journal of Economics* 126 (2011): 1071–1131.

85. Charles H. Hughes, "Considerations of the Medico-Legal Aspects of Aphasia," *Alienist and Neurologist* 24, no. 3 (1903): 301–16, 301, 307. See also Langdon, *Aphasias*, 15, 42.

86. Frederick Bateman, *On Aphasia, or Loss of Speech, and the Localization of the Faculty of Articulate Language*, 2nd ed. (London, 1890), 314.

87. Maudsley, *Responsibility in Mental Disease*, 266–67.

88. Joseph Lefort, "Remarques sur l'interdiction des aphasiques," *Annales d'hygiène publique et de médecine légale*, série 2, no. 38 (1872): 417–34, 417.

89. Lefort, "Remarques sur l'interdiction des aphasiques," 423.

90. Lefort, "Remarques sur l'interdiction des aphasiques," 425.

91. Lefort, "Remarques sur l'interdiction des aphasiques," 431.

92. Lefort, "Remarques sur l'interdiction des aphasiques," 433.

93. Kussmaul, *Die Störungen der Sprache*, 22, 170.

94. Jean-Théophile Gallard, "L'aphasie au point de vue de la médecine légale," *Journal de médecine et de chirurgie pratiques* 48, 3e série (1877): 377–80, 378.

95. "The Jurisprudence of Aphasia," *British Medical Journal*, September 1877, 386. This is likely the text that introduced Gallard's lecture to English-speaking clinicians.

96. William O'Neill, "A Case of Aphasia," *British Medical Journal*, October 1877, 475–76.

97. Legrand du Saulle, *Étude médico-légale sur les testaments*, 587.

98. Emil Frischauer, "Die Testierfähigkeit aphasischer Personen nach österreichischen Rechte," *Wiener medizinische Blätter* 5 (1882): 1260–63, 1260–61.

99. F. Jolly, "Ueber den Einfluss der Aphasie auf die Fähigkeit zur Testamentserrichtung," *Archiv für Psychiatrie und Nervenkrankheiten* 13 (1882): 325–40, 336.

100. Désiré Bernard, *De l'aphasie et de ses diverses forms* (Paris, 1889), 256.

101. William Gowers, *Lectures on the Diagnosis of Diseases of the Brain, Delivered at University College London* (London, 1885), 139; Gowers, *A Manual of Diseases of the Nervous System* (London, 1886–1888), 2:116.

102. George H. Savage, *Insanity and Allied Neuroses* (London, 1884), 352 (emphasis added).

103. Charles K. Mills, *Aphasia and Other Affections of Speech, in Some of Their Medico-Legal Relations, Studied Largely from the Standpoint of Localization* (Milwaukee, 1891), 27, 80.

104. Moses Allen Starr, "The Pathology of Sensory Aphasia with an Analysis of Fifty Cases in Which Broca's Center Was Not Diseased," *Brain* 12 (July 1889): 82–99.

105. Byrom Bramwell, "Remarks on Aphasia and Will-Making I," *British Medical Journal* 1, no. 1898 (May 1897): 1205–10, 1207; Bramwell, "Remarks on Aphasia and Will-Making II," *British Medical Journal* 1, no. 1899 (May1897): 1272–75.

106. Bramwell, "Remarks on Aphasia and Will-Making I," 1210, 1208. See also Henry Charlton Bastian, *Treatise on Aphasia and Other Speech Defects* (London, 1898), 335.

107. Laura Salisbury, "Sounds of Silence: Aphasiology and the Subject of Modernity," in *Neurology and Modernity: A Cultural History of Nervous Systems, 1800–1950*, ed. Laura Salisbury and Andrew Shail (London: Palgrave Macmillan, 2010), 204–30.

108. Jackson, "Observations on the Physiology of Language," 34; Jackson, "On Affections of Speech from Disease of the Brain," 199.

109. William T. Gairdner, "A Discussion on Aphasia in Relation to Testamentary Capacity," *British Medical Journal* 2 (September 3, 1898): 581–83, 582 (emphasis added); see also William Elder, "Reply to Gairdner," *British Medical Journal* 2 (September 3, 1898): 583–85, 583.

110. Geroulanos and Meyers, *The Human Body in the Age of Catastrophe*, 112–37.

111. Redfield, *Law of Wills*, 82; Isaac Redfield, "Senile Dementia," *American Law Register* 12, no. 8, n.s. 3 (June 1864): 449–62; George Rosen, "Psychopathology of Ageing: Cross-Cultural and Historical Approaches," in *Madness in Society: Chapters in the Historical Sociology of Mental Illness* (Chicago: University of Chicago Press, 1968).

112. Redfield, *Law of Wills*, 96.

113. Redfield, *Law of Wills*, 113.

114. Ernest Émile Rousseau, *De l'aphasie dans ses rapports avec l'aliénation mentale: Extrait du Bulletin de la Société médicale de l'Yonne* (Paris, 1883), 1, 21–22.

115. Paul Garnier, "Aphasie et folie: Coexistence d'une psychose systématique avec la cécité et surdité verbales," *Archives générale de médecine* 1, 7e série, v. 20 (1889): 139–51, 309–35, 309.

116. Legrand du Saulle, *Étude médico-légale sur les testaments*, 80.

117. G. Stanley Hall, *Senescence: The Last Half of Life* (New York: D. Appleton and Company, 1922), 202.

118. Alexis Legroux, *De l'aphasie* (Paris, 1875), 70–71.

119. Eugène Billod, "Contribution à l'étude de l'aphasie," *Annales médico-psychologiques* 17, 5e série (1877): 321–45, 330.

120. C. H. Hughes, "The Medico-Legal Aspect of Cerebral Localization and Aphasia," *Alienist and Neurologist* 1, no. 2 (1880): 142–58; no. 3 (1880): 315–22, 321.

121. Maurice Brissot, *L'aphasie dans ses rapports avec la démence et les vésanies: Étude historique, clinique & diagnostique considérations médico-légales* (Paris, 1910), 89.

122. Brissot, *L'aphasie*, 126–27.

123. Brissot, *L'aphasie*, 222.

124. R. Percy Smith, "Aphasia in Relation to Mental Disease," *Journal of Mental Science* 64, no. 264 (1918): 1–18, 16.

125. George M. Beard, *Legal Responsibility in Old Age: Based on Researches into the Relation of Age to Work* (New York, 1874), 10–11; Margaret Lock, *The Alzheimer Conundrum: Entanglements of Dementia and Aging* (Princeton: Princeton University Press, 2013), 42–47.

126. Mercier, *Sanity and Insanity*, 370.

127. Jean-Martin Charcot, *Clinical Lectures on the Diseases of Old Age*, trans. Leigh H. Hunt (New York, 1881), 20.

128. William L. Russell, "Senility and Senile Dementia," *American Journal of Insanity* 58, no. 4 (1902): 625–33, 626.

129. Russell, "Senility and Senile Dementia," 631.

130. C. H. Hughes, "Normal Senility and Dementia Senilis," *Alienist and Neurologist* 30, no. 1 (1909): 63–77, 66.

Chapter Five

1. Benjamin F. Martin, "The Courts, the Magistrature, and Promotions in Third Republic France, 1871–1914," *American Historical Review* 87, no. 4 (1982): 977–1009.

2. "La 'Sinistrose,'" *Le Journal*, December, 29, 1907, 5; "La 'Sinistrose,'" *La Lanterne*, December, 31, 1907, 2; "La 'Sinistrose,'" *Le Temps*, December 30, 1907, 3. I have translated *sinistrose* as "sinistrosis/sinistrotic" in the same way as *neurose* is translated as "neurosis/neurotic." The term's etymology is somewhat linked to the archaic English term "sinistrouse," which ambiguously denotes a sufferer of misfortune as well as a dishonest person.

3. François Ewald, *L'état providence* (Paris: Grasset, 1986); Ewald, *Histoire de l'état providence: Les origines de la solidarité* (Paris: Grasset, 1996); translated as François Ewald, *The Birth of Solidarity: The History of the French Welfare State*, ed. Melinda Cooper; trans. Timothy Scott Johnson (Durham: Duke University Press, 2020); Paul V. Dutton, *Origins of the French Welfare State: The Struggle for Social Reform in France, 1914–1947* (Cambridge: Cambridge University Press, 2002).

4. Laurent Tatu and Julien Bogousslavsky, "The Impossible Succession of Charcot: The Quest for a Suitable Heir," *European Neurology* 65 (2011): 193–97; Paula Teixeiria Marques et al., "Édouard Brissaud: Distinguished Neurologist and Charcot's Pupil," *Arquivos de neuro-psiquiatria* 76, no. 7 (2018): 490–93; Julien Bogousslavsky, "Marcel Proust's Lifelong Tour of the Parisian Neurological Intelligentsia: From Brissaud and Dejerine to Sollier and Babinski," *European Neurology* 57 (2007): 129–36; Jacques Poirier, "Édouard Brissaud, neurologue méconnu et comédien dans l'âme," *Bulletin de l'Académie nationale de médecine* 194, no. 1 (2010): 163–75.

5. "Les accidents du travail," *Le Journal*, January 5, 1908, 5; "La 'Sinistrose,'" *Le Journal*, December 9, 1908, 2; "Accidents du travail: Sinistrose, suite indirecte de l'accident, refus de rente," *Recueil périodique des assurances: Doctrine, jurisprudence, législation* 29 (1911): 63–65.

6. Eugène Quillent, "La 'Sinistrose,'" *L'Humanité*, January 13, 1908, 4.

7. Coron, "Neurasthénie: Causes et remèdes," *L'Humanité*, February 9, 1936, 4.

8. Édouard Brissaud, "La sinistrose," *Le concours médical* 30, no. 1 (February 9, 1908): 114–17, 116.

9. Édouard Brissaud, "Les troubles nerveux post-traumatiques, conférence faite à l'hôpital des 'accidents du travail' et communiquée au Congrès international de Rome des accidents du travail (May, 1909)," *Revue clinique médico-chirurgicale "accidents du travail"* 6 (June 1, 1909): 121–38, 138.

10. Jacques Poirier, "Édouard Brissaud (1852–1909)," *Journal of Neurology* 258, no. 5 (2011): 951–52; Christopher G. Goetz, "The Salpêtrière in the Wake of Charcot's Death," *Archive of Neurology* 45 (1988): 444–47.

11. "Un cas de sinistrose consecutif de la catastrophe de courrieres deboute," *Revue de médecine légale* 15, no. 3 (1908): 174–76; Maxime Laignel-Lavastine, "Psychologie de l'accident du travail," *Le Progrès de Bell-Abbès*, December 6, 1911, 2–3.

12. Coureménos, "La 'Sinistrose' du professeur Brissaud," *Revue clinique médico-chirurgicale "accidents du travail"* 1 (May 1, 1908): 24.

13. L. Perrier, "La sinistrose: Une nouvelle maladie de la volonté," *Foi et vie: Revue de quinzaine, religieuse, morale, littéraire, sociale*, April 5, 1910, 192–96, 193.

14. A. Rémond de Metz and Sauvage, "Sinistrose et psychose," *Annales médico-psychologiques* 2 (1912): 336–43, 336.

15. M. Laffont, *Du rôle du médecin dans les accidents du travail* (Toulouse: Ch. Dirion, 1911), 8–9.

16. Laffont, *Du rôle du médecin*, 54.

17. "Chronique des compagnies d'assurance," *L'Argus*, December 13, 1908, 788–89, 788; reprinted as "La 'Sinistrose,'" *Revue de la prévoyance et de la mutualité* 18 (1909): 208–9.

18. E. P. Hennock, *The Origin of the Welfare State in England and Germany, 1850–1914: Social Policies Compared* (Cambridge: Cambridge University Press, 2007), 71–119; Dutton, *Origins of the French Welfare State*.

19. "Malingering et sinistrose," *L'Intransigeant*, September, 30, 1913, 3.

20. Vesale, "La sinistrose," *L'action française*, May, 31, 1908, 3.

21. Paul Zeys, *La valeur du corps humain devant les tribunaux et les lois sur les accidents du travail en France* (Paris: Société du "Recueil Sirey," 1912), 220–23; J. Eissen, "Un cas de sinistrose," *Annales médico-psychologiques* 11 (1919): 506–12.

22. "La sinistrose," *Paris-midi*, June 15, 1917, 4.

23. R. Benon, "La sinistrose," *Paris médical* 37 (October 16, 1920): 285–88, 285.

24. T. Vasilu and J. Stanesco, "Le délire de revendication sinistrosique," *Annales de médecine légale, de criminologie et de police scientifique* 16, no. 1 (1936): 181; "Les compagnies s'organisent contre la 'sinistrose,'" *L'Argus*, February 8, 1931, 187; Roux, "Une affaire de fraude médical," *L'action française*, January 14, 1933, 3; Brillié, "Tribunal civil d'Auxerre," *Bulletin de la jurisprudence conseils de prud'hommes des tribunaux de commerce et des loyers* 1 (Jan. 1924): 23.

25. Mark S. Micale, "On the 'Disappearance' of Hysteria: A Study of the Clinical Deconstruction of a Diagnosis," *Isis* 84, no. 3 (1993): 496–526.

26. Hennock, *Origin of the Welfare State in England and Germany*, 71; Allan Mitchell, *The Divided Path: The German Influence on Social Reform in France after 1870* (Chapel Hill: University of North Carolina Press, 1991), 223–51; Philip Nord, "The Welfare State in France, 1870–1914," *French Historical Studies* 18, no. 3 (1994): 821–38.

27. Rhodri Hayward, "Medicine and the Mind," in *The Oxford Handbook of the History of Medicine*, ed. M. Jackson (New York: Oxford University Press, 2011), 532. See also Rhodri Hayward, "The Pursuit of Serenity: Psychological Knowledge and the Making of the British Welfare State," in *History and Psyche: Culture, Psychoanalysis, and the Past*, ed. Sally Alexander and Barbara Taylor (New York: Palgrave Macmillan, 2012).

28. Ewald, *L'état providence*; Ewald, *Histoire de l'état providence*; Dutton, *Origins of the French Welfare State*; Hennock, *Origin of the Welfare State in England and Germany*; Greg Eghigian, *Making Security Social: Disability, Insurance, and the Birth of the Social Entitlement State in Germany* (Ann Arbor: University of Michigan Press, 2000); Killen, *Berlin Electropolis*; Greg Eghigian, "German Welfare State as Discourse of Trauma," in *Traumatic Pasts: History, Psychiatry, and Trauma in the Modern Age, 1870–1930*, ed. Mark S. Micale and Paul Lerner (Cambridge: Cambridge University Press, 2001).

29. Foucault, *Psychiatric Power*, 135, 308–23.

30. Ruth Leys, *Trauma: A Genealogy* (Chicago: University of Chicago Press, 2000), 153.

31. Charles Vallon and Georges Genil-Perrin, *La psychiatrie médico-légale dans l'oeuvre de Zacchias (1584–1659)* (Paris: O. Doin, 1912), 46. Also published as Charles Vallon and Georges Genil-Perrin, "La psychiatrie médico-légale dans l'oeuvre de Zacchias," *Revue de psychiatrie et de psychologie expérimentale* 16, 8e série, no. 2 (1912): 46–83; no. 3: 90–106.

32. Paolo Zacchia (Paul Zacchias), *Quaestiones medico-legales, in quibus eae materiae medicae, quae ad legales facultates videntur pertinere, proponuntur, pertractantur, resolvuntur* (Amsterdam, 1651), 160; É. Mahier, *Les questions médico-légales de Paul Zacchias, médecin romain:*

Études bibliographiques (Paris, 1872), 30; Charles Vallon and Georges Genil-Perrin, "Le premier livre sur la simulation des maladies (J.-B. Silvaticus, 1550–1621)," *Archives d'anthropologie criminelle de médecine légale et de psychologie normale et pathologique* 28 (1913): 907–25, 918; Gabriel Tourdes, "Simulation," *Dictionnaire encyclopédique des sciences médicales* (Paris, 1881), 9:681–735.

33. William Guy, *Principles of Forensic Medicine* (London, 1844), 187; Hector Gavin, *On Feigned and Factitious Diseases* (London 1843), 122; Ray, *Treatise on the Medical Jurisprudence of Insanity*, 338.

34. André Barthélemy Dehaussy Robécourt, *Dissertation sur une nouvelle exposition de la doctrine des maladies simulées, et des moyens de les découvrir* (Paris, 1805), 9.

35. Guy, *Principles of Forensic Medicine*, 185.

36. Hubert Coustan, *De la simulation et de l'évaluation des infirmités dans les accidents du travail* (Montpellier: Delord-Boehm et Martial, 1902), 9.

37. Theodric Beck, *Elements of Medical Jurisprudence* (Albany, NY, 1823), 2.

38. George Nesse Hill, *An Essay on the Prevention and Cure of Insanity; with Observations on the Rules for the Detection of Pretenders to Madness* (London, 1814), 393.

39. Hill, *Essay on the Prevention and Cure of Insanity*, 397.

40. Dea H. Boster, "An 'Epileptick' Bondswoman: Fits, Slavery, and Power in the Antebellum South," *Bulletin of the History of Medicine* 83, no. 2 (2009): 271–301; Walter Johnson, *Soul by Soul: Life inside the Antebellum Slave Market* (Cambridge, MA: Harvard University Press, 1999), 183–85.

41. Thomas Blatchford, *An Inaugural Dissertation on Feigned Diseases* (New York, 1817), 47.

42. Blatchford, *Inaugural Dissertation*, 54; Wendy Gonaver, *The Peculiar Institution and the Making of Modern Psychiatry, 1840–1880* (Chapel Hill: University of North Carolina Press, 2019).

43. Ray, *Treatise on the Medical Jurisprudence of Insanity*, 341. See also Henry Wentworth Acland, "Feigned Insanity, How Most Usually Simulated, and How Best Detected" (unpublished, 1844), 5–14.

44. Gavin, *On Feigned and Factitious Diseases*, 130–48.

45. John Charles Bucknill, "'Feigned Insanity,' extracted from 'The Diagnosis of Insanity,'" *American Journal of Insanity (Asylum Journal of Mental Science)* 13, no. 4 (1857): 354–67, 354, 358.

46. Bénédict Augustin Morel, *Rapport médical sur un cas de simulation de folie* (Paris, 1857), 11.

47. William W. Keen, Silas Weir Mitchell, and George R. Morehouse, "On Malingering, Especially in Regards to Simulation of Diseases of the Nervous System," *American Journal of the Medical Sciences* 96 (1864): 367–94, 377.

48. Keen et al., "On Malingering," 384.

49. Jamie L. Bronstein, *Caught in the Machinery: Workplace Accidents and Injured Workers in Nineteenth-Century Britain* (Stanford: Stanford University Press, 2008); Elisabeth A. Cawthon, *Job Accidents and the Law in England's Early Railway Age: Origin of Employer Liability and Workmen's Compensation* (Lewiston, NY: Edwin Mellen, 1997); Vincent Pierre Comiti, "Les maladies et le travail lors de la révolution industrielle française," *History and Philosophy of the Life Sciences* 2, no. 2 (1980): 215–39.

50. Roger Cooter, "The Moment of the Accident: Culture, Militarism, and Modernity in Late-Victorian Britain," in *Accidents in History: Injuries, Fatalities, and Social Relations*, ed. Roger Cooter and Bill Luckin (Amsterdam: Rodopi, 1997).

51. Émile Cheysson, *Les accidents du travail: Observations présentées devant la société d'économie sociale, les 14 février et 14 mars 1898* (Paris, 1898), 9.

52. Lerner, *Hysterical Men*, 23.

53. Eric Michael Caplan, "Trains, Brains, and Sprains: Railway Spine and the Origins of

Psychoneuroses," *Bulletin of the History of Medicine* 69, no. 3 (1995): 387–419. Ralph Harrington, "The Railway Accident: Trains, Trauma, and Technological Crises in Nineteenth-Century Britain," Eric Caplan, "Trains and Trauma in the American Gilded Age," Wolfgang Schäffner, "Events, Series, Trauma: The Probabilistic Revolution of the Mind in the Late Nineteenth and Early Twentieth Century," in Micale and Lerner, *Traumatic Pasts*; Wolfgang Schivelbusch, *The Railway Journey: The Industrialization of Time and Space in the 19th Century* (Berkeley: University of California Press, 1986), 113–49; Frank McKenna, *The Railway Workers, 1840–1970* (London: Faber and Faber, 1980); Rande W. Kostal, *Law and English Railway Capitalism, 1825–1875* (New York: Oxford University Press, 1998), chap. 7.

54. John Eric Erichsen, *On Railway and Other Injuries of the Nervous System* (London, 1866).

55. James Ogden Fletcher, *Railways in Their Medical Aspects* (London, 1867), 49, 58.

56. Fletcher, *Railways in Their Medical Aspects*, 63.

57. Fletcher, *Railways in Their Medical Aspects*, 98.

58. Fletcher, *Railways in Their Medical Aspects*, 64–65.

59. John Eric Erichsen, *On Concussion of the Spine: Nervous Shock and Other Obscure Injuries of the Nervous System in Their Clinical and Medico-Legal Aspects* (New York: William Wood, 1875), 2.

60. Deborah R. Coen, *The Earthquake Observers: Disaster Science from Lisbon to Richter* (Chicago: University of Chicago Press, 2013), 126–29.

61. Georges Guinon, *Les agents provocateurs de l'hystérie* (Paris, 1889). Charcot himself first described trauma as one among the various *agents provocateurs* of hysteria. Charcot, *Leçons* (1887), 8.

62. Strethill H. Wright, "Asthenic Insanity," *Edinburgh Medical Journal* 18 (1872): 237–49.

63. Alexandre Millerand, "Rapport au président de la république Française," *Journal officiel de la république française* 33, no. 347 (December 22, 1901): 7967–79, 7974.

64. Fletcher, *Railways in Their Medical Aspects*, 65.

65. Fletcher, *Railways in Their Medical Aspects*, 104 (emphasis added).

66. Herbert W. Page, *Railway Injuries: With Special Reference to Those of the Back and Nervous System in Their Medico-Legal and Clinical Aspects* (New York, 1892), 139; Schmiedebach, "Post-traumatic Neurosis," 40; Caplan, "Trains, Brains, and Sprains," 395; Ralph Harrington, "On the Tracks of Trauma: Railway Spine Reconsidered," *Journal for the Social History of Medicine* 16, no. 2 (2003): 209–23.

67. Kostal, *Law and English Railway Capitalism*, 276.

68. Lerner, *Hysterical Men*, 23–24.

69. H. Raynar Wilson, *Railway Accidents: Legislation and Statistics, 1825 to 1924* (London: Raynar Wilson Company, 1925), 29.

70. Kostal, *Law and English Railway Capitalism*, 303.

71. Karl Figlio, "What Is an Accident?" in *The Social History of Occupational Health*, ed. Paul Weindling (London: Croom Helm, 1985).

72. Schäffner, "Events, Series, Trauma."

73. Fletcher, *Railways in Their Medical Aspects*, 104; Johann Ludwig Casper, *A Handbook of the Practice of Forensic Medicine, Based upon Personal Experience*, vol. 4, *Biological Division*, trans. George William Balfour (London, 1865), 79.

74. Forbes Winslow, *On Obscure Diseases of the Brain and Disorders of the Mind* (London, 1860), 178.

75. Morel, *Traité des maladies mentales*, 241, 245.

76. Jules Luys, *Études de physiologie et de pathologie cérébrales* (Paris, 1874), 89.

77. Luys, *Études de physiologie*, 100–101.

78. Herbert Spencer, "The Comparative Psychology of Man," in Daniel N. Robinson, ed., *Significant Contributions to the History of Psychology, 1750-1920*, Series D (Washington, DC: University Publications of America, 1977), 7–20; Gabriel Tarde, *Les lois de l'imitation: Étude sociologique* (Paris, 1890); Bruno Karsenti, "Imitation: Returning to the Tarde-Durkheim Debate," in *The Social after Gabriel Tarde: Debates and Assessments*, ed. Matei Candea (London: Routledge, 2010), 44–61; C. Lloyd Morgan, *Habit and Instinct* (London, 1896), chap. 8; Joseph John Murphy, *Habit and Intelligence: A Series of Essays on the Laws of Life and Mind* (London, 1879), 260.

79. James Paget, "Nervous Mimicry," in *Clinical Lectures and Essays*, ed. Howard Marsh (London, 1875); Shirly Roberts, *Sir James Paget: The Rise of Clinical Surgery* (London: Royal Society of Medicine Services, 1989).

80. James Paget, "On the Relation between the Symmetry and the Diseases of the Body," *Medico-Chirurgical Transactions* 25 (1842): 30–41.

81. Paget, "Nervous Mimicry," 172.

82. Blatchford, *Inaugural Dissertation*, 55.

83. Thomas Mayo, "Case of Double Consciousness," *London Medical Gazette, or Journal of Practical Medicine*, n.s., 1 (1845): 1202–3, 1203.

84. Falret, "Folie raisonnante," 501.

85. Edwin Walker, "An Unusual Hysterical Symptom-Group," *Archives of Medicine* 10 (1883): 85–88, 85.

86. Lim Boon Keng, "The Nature of Hysteria," *Edinburgh Medical Journal* 37 (1892): 1017–20, 1020.

87. Paget, "Nervous Mimicry," 172–74.

88. Paget, "Nervous Mimicry," 192.

89. Paget, "Nervous Mimicry," 183. There was even a "family-relationship" linking nervous mimicry to insanity itself (186).

90. Paget, "Nervous Mimicry," 182.

91. Mark S. Micale, *Approaching Hysteria: Disease and Its Interpretations* (Princeton: Princeton University Press, 1995).

92. Fletcher, *Railways in Their Medical Aspects*, 106–7.

93. Erichsen, *On Concussion*, 195, 198, 207.

94. Harrington, "On the Tracks of Trauma."

95. Lerner, *Hysterical Men*, 24–25; Caplan, "Trains and Trauma," 61.

96. Edward Shorter, *From Paralysis to Fatigue: A History of Psychosomatic Illness in the Modern Era* (New York: The Free Press, 1992), 95–200.

97. For example, Benjamin Brodie, *Mind and Matter: or Physiological Inquiries; In a Series of Essays, Intended to Illustrate the Mutual Relations of the Physical Organization and the Mental Faculties* (New York, 1857); Daniel Hack Tuke, *Illustrations of the Influence of the Mind upon the Body in Health and Disease, Designed to Elucidate the Action of the Imagination* (London, 1872); Benjamin W. Richardson, "Physical Disease from Mental Strain," *American Journal of Insanity* 26 (1870): 449–69; Mitchell, *Lectures on Diseases of the Nervous System*.

98. Herbert W. Page, *Injuries of the Spine and Spinal Cord without Apparent Mechanical Lesion, and Nervous Shock in Their Surgical and Medico-Legal Aspects* (London, 1883).

99. Page, *Injuries of the Spine*, 148, 197.

100. Page, *Injuries of the Spine*, 198.

101. James Sully, *Illusions: A Psychological Study* (London, 1881), 123; Lyubov G. Gurjeva, "James Sully and Scientific Psychology, 1870–1910," in *Psychology in Britain: Historical Essays and Personal Reflections*, ed. G. C. Bunn, A. D. Lovie, and G. D. Richards (Leicester: The British Psychological Society, 2001), 72–94.

102. Herbert W. Page, "On the Mental Aspects of Some Traumatic Neuroses" (1895), in *Clinical Papers on Surgical Subjects* (London, 1897), 12–13, 15.

103. Page, "On the Mental Aspects," 15.

104. Page, "On the Mental Aspects," 26.

105. Page, "On the Mental Aspects," 32–33.

106. Page, *Injuries of the Spine*, 235.

107. Page, *Injuries of the Spine*, 251.

108. George M. Beard, *Our Home Physician* (New York, 1869), 741; Beard, *Cases of Hysteria, Neurasthenia, Spinal Irritation and Allied Affections; with Remarks* (Chicago, 1874), 8–9; Beard, *A Practical Treatise on Nervous Exhaustion (Neurasthenia), Its Symptoms, Nature, Sequences, Treatment* (New York, 1880), 86–117; Rudolf Arndt, "Neurasthenia," in Tuke, *Dictionary of Psychological Medicine*, 2:848; Adrien Proust and Gilbert Ballet, *L'hygiène du neurasthénique* (Paris, 1897), 108–10; Killen, *Berlin Electropolis*, chap. 2.

109. Leys, *Trauma*, 301–7.

110. Charcot, *Leçons* (1887), 336–39, 340, 351–52.

111. Charcot, *Leçons* (1887), 392n1; Page, *Injuries of the Spine*, 189. Charcot cites Page's second edition (1885) of *Injuries of the Spine* (207). Charcot also considered the possibility that hypnotically suggestible states also gave rise to secondary personalities, or state of ambulatory automatisms, that displayed a degree of mimetic behavior. Several articles on the subject are among his archival papers (La bibliothèque Charcot, Box 9, "automatisme ambulatoire"), including A. Pitres, "Des variations de la personnalité dans les états hypnotiques," *Gazette des hôpitaux*, 1890, 1318–20; L. Alcindor, "Un cas d'hystérie avec automatisme ambulatoire et tremblement," *Gazette des hôpitaux* 28 (1889): 253–54; M. Chantemesse, "Petit mal comitial: Grandes et petites attaques d'automatisme ambulatoire," July 1890; Henri Colin, "Deux cas d'automatisme ambulatoire: 1ᵉ automatisme d'origine alcoolique; 2ᵉ automatisme hystérique," *Gazette des hôpitaux*, August 1892, 794–97, 852–54.

112. Jean-Martin Charcot and Pierre Marie, "Hysteria," in Tuke, *Dictionary of Psychological Medicine*, 1:630; Charcot, *Leçons* (1887), 432–33.

113. Sigmund Freud, "Über Kriegneurosen, Elektrotherapie, und Psychoanalyse: Ein Auszug aus dem Protokoll des Untersuchungsverfahrens gegen Wagner-Jauregg im Oktober 1920," *Psyche: Zeitschrift für Psychoanalyse und ihre Anwendungen* 16, no. 12 (1972): 939–51, 947. See also Sigmund Freud, "Letter to Sándor Ferenczi 11 October 1920," in *The Correspondences of Sigmund Freud and Sándor Ferenczi*, vol. 3, *1920–1933*, ed. Ernst Falzeder (Cambridge, MA: Harvard University Press, 2000), 35. See also K. R. Eissler, *Freud as an Expert Witness: The Discussion of War Neuroses between Freud and Wagner-Jauregg*, trans. Christine Trollope (Madison: International Universities Press, 1986).

114. Christopher G. Goetz, Michel Bonduelle, and Toby Gelfand, *Charcot: Constructing Neurology* (New York: Oxford University Press, 1995); Mark S. Micale, "Charcot and the Idea of Hysteria in the Male: Gender, Mental Science, and Medical Diagnosis in the Late Nineteenth-Century France," *Medical History* 34 (1990): 363–441.

115. Charcot, *Leçons* (1887), 98, 250–52. Among his archival papers, Charcot also had notes on British surgeon William Thorburn's *Contribution to the Surgery of the Spinal Cord* (London, 1889) (La bibliothèque Charcot, Box 9, "automatisme ambulatoire"). Mark S. Micale, "Charcot

and the Idea of Hysteria in the Male: Gender, Mental Science, and Medical Diagnosis in the Late Nineteenth-Century France," *Medical History 34* (1990): 363–441; Kenneth Levin, "Freud's Paper 'On Male Hysteria' and the Conflict between Anatomical and Physiological Models," *Bulletin of the History of Medicine* 19, no. 4 (1948): 377–97, 382; Micale, "Jean-Martin Charcot and *les névroses traumatiques,*" in Micale and Lerner, *Traumatic Pasts*, 401.

116. Fletcher, *Railways in Their Medical Aspects*, 106–7. See also G. L. Walton, "Two Cases of Hysteria," *Archives of Medicine* 10 (1883): 88–95, 89.

117. James J. Putnam, "The Medico-Legal Significance of Hemianesthesia after Concussion Accidents," *American Journal of Neurology and Psychiatry* 3 (1884): 507–16, 507.

118. Sigmund Freud, "Hysteria" (1888), in *The Standard Edition of the Complete Works of Sigmund Freud*, vol. 1, ed. James Strachey (London: Hogarth Press, 1966), 37–39.

119. Krafft-Ebing, *Lehrbuch der gerichtlichen Psychopathologie*, 235.

120. Paul Möbius, "Bemerkungen über Simulation bei Unfall-Nervenkranken," *Münchener medizinische Wochenschrift* 49 (1890): 887–88, 887; reprinted in Paul Möbius, *Neurologische Beiträge* (Leipzig, 1894). See Francis Schiller, *A Möbius Strip: Fin-de-Siècle Neuropsychiatry and Paul Möbius* (Berkeley: University of California Press, 1982).

121. Hermann Oppenheim, *Die traumatischen Neurosen nach den in der Nervenklinik der Charité in den letzten 5 Jahren gesammelten Beobachtungen* (Berlin, 1889).

122. Paul Lerner, "From Traumatic Neurosis to Male Hysteria: The Decline and Fall of Hermann Oppenheim, 1889–1919," in Micale and Lerner, *Traumatic Pasts*; Heinz-Peter Schmiedebach, "Post-Traumatic Neurosis in Nineteenth-Century Germany: A Disease in Political, Juridical, and Professional Context," *History of Psychiatry* 10 (1999): 27–57; Killen, *Berlin Electropolis*, 91–98; Levin, "Freud's Paper," 396; Bernd Holdorff, "The Fight for 'Traumatic Neurosis,' 1889–1916: Hermann Oppenheim and His Opponents in Berlin," *History of Psychiatry* 22, no. 4 (2011): 465–76.

123. Herman Oppenheim, *Lehrbuch der Nervenkrankheiten für Ärzte und Studierende*, 3rd ed. (Berlin: S. Karger, 1902 [1894]), 1002.

124. Oppenheim, *Lehrbuch der Nervenkrankheiten*, 1007.

125. Axel V. Neel, "Über traumatische Neurosen, deren späteren Verlauf und ihr Verhältnis zur Entschädigungsfrage," *Zeitschrift für die gesamte Neurologie und Psychiatrie* 30, no. 1 (1915): 379–425, 412; John Jenks Thomas, "Malingering and the Feigned Disorders," in *Legal Medicine and Toxicology by Many Specialists* vol. 1, ed. Frederick Peterson, Walter S. Haines, and Ralph W. Webster (Philadelphia and London: W.B. Saunders, 1903), 697; Lerner, *Hysterical Men*, 137–39.

126. Victor Parant, *Relation d'un cas de simulation de la folie, applications médico-légales* (Lille, 1885), 3; Valentin Magnan, *Recherches sur les centres nerveux: Alcoolisme, folie des héréditaires, dégénérés, paralysie générale, médecine légale*, 2nd ed. (Paris, 1893 [1876]), 544–55.

127. George Wilson, "On Feigned Diseases, Their Detection and Management," *Edinburgh Medical Journal* 17 (1871): 329–41, 391–403, 516–33; 391; Stephen Watson, "Malingerers, the 'Weak-minded' Criminal and the 'Moral Imbecile': How the English Prison Medical Officer Became an Expert in Mental Deficiency, 1880–1930," in *Legal Medicine in History*, ed. Michael Clark and Catherine Crawford (Cambridge: Cambridge University Press, 1994).

128. Ambroise Tardieu, *Étude médico-légale sur les blessures* (Paris, 1879), 457–58.

129. See, for example, François Guermonprez, *Simulation des douleurs consécutives aux traumatismes: Diagnostic par les courants induits et interrompus* (Paris, 1881). Cf. Dubiau, *Rapport médico-légal sur un accident de chemin de fer, ayant entrainé un accès d'aliénation mentale* (Paris, 1875); W. S. Chipley, "Feigned Insanity, Motives, Special Tests," *American Journal of Insanity* 22 (1865): 1–24.

130. Tourdes, "Simulation," 681.

131. For example, Eduard Heller, *Simulationen und ihre Behandlung: Für Militair-, Gerichts- und Kassen-Aerzte bearbeitet* (Leipzig, 1890).

132. Jean Crocq, *Les névroses traumatiques: Étude pathogénique et clinique* (Paris, 1896), 141–42.

133. Henri Aucopt, *De l'hystérie traumatique et de la simulation* (Paris: Jules Rousset, 1903), 9.

134. Lerner, *Hysterical Men*, 27–28.

135. Georges Gilles de la Tourette, *Traité clinique et thérapeutique de l'hystérie d'après l'enseignement de la Salpêtrière* (Paris, 1891), 68, 492–93.

136. Gilles de la Tourette, *Traité clinique et thérapeutique*, 77.

137. Gilles de la Tourette, *Traité clinique et thérapeutique*, 520–22. See also Georges Gilles de la Tourette, *L'hypnotisme et les états analogues au point de vue médico-légale* (Paris, 1887).

138. Gilles de la Tourette, *Traité clinique et thérapeutique*, 527–28 (emphasis in original).

139. Micale, "On the 'Disappearance' of Hysteria"; Goetz et al., *Charcot*, 322.

140. Joseph Babinski, *Ma conception de l'hystérie et de l'hypnotisme (pithiatisme)* (Chartres: Imprimerie Durand, 1906), 13.

141. Babinski, *Ma conception de l'hystérie*, 16, 31.

142. Joseph Babinski, "Hystérie-Pithiatisme" (1901)," in *Oeuvre scientifique* (Paris: Masson, 1934), 482.

143. Babinski, "Hystérie-Pithiatisme," 490, 496. Joseph Babinski and Jules Froment, *Hystérie-pithiatisme et troubles nerveux d'ordre réflexe en neurologie de guerre* (Paris: Masson, 1918), 217.

144. Léon Imbert, Constantin Oddo, and P. Chavernac, *Accidents du travail: Guide pour l'évaluation des incapacités* (Paris: Masson, 1913), 344–45.

145. Antoine Porot, ed., *Manuel alphabétique de psychiatrie clinique et thérapeutique* (Paris, Presses universitaires de France, 1952), 426; Porot, "Notes de psychiatrie musulmane," *Annales médico-psychologiques* 9 (1918): 377–84, 381; Antoine Porot and Angelo Hesnard, *Psychiatrie de guerre: Étude clinique* (Paris: Alcan, 1919), 24–25; Keller, *Colonial Madness*, 132.

146. Porot, *Manuel alphabétique*, 496–97.

147. Frantz Fanon, *The Wretched of the Earth*, trans. Richard Philcox (New York: Grove Press, 2004), 225n40.

148. Armand Laurent, *Étude médico-légale sur la simulation de la folie: Considerations cliniques et pratiques* (Paris, 1866), 374–75. See also Gavin, *On Feigned and Factitious Diseases*, 125–26.

149. Morton Prince, "Association Neuroses: A Study of the Pathology of Hysterical Joint Affections, Neurasthenia, and Allied Forms of Neuro-Mimesis (1890/91)," in *Psychotherapy and Multiple Personality: Selected Essays*, ed. Nathan G. Hale Jr. (Cambridge, MA: Harvard University Press, 1975), 78. See Leys, *Trauma*, chap. 2.

150. Emil Kraepelin, *Psychiatrie: Ein Lehrbuch für Studierende und Ärzte*, vol. 2, *Klinische Psychiatrie*, 7th ed. (Leipzig: Barth, 1904), 831–36. For case studies, see Georg Ilberg, "Ein pathologischer Lügner und Schwindler," *Zeitschrift für die gesamte Neurologie und Psychiatrie* 15, no. 1 (1913): 1–12; and M. H. Göring, "Ein hysterischer Schwindler," *Zeitschrift für die gesamte Neurologie und Psychiatrie* 1, no. 1 (1910): 251–57.

151. Otto Binswanger, *Die Hysterie* (Vienna: Hölder, 1904), 844–45.

152. Eugen Bleuler, *Lehrbuch der Psychiatrie*, 2nd ed. (Berlin: Verlag von Julius Springer, 1918 [1916]), 137.

153. Ernst Sträussler, "Zur Frage der Simulation von Geistesstörung," *Zeitschrift für die gesamte Neurologie und Psychiatrie* 46, no. 1 (1919): 207–22, 221.

154. French physicians adopted German nosographies as well. See, for example, Gabriel

Dromard, *La mimique chez les aliénés* (Paris: Alcan, 1909), 262; and J. Caillet, *Contribution à l'étude de la simulation des troubles mentaux chez les criminels, ses rapports avec la dégénérescence* (Bordeaux: Imprimerie Commerciale, 1908).

155. Ernst Kretschmer, "Hysterische Erkrankung und hysterische Gewöhnung," *Zeitschrift für die gesamte Neurologie und Psychiatrie* 37, no. 1 (1917): 64–91, 65; Ernst Kretschmer, "Die Gesetze der willkürlichen Reflexverstärkung in ihrer Bedeutung für das Hysterie- und Simulationsproblem," *Zeitschrift für die gesamte Neurologie und Psychiatrie* 41, no. 1 (1918): 354–85, 380. See also W. Mayer, "Über Simulation und Hysterie," *Zeitschrift für die gesamte Neurologie und Psychiatrie* 39, no. 1 (1918): 315–28; Erwin Ackerknecht, *A Short History of Psychiatry* (New York: Hafner, 1968), 80.

156. Charles S. Myers, *Contributions to the Study of Shell Shock: Being an Account of Certain Disorders of Speech, with Special Reference to Their Causation and Their Relation to Malingering*, pt. 4 (repr.; London: The Lancet, 1916), 18.

157. Frederick Parkes Weber, "Association of Hysteria with Malingering, and on the Phylogenetic Aspect of Hysteria as Pathological Exaggeration (or Disorder) of Tertiary (Nervous) Sex Characteristics," *Proceedings of the Royal Society of Medicine* 5 (1911): 26–36, 28–29. See also Frederick Parkes Weber, "Two Strange Cases of Functional Disorder with Remarks on the Association of Hysteria and Malingering," *International Clinics* 1, series 22 (1912): 125–38; and Frederick Parkes Weber, "Possible Pitfalls in Life Assurance Examination, and Remarks on Malingering," *British Medical Journal* (Feb. 9, 1918): 167 (Wellcome Library Archives: PP/FPW/B.267: Box 135 [B/278 B/4], "psychoneuroses, malingering etc. 1910–1949").

158. William Norwood East, *An Introduction to Forensic Psychiatry in the Criminal Courts* (London: Churchill, 1927), 364–65. For the history of factitious disorders, see Richard A. A. Kanaan and Simon C. Wessely, "The Origins of Factitious Disorder," *History of the Human Sciences* 23, no. 2 (2010): 68–85, 77; Ryan Ross, "Between Shell Shock and PTSD? Accident Neurosis and Its Sequelae in Post-War Britain," *Social History of Medicine* 32, no. 3 (2018): 565–85; Allan Young, "Our Traumatic Neurosis and Its Brain," *Science in Context* 14, no. 4 (2001): 661–83; Chris Millard, "Concepts, Diagnosis, and the History of Medicine: Historicizing Ian Hacking and Munchausen Syndrome," *Social History of Medicine* 30, no. 3 (2016): 567–89; Cecil R. Reynolds, "Common Sense, Clinicians, and Actuarialism in the Detection of Malingering during Head Injury Litigation," in *Detection of Malingering during Head Injury Litigation*, ed. Cecil R. Reynolds (New York: Plenum, 1998), 263; Michael Pettit, *The Science of Deception: Psychology and Commerce in America* (Chicago: University of Chicago Press, 2013), 157–93.

159. Albert Mairet, *La simulation de la folie* (Montpellier: Coulet, 1908), 71.

160. Mairet, *La simulation*, 318.

161. Henri Giraud, *Étude sur les blessures simulées dans l'industrie* (Lille, 1895), 69.

162. René Sand, *La simulation et l'interprétation des accidents du travail* (Paris: Maloine, 1907), 12; Henry E. Sigerist, "From Bismarck to Beveridge: Developments and Trends in Social Security Legislation," *Bulletin of the History of Medicine* 13, no. 4 (1943): 365–88; Alfons Labisch, "From Traditional Individualism to Collective Professionalism: State, Patient, Compulsory Health Insurance, and the Panel Doctor Question in Germany, 1883–1931," in *Medicine and Modernity: Public Health and Medical Care in Nineteenth- and Twentieth-Century Germany*, ed. Manfred Berg and Geoffrey Cock (Cambridge: Cambridge University Press, 1997), 39.

163. Sand, *La simulation*, 12. The one exception is the United States, which instituted patchwork compensation systems only at the state level and not thoroughly until the 1920s, even though it drew from European models. See Rodgers, *Atlantic Crossings*, 209–66; Price V. Fishback and Shawn Everette Kantor, *A Prelude to the Welfare State: The Origins of Workers' Compen-*

sation (Chicago: University of Chicago Press, 2000); John Fabian Witt, *The Accidental Republic: Crippled Workingmen, Destitute Widows, and the Remaking of American Law* (Cambridge, MA: Harvard University Press, 2004); Nate Holdren, *Injury Impoverished: Workplace Accidents, Capitalism, and Law in the Progressive Era* (Cambridge: Cambridge University Press, 2020).

164. Sand, *La simulation*, 16.

165. Sand, *La simulation*, 18.

166. Sand, *La simulation*, 25–26.

167. Ludwig Becker, ed., *Die Simulation von Krankheiten und ihre Beurteilung* (Leipzig: Thieme, 1908), 9.

168. Becker, *Die Simulation*, 103, 107.

169. Roger Cooter, "Malingering in Modernity: Psychological Scripts and Adversarial Encounters during the First World War," in *War, Medicine and Modernity*, ed. Roger Cooter, Mark Harrison, and Steve Sturdy (Stroud: Sutton, 1998), 125–48; Peter Leese, *Shell Shock: Traumatic Neurosis and the British Soldiers of the First World War* (New York: Palgrave, 2002).

170. Lerner, *Hysterical Men*, 32–39.

171. Killen, *Berlin Electropolis*, 91.

172. Paul Lerner, "Rationalizing the Therapeutic Arsenal: German Neuropsychiatry in World War I," in Cooter, Harrison, and Sturdy, *Medicine and Modernity*, 121–48; Lerner, *Hysterical Men*, 32–36; Hermann Oppenheim, "Der Krieg und die traumatischen Neurosen," *Berliner klinische Wochenschrift* 52, no. 11 (1915): 257–61. See also Lerner, "From Traumatic Neurosis to Male Hysteria," 150.

173. Eghigian, "German Welfare State as Discourse of Trauma," 92–112; Killen, *Berlin Electropolis*, chap. 3; Lerner, *Hysterical Men*, 34. For more on worker compensation histories, see P. W. J. Bartrip and S. B. Burman, *The Wounded Soldiers of Industry: Industrial Compensation Policy, 1833–1897* (Oxford: Clarendon, 1983); Dietrich Milles, "From Workers' Diseases to Occupational Diseases: The Impact of Experts' Concepts on Workers' Attitudes, in *The Social History of Occupational Health*, ed. Paul Weindling (London: Croom Helm, 1985).

174. Eghigian, "German Welfare State," 110.

175. Binswanger, *Hysterie*, 845. See also Doris Kaufmann, "Science as Cultural Practice: Psychiatry in the First World War and Weimar Germany," *Journal of Contemporary History* 34, no. 1 (1999): 125–44.

176. Coustan, *De la simulation*, 64.

177. "Accidents du travail—la simulation," *L'agent d'assurance* 21, no. 13 (July 5, 1903): 136–37.

178. Émile Forgue and Émile Jeanbrau, "Simulation," in *Nouvelle pratiques médico-chirurgicale illustrée*, vol. 7 (Paris: Masson, 1911), 939–46.

179. "La simulation en matière d'accident du travail," *L'agent d'assurances* 21, no. 19 (October 5, 1903), 186.

180. Mathew Thomson, "Neurasthenia in Britain: An Overview," in *Cultures of Neurasthenia from Beard to the First World War*, ed. Marijke Gijswijt-Hofstra and Roy Porter (New York: Rodopi, 2001), 85.

181. Archibald M'Kendrick, *Malingering and Its Detection under the Workmen's Compensation and Other Acts* (Edinburgh: E. & S. Livingstone, 1912), 10–11, 23.

182. John Collie, *Medical Evidence and the Laws Related to Compensation for Injury* (London: The Academical Press, 1909), 4, 6.

183. John Collie, *Malingering and Feigned Sickness* (London: E. Arnold, 1913), v.

184. Collie, *Malingering and Feigned Sickness*, 8–9.

185. Collie, *Malingering and Feigned Sickness*, 11.

186. Collie, *Malingering and Feigned Sickness*, 101.

187. Erichsen, *On Concussion*, 198.

188. Page, *Injuries of the Spine*, 280 (emphasis added).

189. Page, *Injuries of the Spine*, 281–82.

190. Herbert W. Page, "On the Abuse of Bromide of Potassium in the Treatment of Traumatic Neurasthenia" (1885), in *Clinical Papers*, 44.

191. Putnam, "Medico-Legal Significance," 511; Charcot, *Leçons* (1887), 339; Thorburn, *Contribution to the Surgery of the Spinal Cord*, 222.

192. Pearce Bailey, *Accident and Injury: Their Relations to Disease of the Nervous System* (New York, 1898), 267–68.

193. Allan McLane Hamilton, *Railway and Other Accidents with Relation to Injury and Disease of the Nervous System: A Book for Court Use* (New York: William Wood and Company, 1904), 9.

194. Oppenheim, *Lehrbuch der Nervenkrankheiten*, 1007–8.

195. Alfred E. Hoche, "Geisteskrankheit und Kultur" (1910), in *Aus der Werkstatt* (Munich: Lehmann, 1937), 1–24, 15–16. See also Lerner, "Traumatic Neurosis," 151.

196. Hoche, "Geisteskrankheit," 16; Lerner, *Hysterical Men*, 34.

197. Killen, *Berlin Electropolis*, 103.

198. A. Beaumont, "La loi du 9 avril 1898 et la sinistrose," *Recueil spécial des accidents du travail*, 8e année, no. 1 (1907): 338–43, 339–40. For additional association of sinistrosis with "what the Germans called 'pension hysteria,'" see Imbert et al., *Accidents du travail*, 857, 886.

199. Émile Forgue and Émile Jeanbrau, *Guide pratique du médecin dans les accidents du travail leurs suites médicales et judiciaires*, 2nd ed. (Paris: Masson, 1909), 948, 454; reprinted in part in Forgue and Jeanbrau, "Sinistrose," 947–50.

200. A. Bassett Jones and Llewellyn J. Llewellyn, *Malingering or the Simulation of Disease* (London: W. Heinemann, 1917), 28–33, 46–47.

201. Charles Juillard, "Simulation et abus dans les assurances ouvrières au point de vue médical," *Revue de médecine légale* 15, no. 3 (1908): 321–27, 355–62, 322, 360.

202. G. Oddo, "Les névroses et les accidents du travail," *Paris médical* 9 (1913): 281–87, 284; reprinted in G. Oddo, "Les névroses et les accidents du travail," *Le poitou médical: Revue mensuelle* (March 1913): 58–67.

203. Maxime Laignel-Lavastine, "Les troubles psychiques dans les accidents du travail," *Annales de hygiène publique et de médecine légale*, série 4, no. 17 (1912): 5–21, 7.

204. Laignel-Lavastine, "Les troubles psychiques dans les accidents du travail," 6.

205. Joanny Roux, *Les névroses post-traumatiques: Hystérie, neurasthénie, sinistrose* (Paris: J.-B. Baillière, 1913), 80.

206. Roux, *Les névroses post-traumatiques*, 91.

207. E. Dupré, "L'oeuvre psychiatrique et médico-légale du Professeur Brissaud," *Revue clinique médico-chirurgicale "accidents du travail"* 4 (April 1, 1910): 73–90.

208. The case was reported in *Revue neurologique* (1914–1915): 1112–13, citing another article: Gustave Roussy and J. Boisseau, "Les 'sinistrose de guerre': Accidents nerveux par éclatement d'obus à distance," *Réunion médico-chirurgicale de la Xe Armée*, October 10, 1915.

209. Maxime Laignel-Lavastine and Paul Courbon, "La sinistrose de guerre," *Revue neurologique* 1 (1918): 322–27, 322.

210. Laignel-Lavastine and Courbon, "La sinistrose de guerre," 323.

211. Laignel-Lavastine and Courbon, "La sinistrose de guerre," 324.

212. E. de Giscarde, *La fraude, la sinistrose et les médecins marrons dans les accidents du tra-*

vail (Paris: Éditions de "L'Usine," 1921); Cristina Ferreira, "Retour sur la sinistrose, dite névrose de revendication," *Carnets de bord* 13 (2007): 78–87.

213. Simmel, *Georg Simmel: On Individuality and Social Forms*, 143.

214. Cheysson, *Les accidents du travail*, 10–11; Anson Rabinbach, "Social Knowledge and the Politics of Industrial Accidents," in *The Eclipse of the Utopias of Labor* (New York: Fordham University Press, 2018). See also Raymond Saleilles, *Les accidents de travail et la responsabilité civile* (Paris, 1897), 5–6; Marco Sabbioneti, "Raymond Saleilles (1855–1912)," in *Great Christian Jurists in French History*, ed. Olivier Descamps and Rafael Domingo (Cambridge: Cambridge University Press, 2019), 324–41; Adolphe Prins, *La défense sociale et la transformation du droit pénal* (Brussels: Misch et Thron, 1910), 57–58.

215. Ewald, *Birth of Solidarity*, 187–222.

Chapter Six

1. John Forrester, "If *p*, Then What? Thinking in Cases," *History of the Human Sciences* 9, no. 3 (1996): 1–25, 4.

2. Donna M. Lucey, *Archie and Amélie: Love and Madness in the Gilded Age* (New York: Harmony Books, 2006), 24, 47; Lately Thomas, *A Pride of Lions: The Astor Orphans, The Chanler Chronicles* (New York: W. Morrow, 1971).

3. Sven Beckert, *The Monied Metropolis: New York City and the Consolidation of the American Bourgeoisie, 1850–1896* (Cambridge: Cambridge University Press, 2001), 334.

4. Knight, *Risk, Uncertainty, and Profit*, 224–28.

5. Mercier, *The Nervous System and the Mind*, 189–91.

6. For example, "Eccentric Astor Millionaire Slays a Virginia Wife Beater," *New York World*, March 16, 1909; "Here to Push Rural Movies," *Nashville Banner*, March 24, 1922. (John Armstrong Chaloner Papers, 1876–1933, Box 7, "assorted newspaper clippings, 1888–1933," David M. Rubenstein Rare Book & Manuscript Library, Duke University).

7. Michel Foucault, "Technologies of the Self," in *The Essential Works of Foucault, 1954–1984*, vol. 1, *Ethics, Subjectivity, and Truth*, ed. Paul Rabinow (New York: New Press, 1994).

8. "Chanler Would Change Name," *New York Tribune*, May 23, 1908 (John Armstrong Chaloner Papers, 1876–1933, Box 7, "assorted newspaper clippings, 1888–1933," David M. Rubenstein Rare Book & Manuscript Library, Duke University); Lucey, *Archie and Amélie*, 249.

9. See for example, Lucey, *Archie and Amélie*; Carole Haber, "Who's Looney Now? The Insanity Case of John Armstrong Chaloner," *Bulletin of the History of Medicine* 60, no. 2 (1986): 17–193; Thomas, *A Pride of Lions*; J. Bryan III, "Johnny Jackanapes, the Merry-Andrew of the Merry Mills: A Brief Biography of John Armstrong Chaloner," *Virginia Magazine of History and Biography* 73, no. 1 (1965): 3–21.

10. John Armstrong Chaloner, *Four Years behind the Bars of "Bloomingdale" or the Bankruptcy of Law in New York* (Roanoke Rapids, NC: Palmetto, 1906).

11. Georges Canguilhem, "The Death of Man, or Exhaustion of the Cogito," trans. Catherine Porter, in *The Cambridge Companion to Foucault*, ed. Gary Gutting (Cambridge: Cambridge University Press, 1994), 86.

12. Beckert, *Monied Metropolis*, 238.

13. Amélie Rives, *The Quick or the Dead? A Study* (Philadelphia, 1888).

14. Quoted in Chaloner, *Four Years*, 118–24, 123.

15. Lucey, *Archie and Amélie*, 174, 212.

16. F. B. Arendell, "Rapids of the Roanoke," *News and Observer*, Raleigh, NC, Sunday,

May 23, 1897, 2; Robert B. Robinson III, ed., *Roanoke Rapids: The First Hundred Years, 1897-1997* (Lawrenceville, VA: Brunswick, 1997), 106-7.

17. Leland M. Roth, "Three Industrial Towns by McKim, Mead & White," *Journal of the Society of Architectural Historians* 38, no. 4 (1979): 317-47, 329.

18. Beckert, *Monied Metropolis*, 22-24, 85-89.

19. Mark Carlson, "The Panic of 1893," in *Routledge Handbook of Major Events in Economic History*, ed. Randall E. Parker and Robert Whaples (New York: Routledge, 2013).

20. "Roanoke Rapids," *Wilmington Messenger*, Tuesday, September 10, 1895 (John Armstrong Chaloner Papers, 1876-1933, Box 7, "assorted newspaper clippings, 1888-1933," David M. Rubenstein Rare Book & Manuscript Library, Duke University).

21. F. B. Arendell, "Rapids of the Roanoke," *News and Observer*, Sunday, May 23, 1897, 2.

22. F. B. Arendell, "Lowell of the Future," *News and Observer*, Raleigh, NC, Sunday morning, March 6, 1898, 1.

23. Lucey, *Archie and Amélie*, 175-79; Beckert, *Monied Metropolis*, 28.

24. Lucy, *Archie and Amélie*, 173-74.

25. In Department H ("Manufacturers"); Exhibition no. 693; classification no. Gal. G-5 659; classification heading: "Clothing and Costumes;" sub-heading: "Sewing machines for domestic purposes." James Allison, *World's Columbian Exposition, 1893, Official Catalogue, Part VIII: Manufactures and Liberal Arts Building, Leather and Shoe Building* (Chicago, 1893), 38.

26. Andrew Godley, "Selling the Sewing Machine around the World: Singer's International Marketing Strategies, 1850-1920," *Enterprise & Society* 7, no. 2 (2006): 266-314.

27. Letter to Rives, August 20, 1896 (John Armstrong Chaloner Papers, 1876-1933, Box 1, "Correspondence, 1876-1904," David M. Rubenstein Rare Book & Manuscript Library, Duke University) (emphasis added).

28. Lucey, *Archie and Amélie*, 46, 146; Thomas, *Pride*, 32.

29. Letter to Rives, August 20, 1896.

30. Gary R. Freeze, "Poor Girls Who Might Otherwise Be Wretched: The Origins of Paternalism in North Carolina's Mills, 1836-1880," in *Hanging by a Thread: Social Change in Southern Textiles*, ed. Jeffrey Leiter, Michael D. Schulman, and Rhonda Zingraff (Ithaca: Cornell University Press, 1991), 21-32; Mary H. Blewett, "Passionate Voices and Cool Calculations: The Emotional Landscape of the Nineteenth-Century New England Textile Industry," in *An Emotional History of the United States*, ed. Peter N. Stearns and Jan Lewis (New York: New York University Press, 1998).

31. Letter to Rives, August 20, 1896.

32. Transcript of deposition, 1911; "Papers on Insanity Case, 64p," MSS 38-394 c., Box 1 (University of Virginia Special Collections); "N.D. Transcript of testimony in unidentified court proceedings," MS 38-394 A, Box 5 (University of Virginia Special Collections).

33. J. A. Chaloner, *The X-Faculty, or the Pythagorean Triangle of Psychology*, 2nd ed. (Roanoke Rapids, N.C.: Palmetto, 1911), 7 (emphasis in original).

34. Transcript of deposition, 1911; "Papers on Insanity Case, 64p."

35. John Armstrong Chanler, "The X Faculty" (John Armstrong Chaloner Papers, 1876-1933, Box 6, "assorted manuscript drafts and notes, 1905," David M. Rubenstein Rare Book & Manuscript Library, Duke University).

36. Chanler, "The X Faculty," 1.

37. Max Dessoir, *Das Doppel-Ich* (Leipzig, 1890); Frank Podmore, *Studies in Psychical Research* (London, 1897), 279-86; Ellenberger, *The Discovery of the Unconscious*, 145; Courtenay Raia, *The New Prometheans: Faith, Science, and the Supernatural Mind in the Victorian Fin de*

Siècle (Chicago: University of Chicago Press, 2019), chap. 4; Alan Gauld, *A History of Hypnotism* (Cambridge: Cambridge University Press, 1992), 385–418; Ann Taves, *Fits, Trances, and Visions: Experiencing Religion and Explaining Experience from Wesley to James* (Princeton: Princeton University Press, 1999), 207–13, 254–60; Oppenheim, *The Other World*, 111–58, 205–66.

38. Letter from Chaloner to County Judge Taylor, Kings County Court, Brooklyn, NY, "1924 March 22," MS 38–394 A, Box 3 (University of Virginia Special Collections).

39. Chanler, "The X Faculty," 12.

40. Chanler, "The X Faculty," 8.

41. Chanler, "The X Faculty," 12.

42. Chanler, "The X Faculty," 10.

43. Chanler, "The X Faculty," 11.

44. Chanler, "The X Faculty," 13.

45. Chanler, "The X Faculty," 29; Bacopoulos-Viau, "Automatism, Surrealism, and the Making of French Psychopathology."

46. Chanler, "The X Faculty," 14.

47. Chanler, "The X Faculty," 15.

48. "John Armstrong Chanler Controlled by Fourth Dimension," *New York Herald*, Sunday, October 28, 1906, 8 (John Armstrong Chaloner Papers, 1876–1933, Box 7, "assorted newspaper clippings, 1888–1933," David M. Rubenstein Rare Book & Manuscript Library, Duke University).

49. Chanler, "The X Faculty," 29; "Chaloner Won $600 on 'X-Faculty' Tip," *Chicago Examiner*, October 7, 1911; "Planchette Tipped Chaloner to $600," *SF Examiner* October 7, 1911; "Chaloner Claims 'X-Faculty,'" *Providence Journal*, October 7, 1911; "Chaloner's 'X-Faculty' Useful in Stock Market," *Boston Advertiser*, October 7, 1911; "Occult Tips on Market," *Philadelphia Record*, October 7, 1911; "Chaloner's 'X Force' Beat Stock Market," *New York City World*, October 7, 1911 (John Armstrong Chaloner Papers, 1876–1933, Box 7, "assorted newspaper clippings, 1888–1933," David M. Rubenstein Rare Book & Manuscript Library, Duke University).

50. Chanler, "The X Faculty," 15–16.

51. Chanler, "The X Faculty," 18.

52. Chanler, "The X Faculty," 2.

53. Chanler, "The X Faculty," 30–31.

54. Chanler, "The X Faculty," 35; Chaloner, *Four Years*, 152. See also transcript of deposition, 1911; "Papers on Insanity Case, 64p."

55. Chanler, "The X Faculty," 37–38.

56. Chanler, "The X Faculty," 41.

57. Chanler, "The X Faculty," 47.

58. Chanler, "The X Faculty," 47.

59. Chanler, "The X Faculty," 64.

60. Herman Schubert, "The Squaring of the Circle: An Historical Sketch of the Problem from the Earliest Times to the Present Day," *The Monist* 1 (1890): 197–228.

61. Chanler, "The X Faculty," 65.

62. Chanler, "The X Faculty," 74.

63. Chanler, "The X Faculty," 77.

64. John Armstrong Chaloner, *Scorpio (Sonnets)* (Roanoke Rapids, NC: Palmetto, 1907); Chaloner, *Scorpio No. I* (Roanoke Rapids, NC: Palmetto, 1913); Chaloner, *Scorpio No. II* (Roanoke Rapids, NC: Palmetto, 1913); Chaloner, *Pieces of Eight: A Sequence of Twenty-Four War-Sonnets* (Roanoke Rapids, NC: Palmetto, 1914); Chaloner, *The Swan-Song of "Who's Looney Now?"* (Roa-

noke Rapids, NC: Palmetto, 1914); Chaloner, *"Saul": A Tragedy in Three Acts* (Roanoke Rapids, NC: Palmetto, 1915); Chaloner, *Sequence of Thirty-One Sonnets Entitled "'English Bards and Scotch Reviewers' Up-to-Date"* (Roanoke Rapids, NC: Palmetto, 1915); Chaloner, *Jupiter Tonans: A Sequence of Seven Sonnets* (Roanoke Rapids, NC: Palmetto, 1916).

65. Maudsley, *Pathology of Mind*, 297; Maudsley, *Responsibility in Mental Disease*, 55–56.

66. George H. Savage, "Mental Diseases: Introduction," in Allbut and Rolleston, *System of Medicine by Many Writers*, 842–74, 826–27.

67. Chaloner, *Four Years*, 69.

68. Chaloner, *Four Years*, 70.

69. "John Armstrong Chanler Controlled by Fourth Dimension," 8.

70. "John Armstrong Chaloner Thinks His Resemblance to Napoleon Will Aid in Proving His Sanity," *News Leader*, June 8, 1912, 1.

71. Darrin M. McMahon, *Divine Fury: A History of Genius* (New York: Basic Books, 2013), 115–24.

72. Mercier, *Human Temperaments*, 3rd ed., 101–2.

73. Laure Murat, *The Man Who Thought He Was Napoleon: Toward a Political History of Madness*, trans. Deke Dusinberre (Chicago: University of Chicago Press, 2014), 132.

74. Chaloner, *Four Years*, 79, 249–50; Lucey, *Archie and Amélie*, 208–13.

75. Pliny Earle, *History, Description, and Statistics of the Bloomingdale Asylum for the Insane* (New York, 1848), 43.

76. The Society of the New York Hospital, *One Hundred and Twenty-Seventh Annual Report for the Year 1897* (New York, 1898); Patient Record for John A. Chanler, *Bloomingdale Asylum, Discharges 1891–1920* (Medical Center Archives, Weil Cornell Medical College).

77. Chaloner, *Four Years*, 156.

78. Chaloner, *Four Years*, 133.

79. Chaloner, *Four Years*, 243, 57.

80. Karen Halttunen, "Gothic Mystery and the Birth of the Asylum: The Cultural Construction of Deviance in Early-Nineteenth-Century America," in *Moral Problems in American Life: New Perspectives in Cultural History*, ed. Karen Halttunen and Lewis Perry (Ithaca: Cornell University Press, 1998), 41–57.

81. Chaloner, *Four Years*, 249–50.

82. Letter written to Dr. H. B. Birce (sp.), "Correspondences 1894–1901," MSS 38–394 f (University of Virginia Special Collections).

83. Chaloner, *Four Years*, 186.

84. Chaloner, *Four Years*, 148.

85. Transcript of deposition, 1911; "Papers on Insanity Case, 64p."

86. Chaloner, *Four Years*, 161.

87. "John A. Chanler Escapes," *New York Times*, December 5, 1900, 2.

88. "How Mr. Chanler Escaped," *New York Times*, December 13, 1900, 1.

89. The Society of the New York Hospital, *Annual Report for the Year 1900*, 24; Patient Record for John A. Chanler.

90. "John Armstrong Chanler's Case," *The Times*, Richmond, VA, Sunday, February 3, 1901, 1.

91. J. Madison Taylor, "Personal Glimpses of S. Weir Mitchell: Incidents in the Life of the Physician, Scientist, Novelist, and Poet Narrated by His Associate and Assistant," *Annals of Medical History* 1, n.s. (1929): 583–98.

92. Lucey, *Archie and Amélie*, 165–66; Silas Weir Mitchell, *Fat and Blood: And How to Make*

Them (Philadelphia, 1877), 36–50. See also Nancy Cervetti, *S. Weir Mitchell, 1829–1914: Philadelphia's Literary Physician* (University Park: Pennsylvania State University Press, 2012), 108–13, 244–47.

93. Nancy Frances Kane, *Textiles in Transition: Technology, Wages, and Industry Relocation in the U.S. Textile Industry, 1880–1930* (New York: Greenwood, 1988), 1–45.

94. Cathy L. McHugh, *Mill Family: The Labor System in the Southern Cotton Textile Industry, 1880–1915* (New York: Oxford University Press, 1988), 5.

95. Beth English, "Beginnings of the Global Economy: Capital Mobility and the 1890s U.S. Textile Industry," in *Global Perspectives on Industrial Transformation in the American South*, ed. Susanna Delfino and Michel Gillespie (Columbia: University of Missouri Press, 2005), 175–98.

96. Patrick J. Hearden, *Independence and Empire: The New South's Cotton Mill Campaign, 1865–1901* (DeKalb: Northern Illinois University Press, 1982), 147.

97. Mildred Gwin Andrews, *The Men and the Mills: A History of the Southern Textile Industry* (Macon, GA: Mercer, 1987), 57–77.

98. Court proceedings for the November 1901 trial in Albemarle County, Virginia, are reprinted in Chaloner, *Four Years*, which is what I will hereafter cite.

99. "John Armonstrong Chanler Free," *The Times*, Richmond, VA, Thursday, November 7, 1901, 2.

100. Chaloner, *Four Years*, 56, 63–64.

101. Chaloner, *Four Years*, 98–102, 98–99.

102. Haber, "Who's Looney Now?," 182–83; Bonnie Ellen Blustein, "'A Hollow Square of Psychological Science': American Neurologists and Psychiatrists in Conflict," in Scull, *Madhouses, Mad-Doctors, and Madmen*, 241–70; Edward M. Brown, "Neurology's Influence on American Psychiatry: 1865–1915," in *History of Psychiatry and Medical Psychology, with an Epilogue on Psychiatry and the Mind-Body Relation*, ed. Edwin R. Wallace and John Gach (New York: Springer, 2008); Gerald N. Grob, *Mental Illness and American Society, 1875–1940* (Princeton: Princeton University Press, 1983), 49–71.

103. Horatio Curtis Wood, *Nervous Diseases and Their Diagnosis: A Treatise upon the Phenomena Produced by Disease of the Nervous System, with Especial Reference to the Recognition of their Causes* (Philadelphia, 1887), 489–90; Horatio Curtis Wood, *Insanity in Its Relations to Law: An Address Delivered before the Medical Society of the State of Pennsylvania, June, 1888* (Philadelphia, 1888), 8.

104. Chaloner, *Four Years*, 67–68.

105. Joseph Jastrow, "Joseph Jastrow," in *A History of Psychology in Autobiography*, vol. 1, ed. Carl Murchison (Worcester, MA: Clark University Press, 1930), 144; Théodule Ribot et al., "A Symposium on the Subconscious," *Journal of Abnormal Psychology* 2, nos. 1–2 (1907): 22–43, 58–92, 34.

106. Joseph Jastrow, *The Subconscious* (Boston: Houghton, Mifflin and Company, 1905), 299.

107. Chaloner, *Four Years*, 75–80, 77, 79.

108. Andreas Sommer, "Psychical Research and the Origins of American Psychology: Hugo Münsterberg, William James, and Eusapia Palladino," *History of the Human Sciences* 25, no. 2 (2012): 23–44.

109. "Chaloner Claims the Discovery of New Force," *New York Herald*, October, 4 1911 (John Armstrong Chaloner Papers, 1876–1933, Box 7, "assorted newspaper clippings, 1888–1933," David M. Rubenstein Rare Book & Manuscript Library, Duke University).

110. Letter from Jastrow to Chaloner, Jan. 24, 1911, "Psychical Correspondences," MSS 38-394 f (University of Virginia Special Collections).

111. Thomson Jay Hudson, *The Law of Psychic Phenomena: A Working Hypothesis for the Systematic Study of Hypnotism, Spiritism, Mental Therapeutics, etc.* (Chicago, 1893); Taves, *Fits, Trances, and Visions*, 310–13; "Review of *Laws of Psychic Phenomena*," *Journal of Mental Science* 40 (1894): 113.

112. Chaloner, *Four Years*, 87.

113. Chaloner, *Four Years*, 88.

114. Lucey, *Archie and Amélie*, 247.

115. Chaloner, *Four Years*, 82; Sofie Lachapelle, *Investigating the Supernatural: From Spiritism and Occultism to Psychical Research and Metaphysics in France, 1853–1931* (Baltimore: Johns Hopkins University Press, 2011); Andreas Sommer, "James and Psychical Research in Context," in *The Oxford Handbook of William James*, ed. Alexander Klein (New York: Oxford University Press, 2020).

116. William James, "Person and Personality: From Johnson's Universal Cyclopaedia (1895)," in *Essays in Psychology* (Cambridge, MA: Harvard University Press, 1983), 321.

117. William James, "A Case of Psychic Automatism, including 'Speaking with Tongues,' by Albert Le Baron, Communicated by Professor William James (1896)," in *Essays in Psychical Research* (Cambridge, MA: Harvard University Press, 1986), 164–65.

118. Chaloner, *Four Years*, 83.

119. Transcript of deposition, 1911; "Papers on Insanity Case, 64p."

120. Chaloner, *Four Years*, 84.

121. Amy Dru Stanley, *From Bondage to Contract: Wage Labor, Marriage, and the Market in the Age of Slave Emancipation* (Cambridge: Cambridge University Press, 1998), 264–68; Jeffrey Sklansky, *The Soul's Economy: Market Society and Selfhood in American Thought, 1820–1920* (Chapel Hill: University of North Carolina Press, 2002), 146–51.

122. T. J. Jackson Lears, "'What If History Was a Gambler?,'" in *Moral Problems in American Life*, 309–29; Jonathan Levy, *Freaks of Fortune: The Emerging World of Capitalism and Risk in America* (Cambridge, MA: Harvard University Press, 2012), 255–63.

123. William James, "Notes on Automatic Writing (1889)," in *Essays in Psychical Research*, 45.

124. Frederic W. H. Myers, *Human Personality and Its Survival of Bodily Death* (London: Longmans, Green, and Co., 1903), 1:77.

125. Eugene Taylor, *William James on Exceptional Mental States: The 1896 Lowell Lectures* (New York: Ch. Scribner's Sons, 1982), 149–65.

126. James, "Person and Personality," 319–20 (emphasis in original); Richard M. Gale, *The Divided Self of William James* (Cambridge: Cambridge University Press, 1999).

127. Taylor, *William James on Exceptional Mental States*, 42; Francesca Bordogna, *William James at the Boundaries: Philosophy, Science, and the Geography of Knowledge* (Chicago: University of Chicago Press, 2008), 99–124, 189–217.

128. Lucey, *Archie and Amélie*, 260–65.

129. "Eccentric Astor Millionaire Slays a Virginia Wife Beater," *New York World*, March 16, 1909 (John Armstrong Chaloner Papers, 1876–1933, Box 7, "assorted newspaper clippings, 1888–1933," David M. Rubenstein Rare Book & Manuscript Library, Duke University).

130. Richard Noll, *American Madness: The Rise and Fall of Dementia Praecox* (Cambridge, MA: Harvard University Press, 2011), 153–58.

131. New York Evening Post Co. v. Chaloner, 265 F. 204, No. 120 (United States Court of Appeals for the Second Circuit, Decided February 18, 1920).

132. Chaloner, *Four Years*, 211–20.

133. "John Armstrong Chanler Controlled by Fourth Dimension"; "John A. Chanler's Estate," *Richmond Dispatch*, Tuesday, November 12, 1901 (John Armstrong Chaloner Papers, 1876–1933, Box 7, "assorted newspaper clippings, 1888–1933," David M. Rubenstein Rare Book & Manuscript Library, Duke University).

134. John Armstrong Chaloner, *Robbery under Law, or The Battle of the Millionaires: A Play in Three Acts and Three Scenes* (Roanoke Rapids, NC: Palmetto, 1915), 46.

135. Chaloner, *Robbery under Law*, 44.

136. "Chanler Breaks a Long Silence: Sensational Address at Roanoke Rapids," *News and Observer*, Raleigh, NC, Wednesday morning, October 17, 1906; "Chanler on the War Path," *Roanoke News*, October 18, 1906; "John A. Chanler's Estate" (John Armstrong Chaloner Papers, 1876–1933, Box 7, "assorted newspaper clippings, 1888–1933," David M. Rubenstein Rare Book & Manuscript Library, Duke University).

137. John Armstrong Chaloner, "The Lunacy Law of the World: Being That of Each of the Forty-Eight States and Territories of the United States, with an Examination Thereof and Leading Cases Thereon; Together with That of the Six Great Powers of Europe—Great Britain, France, Italy, Germany, Austria-Hungary, and Russia" (John Armstrong Chaloner Papers, 1876–1933, Box 5, David M. Rubenstein Rare Book & Manuscript Library, Duke University); "John Armstrong Chanler Controlled by Fourth Dimension" (John Armstrong Chaloner Papers, 1876–1933, Box 7, "assorted newspaper clippings, 1888–1933," David M. Rubenstein Rare Book & Manuscript Library, Duke University).

138. Andrew Scull, *Desperate Remedies: Psychiatry's Turbulent Quest to Cure Mental Illness* (Cambridge, MA: Harvard University Press, 2022); Anne Harrington, *Mind Fixers: Psychiatry's Troubled Search for the Biology of Mental Illness* (New York: Norton, 2019); Andrew Scull, *Madhouse: A Tragic Tale of Megalomania and Modern Medicine* (New Haven: Yale University Press, 2005); Paul Lerner, "Rationalizing the Therapeutic Arsenal: German Neuropsychiatry in World War I," in *Medicine and Modernity: Public Health and Medical Care in Nineteenth- and Twentieth-Century Germany*, ed. Manfred Berg and Geoffrey Cock (Cambridge: Cambridge University Press, 1997), 121–48.

139. "1910 Miscellaneous Items," MSS 38–394 A, Box 2 (University of Virginia Special Collections).

140. Richard M. Valelly, *The Two Reconstructions: The Struggle for Black Enfranchisement* (Chicago: University of Chicago Press, 2004).

141. Chanler, "The X Faculty," 54.

142. "1910 Miscellaneous Items"; Cheryl I. Harris, "Whiteness as Property," in *Critical Race Theory: The Key Writings that Formed the Movement*, ed. Kimberlé Crenshaw, Neil Gotanda, Gary Peller, and Kendall Thomas (New York: New Press, 1995).

143. Chaloner, *The X-Faculty*, 6–7.

144. Located in John Armstrong Chaloner Papers, 1876–1933, Box 8, "photographs and postcards, 1902–1932 and undated," David M. Rubenstein Rare Book & Manuscript Library, Duke University.

145. "Notes 1908, April 30," MSS 38–394 c., Box 2 (University of Virginia Special Collections).

146. John Armstrong Chaloner, *Hell, Per a Spirit-Message Therefrom* (Roanoke Rapids, NC: Palmetto, 1917), 16.

147. John Armstrong Chaloner, "Lecture at Christ Church, Broadway at 71st Street, New York City, Sunday, November 21st, 1920" (John Armstrong Chaloner Papers, 1876–1933, Box 7, "Robbery under Law revisions, 1920–1921," David M. Rubenstein Rare Book & Manuscript Library, Duke University).

148. John Armstrong Chaloner, "No date 2 Copies of Article," MS 38-394 A, Box 4 (University of Virginia Special Collections).

149. Chaloner, "Lecture at Christ Church."

150. John Armstrong Chaloner, *A Brief for the Defense of the Unequivocal Divinity of the Founder of Christianity as the Son of Jehovah in the Legal Sense That the Legitimate Only Son of a Kind Is Heir Apparent and of "The Blood Royal"* (New York: Palmetto, 1924); "1914–1923 Religions, Divine Related Items," MS 38-394 A, Box 3 (University of Virginia Special Collections); "Ca. 1924 Proofs & Reviews for Brief for Defense," MS 38-394 A, Box 4 (University of Virginia Special Collections).

151. Chaloner, "Lecture at Christ Church."

152. "1890–1901 J. A. Chaloner Paris Prize Foundation," MSS 38-394 A, Box 2 (University of Virginia Special Collections); letter from Scrugham to Chaloner, July 14, 1925, "1921–1930 Correspondences," MSS 38-394 A, Box 1 (University of Virginia Special Collections); "Nevada Receives Chaloner Gift," *Virginia Daily Progress*, Friday, March 20, 1925; "Graduate Fellowship in Mackay School of Mind Is Announced," *Nevada Statesman*, Friday, March 18, 1925; "Millionaire Chaloner Goes into Movies," *New Orleans States*, August 22, 1920; "Chaloner Opens Film House Named after Himself," *New York Herald*, December 22, 1922 (John Armstrong Chaloner Papers, 1876–1933, Box 7, "assorted newspaper clippings, 1888–1933," David M. Rubenstein Rare Book & Manuscript Library, Duke University); John Armstrong Chaloner, *Movies for the Farmers in Rural Public Schools* (Roanoke Rapids, NC: Palmetto, 1922) (John Armstrong Chaloner Papers, 1876–1933, Box 7, "assorted printed materials and publications, 1901–1931," David M. Rubenstein Rare Book & Manuscript Library, Duke University).

153. "Chaloner Produces 'Telekinesis,' New Astronomical Craps Game," *New York Times*, Jan. 10, 1923 (John Armstrong Chaloner Papers, 1876–1933, Box 7, "assorted newspaper clippings, 1888–1933," David M. Rubenstein Rare Book & Manuscript Library, Duke University).

154. "N.D. Third lecture . . . ," MS 38-394 A, Box 5 (University of Virginia Special Collections); "1915 Review and Printed Copy of *Robbery under Law*," MSS 38-393 A, Box 3 (University of Virginia Special Collections).

155. "Chaloner 'Talks to Dead,'" *New York Times*, July 29, 1929 (John Armstrong Chaloner Papers, 1876–1933, Box 7, "assorted newspaper clippings, 1888–1933," David M. Rubenstein Rare Book & Manuscript Library, Duke University).

156. "'Who's-Looney-Now': Chaloner Sees Millennium Just Ahead," *Washington Post*, August 12, 1934, 6.

157. "Mr. Chaloner Vindicated" (John Armstrong Chaloner Papers, 1876–1933, Box 4, "legal papers, 1891–1919 and undated," David M. Rubenstein Rare Book & Manuscript Library, Duke University); "N.D. Clippings about Chaloner, Amélie Rives & Prince Troubetzkoy," MSS 38-395 A, Box 4 (University of Virginia Special Collections).

158. "A 'Who's Looney Now?' Club," *Greensboro Daily News*, Wednesday, May 14, 1919, 4 (John Armstrong Chaloner Papers, 1876–1933, Box 7, "assorted newspaper clippings, 1888–1933," David M. Rubenstein Rare Book & Manuscript Library, Duke University).

159. "Mr. Chaloner Vindicated."

160. Chanler, *Four Years*, 250; Beckert, *Monied Metropolis*, 281.

161. "Mr. Chaloner Vindicated" (emphasis added).

162. Weber, *Protestant Ethic*, 124; Lawrence A. Scaff, *Max Weber in America* (Princeton: Princeton University Press, 2011), 181–93.

163. Philip Rucker and Carol Leonnig, *A Very Stable Genius: Donald J. Trump's Testing of America* (New York: Penguin, 2020), 6.

Conclusion

1. Karl Marx, *Grundrisse: Foundations of the Critique of Political Economy (Rough Draft)*, trans. Martin Nicolaus (New York: Penguin, 1993), 269; Simon Mohun, "Value, Value-Form and Money," in *Debates in Value Theory*, ed. Simon Mohun (London: St. Martin's Press, 1991).

2. David Harvey, *Marx, Capital, and the Madness of Economic Reason* (London: Profile Books, 2018), 173.

3. Karl Marx, "Economic and Philosophic Manuscripts," in *Selected Writings*, ed. Lawrence H. Simon (Indianapolis: Hackett, 1994), 64–78.

4. Erich Fromm, *Beyond the Chains of Illusion: My Encounter with Marx and Freud* (New York: Continuum, 2009), 35, 38; Jay, *Dialectical Imagination*, 90–92.

5. Fromm, *Beyond the Chains of Illusion*, 45.

6. G. W. F. Hegel, *Hegel's Philosophy of Mind: Part Three of the "Encyclopedia of the Philosophical Sciences" (1830)* (Oxford: Clarendon, 1971), 122–39; Gladys Swain, "De Kant à Hegel: Deux époques de la folie," in *Dialogue avec l'insensé: Essais d'histoire de la psychiatrie* (Paris: Gallimard, 1994), 1–28; Daniel Berthold-Bond, *Hegel's Theory of Madness* (Albany: State University of New York Press, 1995).

7. Hegel, *Hegel's Philosophy of Mind*, 128 (emphasis added).

8. G. W. F. Hegel, *Outlines of the Philosophy of Right*, trans. T. M. Knox (Oxford: Oxford University Press, 2008), 77–78.

9. Jacques Derrida, *Given Time: I. Counterfeit Money*, trans. Peggy Kamuf (Chicago: University of Chicago Press, 1992), 7.

10. Derrida, *Given Time*, 40.

11. Becker, "Economic Approach to Human Behavior," 7–8.

12. Foucault, *History of Madness*, 46.

13. Thomas Kuhn, "Postscript—1969," in *Structure of Scientific Revolutions*.

14. Foucault, *History of Madness*, 378, 503.

15. François Dosse, *Histoire du structuralisme*, vol. 1, *Le champ du signe, 1945–1966* (Paris: Éditions la Découverte, 1991), 130. See also Jean-Christophe Coffin, "'Misery' and 'Revolution': The Organization of French Psychiatry, 1900–1980," in *Psychiatric Cultures Compared: Psychiatry and Mental Health Care in the Twentieth Century; Comparisons and Approaches*, ed. Marijke Gijswijt-Hofstra et al. (Amsterdam: Amsterdam University Press, 2005); Line Joranger, "Individual Perception and Cultural Development: Foucault's 1954 Approach to Mental Illness and Its History," *History of Psychology* 19, no. 1 (2016): 40–51; Saïd Chebili, "Le passage à l'acte: Problématique au coeur du différend entre Michel Foucault et Henri Ey," *L'information psychiatrique* 82, no. 5 (2006): 421–28.

16. Henri Ey, *Consciousness: A Phenomenological Study of Being Conscious and Becoming Conscious*, trans. John H. Flodstrom (Bloomington: Indiana University Press, 1978), 384 (emphasis in original); Henri Ey, *La conscience*, 3rd ed. (Paris: Desclée de Brouwer, 1983), 483.

17. Ey, *Consciousness*, 385; Ey, *La conscience*, 484–85; Ey, "À propos du numéro spécial de *La Nef* sur l'antipsychiatrie," *L'évolution psychiatrique* 37 (1972): 294.

18. Henri Ey, "Commentaires critiques sur 'L'histoire de la folie' de Michel Foucault," *L'évolution psychiatrique* 36, no. 2 (1971): 243–58, 243; Sherry Turkle, "French Anti-psychiatry," in *Critical Psychiatry: The Politics of Mental Health*, ed. David Ingleby (New York: Pantheon, 1980), 150–83; Jacques Postel and David F. Allen, "History and Anti-psychiatry in France," in Micale and Porter, *Discovering the History of Psychiatry*, 384–414; Roudinesco, *Histoire de la psychanalyse en France*, 1:413–38.

19. Freud, "Civilization and Its Discontents," 57–146, 144; Adorno et al., *Authoritarian Personality*; Brandy X. Lee, ed., *The Dangerous Case of Donald Trump: 37 Psychiatrists and Mental Health Experts Assess a President* (New York: Thomas Dunne Books, 2019).

20. Nikolas Rose, "Government and Control," *British Journal of Criminology* 40 (2000): 321–39; Nikolas Rose, "Governing Risky Individuals: The Role of Psychiatry in New Regimes of Control," *Psychiatry, Psychology and Law* 5, no. 2 (1998): 177–95; Andrew Scull, "Psychiatry and Social Control in the Nineteenth and Twentieth Centuries," *History of Psychiatry* 2 (1991): 149–69; Robert Castel, *The Regulation of Madness: The Origins of Incarceration in France*, trans. W. D. Halls (Berkeley: University of California Press, 1988).

21. Martin, *Bipolar Expeditions*, 7.

22. Verhoeven, "An Interview with Erving Goffman, 1980," 217.

23. Bourdieu, "Habitus"; Bourdieu, *Outline of a Theory of Practice*; Bourdieu, *Logic of Practice*.

24. Mead, *Mind, Self, and Society*, 142.

25. Michel Foucault, "The Ethics of the Concern for Self as a Practice of Freedom," in *Essential Works of Foucault, 1954–1984*, 1:291–92; Durkheim, *Rules of Sociological Method*, 6–17; Tarde, *Laws of Imitation*, xiii–xxiv.

26. Julia Ott, "What *Was* the Great Bull Market? Value, Valuation, and Financial History," in *American Capitalism: New Histories*, ed. Sven Beckert and Christine Desan (New York: Columbia University Press, 2018); François Vatin, "Valuation as Evaluating and Valorizing," *Valuation Studies* 1, no. 1 (2013): 31–50; Michèle Lamont, "Toward a Comparative Sociology of Valuation and Evaluation," *Annual Review of Sociology* 38 (2012): 201–21.

27. Weber, *Protestant Ethic*, 116; Peter E. Gordon, "Contesting Secularization: The Idea of a Normative Deficit of Modernity after Max Weber," in *Formations of Belief: Historical Approaches to Religion and the Secular*, ed. Philip Nord, Katja Guenther, and Max Weiss (Princeton: Princeton University Press, 2019).

28. Weber, *Protestant Ethic*, 111.

29. Walter Benjamin, "Critique of Violence," in Bullock and Jennings, *Selected Writings*, vol. 1.

30. Thorstein Veblen, *The Theory of the Leisure Class* (New York: Oxford University Press, 2007 [1899]), 46–69.

31. Nietzsche, *On the Genealogy of Morality*, 39–47.

32. See, for example, Jürgen Habermas, *The Structural Transformation of the Public Sphere: An Inquiry into a Category of Bourgeois Society*, trans. Thomas Burger (Cambridge, MA: MIT Press, 1991 [1962]); Habermas, *Toward a Rational Society: Student Protest, Science, and Politics*, trans. Jeremy J. Shapiro (Boston: Beacon, 1971); Deborah Cook, *Adorno, Habermas, and the Search for a Rational Society* (New York: Routledge, 2004).

33. Adorno et al., *Authoritarian Personality*, 1–27, 744–83, 971–76; Michal Shapira, *The War Inside: Psychoanalysis, Total War, and the Making of the Democratic Self in Postwar Britain* (Cambridge: Cambridge University Press, 2013), 87–111.

34. For an extended discussion of the purported psychopathology of racism, see Sander L. Gilman and James M. Thomas, *Are Racists Crazy? How Prejudice, Racism, and Antisemitism Became Markers of Insanity* (New York: New York University Press, 2016), especially chap. 6.

35. Philip Mirowski, "What Were von Neumann and Morgenstern Trying to Accomplish?," in *Toward a History of Game Theory*, ed. E. Roy Weintraub (Durham: Duke University Press, 1992), 113–47; Paul Erickson, *The World the Game Theorists Made* (Chicago: University of Chicago Press, 2015), 122–62; Erickson et al., *How Reason Almost Lost Its Mind*; Hunter Crowther-Heyck, *Herbert A. Simon: The Bounds of Reason in Modern America* (Baltimore: Johns Hopkins

University Press, 2005), 184–290; Herbert A. Simon, "What Is an 'Explanation' of Behavior?," *Psychological Science* 3, no. 3 (1992): 150–61, 159.

36. Joseph Dumit, "Drugs for Life," *Molecular Interventions* 2, no. 3 (2002): 124–27; Joseph Dumit, *Drugs for Life: How Pharmaceutical Companies Define Our Health* (Durham: Duke University Press, 2012); Brian P. Casey, "Salvation through Reductionism: The National Institute of Mental Health and the Return to Biological Psychiatry," in *The History of the Brain and Mind Sciences: Technique, Technology, Therapy*, ed. Stephen T. Casper and Delia Gavrus (Rochester: University of Rochester Press, 2017), 229–56.

37. Henri Ey, *Défense et illustration de la psychiatrie: La réalité de la maladie mentale* (Paris: Masson, 1978), 81.

38. Milton Friedman, *Capitalism and Freedom* (Chicago: University of Chicago Press, 1962), 39.

39. Dora B. Weiner, "'*Le geste de Pinel*': The History of a Psychiatric Myth," and Patrick Vandermeersch, "'*Les mythes d'origine*' in the History of Psychiatry," in Micale and Porter, *Discovering the History of Psychiatry*.

40. Emile Kraepelin, *Hundert Jahre Psychiatrie: Ein Beitrag zur Geschichte menschlicher Gesittung* (Berlin: J. Springer, 1918), 94–95; originally published as "Hundert Jahre Psychiatrie," *Zeitschrift für die gesamte Neurologie und Psychiatrie* 38, no. 1 (December 1918): 161–275; Goldstein, *Console and Classify*, 71.

41. Frantz Fanon, "Mental Alterations, Character Modifications, Psychic Disorders, and Intellectual Deficit in Spinocerebellar Heredodegeneration: A Case of Friedreich's Ataxia with Delusions of Possession (1951)," in *Alienation and Freedom*, ed. Jean Khalfa and Robert J. C. Young (London: Bloomsbury, 2018), 247. See also Frantz Fanon, "Letter to the Resident Minister (1956)," in *Toward the African Revolution: Political Essays*, trans. Haakon Chevalier (New York: Grove Press, 1967), 53.

42. Frantz Fanon and Charles Geronimi, "Day Hospitalization in Psychiatry: Value and Limits; Part Two: Doctrinal Considerations (1959)," in Khalfa and Young, *Alienation and Freedom*, 497, 501; Camille Robcis, "Frantz Fanon, Institutional Psychotherapy, and the Decolonization of Psychiatry," *Journal of the History of Ideas* 81, no. 2 (2020): 303–25, 321.

43. David Macey, *Frantz Fanon: A Biography* (New York: Verso, 2000), chap. 6; Lewis Gordon, *What Fanon Said: A Philosophical Introduction to His Life and Thought* (New York: Fordham University Press, 2015), chap. 4.

44. Robcis, *Disalienation*, 48–73.

45. Achille Mbembe, "Metamorphic Thought: The *Works* of Frantz Fanon," *African Studies* 71, no. 1 (2012): 19–28, 25.

46. Frantz Fanon, "The Meeting between Society and Psychiatry: Frantz Fanon's Course on Social Psychopathology at the Institut des hautes études in Tunis; Notes Taken by Lilia Ben Salem, Tunis, 1959–1960," in Khalfa and Young, *Alienation and Freedom*, 520–21.

47. Fanon, *Wretched of the Earth*, 44.

48. Marx, "Economic and Philosophic Manuscripts," 78.

49. Herbert Marcuse, *Studies in Critical Philosophy* (New York: Beacon Press, 1973), 13.

50. Marcuse, *Studies in Critical Philosophy*, 25.

51. Fanon, *Wretched*, 44.

52. Fanon, "Meeting between Society and Psychiatry," 530.

53. Fanon, *Wretched*, 51, 62.

54. Fanon, *Wretched*, 2; Achille Mbembe, *Necropolitics* (Durham: Duke University Press, 2019), 118, 129, 141.

55. Frantz Fanon, *A Dying Colonialism*, trans. Haakon Chevalier (New York: Grove Press, 1965), 179.

56. Quoted in Khalfa and Young, *Alienation and Freedom*, 190.

57. Michel Foucault, "Political Spirituality as the Will for Alterity: An Interview with the *Nouvel Observateur*," *Critical Inquiry* 47, no. 1 (2020): 121–34, 124.

58. Michel Foucault, "What Are the Iranians Dreaming [*Rêvent*] About?," in *Foucault and the Iranian Revolution: Gender and the Seductions of Islam*, ed. Janet Afary and Kevin B. Anderson (Chicago: University of Chicago Press, 2005), 207; Sabina Vaccarino Bremner, "Introduction to Michel Foucault's 'Political Spirituality as the Will for Alterity,'" *Critical Inquiry* 47, no. 1 (2020): 115–20; Behrooz Ghamari-Tabrizi, *Foucault in Iran: Islamic Revolution after the Enlightenment* (Minneapolis: University of Minnesota Press, 2016), 55–74; Arash Davari, "A Return to Which Self? Ali Shari'ati and Frantz Fanon on the Political Ethics of Insurrectionary Violence," *Comparative Studies in South Asia, Africa and the Middle East* 34, no. 1 (2014): 86–105.

59. Frantz Fanon, "Letter to Ali Shariati," in Khalfa and Young, *Alienation and Freedom*, 668–69. The authenticity of this letter, however, and whether any letters were in fact exchanged are topics of some debate. See Arash Davari and Siavash Saffari, "Mystical Solidarities: Ali Shariati and the Act of Translation," *Philosophy and Global Affairs* 2, no. 1 (2022): 91–104.

60. Michel Foucault, *The Hermeneutics of the Subject: Lectures at the Collège de France, 1981–1982*, trans. Graham Burchell (New York: Picador, 2005), 25–30; Nima Bassiri, "Michel Foucault and the Practices of 'Spirituality': Self-Transformation in the History of the Human Sciences," in McCallum, *Palgrave Handbook of the History of the Human Science*, 375–400.

61. Foucault, "Political Spirituality," 133.

62. Michel Foucault, "The Subject and Power," in *The Essential Works of Foucault, 1954–1984*, 3:336.

63. On the idea of a spiritual medicine in the work of Fanon and in traditions and contemporary practices of Islamic healing, see Stefania Pandolfo, *Knot of the Soul: Madness, Psychoanalysis, Islam* (Chicago: University of Chicago Press, 2018), 6–30, 242–82.

Index